THE
BEEKEEPER'S
BIBLE

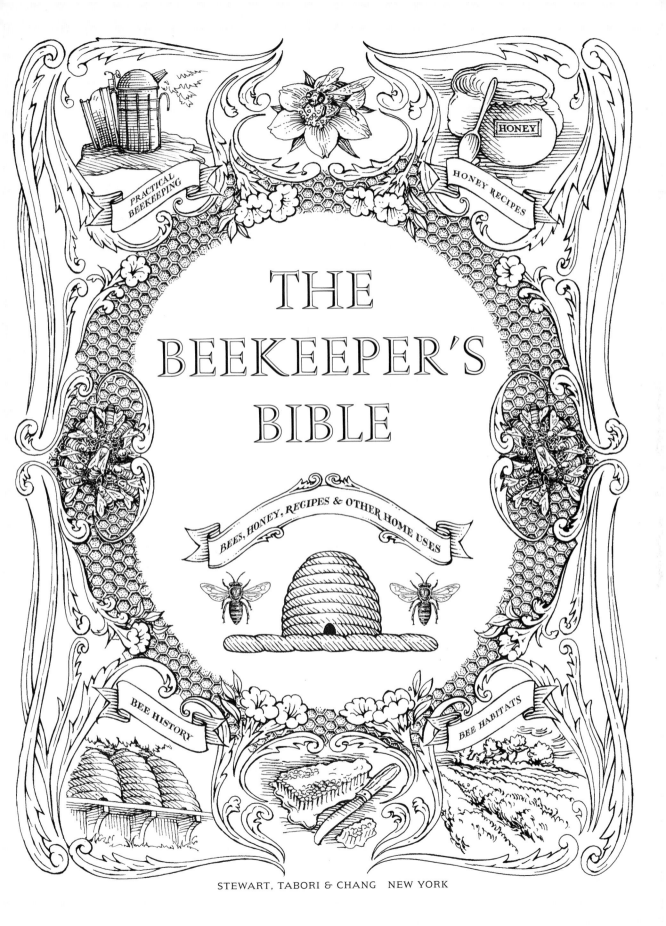

PRACTICAL BEEKEEPING

HONEY RECIPES

THE BEEKEEPER'S BIBLE

BEES, HONEY, RECIPES & OTHER HOME USES

BEE HISTORY

BEE HABITATS

STEWART, TABORI & CHANG   NEW YORK

CONTENTS

11
BEES AND BEEKEEPING HISTORY

79
UNDERSTANDING THE HONEYBEE

153
PRACTICAL BEEKEEPING

267
HONEY AND OTHER BEE PRODUCTS

305
RECIPES AND HOME CRAFTS

398
INDEX

Swarming

S. Europe.

Worker.

Drone

American

Queen.

Indian.

Published by THOMAS VARTY, 31, Strand, London.

Europe. Worker.

Defence made with Bees.

Honey Guide and Ratels.

Published by THOMAS VARTY, 31, Strand, London.

THE BEE.

Introduction

THE HISTORIC RELATIONSHIP between humans and their bees is long and enduring. Honey, beeswax, and mead (the alcoholic drink made from honey) are part of a worldwide industry, yet, in the twenty-first century, the numbers of honeybees are falling at an alarming rate, due to a mysterious condition known as Colony Collapse Disorder, which emerged late in 2006 and for which no one has yet discovered the cause. It is only as more and more of the world's honeybees die that we are now beginning to appreciate not only how fragile their survival really is but also their importance to the agricultural economy globally owing to their pollination of crops. If bees are to survive into the twenty-second century, we must take them seriously now.

The Beekeeper's Bible aims to offer all of us an opportunity to become better acquainted with bees—with their history, evolution, and biology, and with the practical aspects of beekeeping and produce. The book also gives detailed information about bees and bee colonies, bee products and their uses, and provides recipes for making your own products. While it should provide valuable information for those who are already beekeepers, it also aims to interest anyone considering getting their first colony or who is simply curious about the background of one of our most fascinating insects.

J. Stewart Del.

HONEY BEE 1.Worker. 2.Male. 3.Queen. 4.5.COMMON HUMBLE
9 DONOVAN'S HUMBLE BEE. 10.HARRIS' HUMBLE BEE.

PLATE 70

J Bishop Sc

2. LAPIDARY BEE. 6.Male. 7.Female. 8.MOSS or CARDER BEE.

LSE HUMBLE BEES. 11.Apathus Vestalis. 12.Apathus Rupestris.

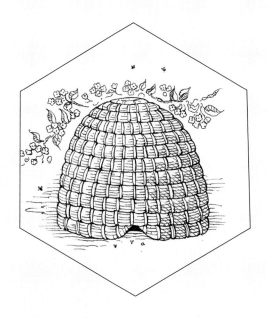

PART ONE

BEES AND BEEKEEPING HISTORY

The Honeybee in Myth and Symbol

Honey-hunting and Early Beekeeping

Scientific Advances in Beekeeping

The Global Spread of the European Honeybee

The Honeybee in Myth and Symbol

THROUGHOUT HISTORY, both bees and the honey they so magically produce have featured in myth and philosophy. The flowing sweetness of honey has, since ancient times, been associated with great eloquence ("honeyed words"), while features of bee society and behavior have long been identified with our own. A dominant female is still a "queen bee" in common parlance, and when we work hard we are "as busy as a bee" or appear as a "hive of industry." Such symbolism has its roots in the ancient cultures of the world.

BEES IN MYTHOLOGY

Our relationship with the honeybee goes back thousands of years, to the dawn of human history. Prehistoric spiritual rock art executed by the San (Bushmen) of southern Africa, whose descendants still live in the Kalahari, abounds with images of bees and their nests, suggesting that bees had a special importance in the Stone Age. Indeed, the bee is part of the creation myth of the San: an exhausted bee, having carried a mantis across a river, left the mantis on a floating flower; before the bee died, however, it planted a seed in the mantis's body. That seed became the first human.

While there is a gap in our archaeological knowledge from the Stone Age to the first beekeepers of ancient Egypt, we can see from the mythology of ancient civilizations how bees held a sacred significance and mythical power, often appearing as a sacred insect that acted as a bridge between the natural world and the underworld. Images of bees appear on some tombs, while others—specifically Mycenaean tombs—were shaped as beehives.

Bees symbolize vital principles and embody the soul. According to the ancient Greeks, the bee was sometimes identified with Demeter, the goddess of the earth and crops, who represented the soul sent to hell. The bee also symbolizes the soul that flies away from the body in the Siberian, Central Asian, and South American traditions.

This simple prehistoric cave drawing in La Cueva de la Araña in eastern Spain shows how early humans plundered wild bees' nests for honey.

OTHER BEE GODS AND GODDESSES

Northern Europe

Nanosvelta, a Roman-Germanic goddess, carries a staff with a beehive on top.

The Poles, Livlanders, and Silesians in eastern Europe had a bee god called Babilos and a goddess called Austeia.

In Russia, images of a bee god, Zosim (believed to have discovered beekeeping), were placed in beehives for protection.

The Mordva, an indigenous Russian people, had a beehive god who was the eldest son of their mother-goddess.

India (Hinduism)

Kama, the Indian god of love, is often depicted with bees (like his Greek counterpart, Eros)—symbolic of the bittersweet nature of love.

Vishnu, Krishna, and Indra together are called the Madhava or Nectar-born Ones: Vishnu is shown as a blue bee on a lotus flower (a symbol of life and resurrection), and Krishna is often depicted with a blue bee on his forehead.

Vishnu is also represented by his three "steps" (with which he strode across the universe), encompassing sunrise, zenith and sunset. In the highest step, which is associated with life after death, there is a spring of mead, the alcoholic drink made from fermented honey and water.

Siva (commonly known as the Destroyer) has another, lesser-known form as Madheri (the Suave One), represented by a bee above a triangle.

Africa

Bees in Egypt were believed to be the tears of Ra, the sun god and the giver of life and resurrection.

The Kung Bushmen in the Kalahari believe bees are the carriers of supernatural power.

Ancient Maya

The ancient Maya of Mesoamerica kept native stingless bees (see page 31) and celebrated the bee god, Ah Mucan Cab, on the fifth month in their 13-month calendar, by downing honey and *balché*, an alcoholic honey drink. They also gave burnt offerings and asked for abundant flowers so the bees could produce plenty of honey.

In western Asia, a great mother–fertility goddess, known variously as Ma, Anaitis, Rhea, Cybele, Istar, Atergatis, Artemis, and Diana, was worshipped. Her cult reached its zenith in the Temple of Artemis at Ephesus (in present-day Turkey), one of the seven wonders of the ancient world. At the temple are two Roman copies of a wooden statue of Artemis, originally created in the eighth century BC and decorated with bees carved in relief. Artemis's priests were known as Essenes or "king bees" and were likened to the bee nymphs or Mellissae that feature in Greek mythology (see page 23). These bee nymphs also gave the gift of prophecy to Artemis's twin, Apollo, whose shrine at Delphi is one of the most significant historic sites of ancient Greece. The priestess at Delphi, through whom Apollo foretold the future, was known as the Delphic Bee.

Bee mythology also persists in Hinduism (see box above), and the qualities represented by bees continue to appear in modern religions (see pages 16 and 18).

MYTHOLOGY OF THE ORIGIN OF BEES

The origin of bees was regarded as a source of fascination to many ancient cultures. In Greek mythology, the god Aristaeus, son of Apollo, is often credited with being the first beekeeper, having been taught to tend bees by Mother Earth Gaia's nymphs. According to the legend, Aristaeus fell in love with Eurydice, wife of Orpheus who, in haste to escape Aristaeus's unwanted attentions, trod on a serpent that bit and killed her. In punishment, Eurydice's nymphs destroyed Aristaeus's precious bees.

In order to recover his bees, Aristaeus had to appease the nymphs by slaughtering four bulls and four heifers, leaving their carcasses for nine days in a leafy grove as a sacrificial offering. Miraculously, at the end of this period, bees swarmed from the carcass and Aristaeus was able to rebuild his hive and pass on his knowledge of bee-keeping to humankind.

This belief that bees were born from the carcasses of dead animals persisted for centuries (see page 53).

BEE SYMBOLISM

The qualities of bees and their community in the hive have offered a source of inspiration since ancient times. Like honey, they are associated with eloquence and the power of the spoken or written word (see pages 20 and 24). The word for "bee" in Hebrew, *dbure*, is related to the word *dbr*, "speech." (The biblical name "Deborah" derives from this too.) According to legend, some bees settled on the lips of Plato when he was a baby, indicating his future brilliance with words. A similar legend attaches to other famous figures of the past, including Sophocles (also known as the "Attic Bee" because of his industriousness), Virgil, and Saint Ambrose, the patron saint of beekeepers. And there

are texts from India in which the bee represents the spirit intoxicated with the "pollen of knowledge."

A model of human society

The workings of the bee community have fascinated writers and philosophers through the ages, serving as a model or illustration for their ideas about human society. Many writers of antiquity who observed bees closely—such as Aristotle in his *Historia Animalium* (see page 33)—drew parallels between apian and human society, and found much to admire in the former. Firstly, there was the bees' orderly industriousness and cooperation. Virgil (70–19 BC) had a notion of a commonwealth of bees, where "All's the state; the state provides for all"—a notion taken up by later writers. In the *Georgics* (see page 50), he describes bees as a model for commerce, with the hive as a busy shop and the inhabitants as trading citizens.

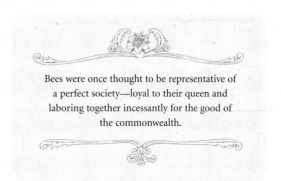

Bees were once thought to be representative of a perfect society—loyal to their queen and laboring together incessantly for the good of the commonwealth.

"Where the bee sucks, there suck I / In a cowslip's bell I lie"
Shakespeare, *The Tempest (V.i)*

The bees, as model citizens, deserved only the best beekeepers. The Roman writer Varro (see page 33) claimed bees "detest the lazy," evidenced by their culling of the drones. And Columella dedicated a

Statue of a fertility goddess, often decorated with bees.

chapter of his *De re rustica* (see page 26) to bee husbandry, insisting that perfect honesty is vital, for bees revolt against "fraudulent management."

Bee society appears in the work of much later writers. William Shakespeare in *Henry V* (I.ii.187–204) offers a fable of bees as a pattern of commonwealth in time of war (he would have had ready access to Virgil's *Georgics* IV). The first use of bee society as a lesson for a young prince was in Seneca's *De Clementia*, written for his pupil Nero. Thomas Elyot also used it for the young Edward VI. Again in the context of war, Leo Tolstoy—a keen beekeeper himself—makes a detailed analogy between Moscow, devoid of inhabitants after its capture by Napoleon, and a "queenless hive" in *War and Peace* (1869):

> In a queenless hive no life is left, though to a superficial glance it seems as much alive as other hives . . . To the beekeeper's tap on the wall of the sick hive, instead of the former instant unanimous humming of tens of thousands of bees . . . the only reply is a disconnected buzzing from different parts of the deserted hive. From the alighting board, instead of the former spirituous fragrant smell of honey and venom, and the warm whiffs of crowded life, comes an odor of emptiness and decay mingling with the smell of honey.

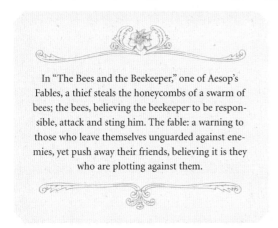

In "The Bees and the Beekeeper," one of Aesop's Fables, a thief steals the honeycombs of a swarm of bees; the bees, believing the beekeeper to be responsible, attack and sting him. The fable: a warning to those who leave themselves unguarded against enemies, yet push away their friends, believing it is they who are plotting against them.

prosperity and the bees lose the hive, thereby illustrating how "private vices" are necessary to produce "public benefit." Regarded as cynical and offensive at the time, the book was nonetheless very influential, offering an early examination of the workings of capitalism not fully developed until the likes of Adam Smith and Karl Marx.

The bee is considered an emblem of Christ: his mildness and mercy on one side (honey) and his justice on the other (the sting).

A political and economic model

In Percy Bysshe Shelley's *A Song: Men of England*, written in 1819, the year of the Peterloo Massacre, in which demonstrators for parliamentary reform had been harshly dispersed by armed cavalrymen, the political lyric pictures capitalist society as divided into two hostile classes: the parasitic class (the drones) and the working class (the worker bees). It is a rallying cry for the English working people to rise up against their political masters and oppressors and to free themselves from economic exploitation.

The hive was used by others as a model for democracy, where the individual bee subsumes its individuality for the good of the colony, and as a symbol of monarchy, tyranny, and dictatorship. The efficiency of the social system depends on the absolute rule of the queen.

The hive was also used as a model for economic principles. In 1714 Dutch-born philosopher Bernard Mandeville (1670–1733) published *The Fable of the Bees: or, Private Vices, Publick Benefits*, consisting of a poem, "The Grumbling Hive: or, Knaves turn'd Honest" and a commentary on it. The poem, a political satire on eighteenth-century England, presents a bee community that is thriving—if corrupt—until a number of the bees decide to turn from corruption and self-interest to honesty and virtue. As a result, there is a sharp drop in

It was said that a swarm of bees settled on baby Ambrose's face, leaving behind a drop of honey and presaging his honeyed tongue.

Bees flying down to three hives, from the twelfth-century Aberdeen bestiary.

A Christian symbol

St. Ambrose, a fourth-century bishop of Milan in Italy, saw the bees' apparent purity, chastity, and industrious behavior as an allegory for the monastic life, and wrote numerous religious texts for the education and guidance of the early Church. While, according to the *Confessions of St. Augustus,* "Ambrose bore his celibacy with hardship," he knew from Virgil and other ancient scholars that sex played no part in the life of the worker bee, and he encouraged Christian monks to think of the bee and its chaste, hard-working life as a model for their own.

An emblem of thrift and industry

Because of their association with selfless hard work, bees and their hives have frequently been adopted as a symbol by various groups. The

BEES IN HERALDRY

The bee is the most popular insect found in heraldry, and even the beehive occurs often as a crest. As a sacred emblem of Egyptian royalty, the bee began to be seen as a symbol of Wisdom. To the Merovingians, therefore, the bee was a most hallowed creature and 300 golden bees were discovered stitched to the cloak of Childeric I (son of Meroveus) when his grave was unearthed in 1653.

The Stuart kings of Britain claimed descent from the Merovingians, and the exiled Stuarts in Europe adopted the bee. Emperor Napoleon, in turn, had bees attached to his own coronation robe in 1804 (claiming this right by virtue of his descent from the natural son of King Charles II). Napoleon's adoption of the bee as his personal badge also gave the bee considerable importance in the French armory. It appeared on the mantle and pavilion around the armorial bearings of the French empire.

The bee is still used to denote industrious activities such as a "spelling bee" or a "quilting bee."

through hard work and an orderly society working for common goals—with its religious beliefs, and chose as its emblem the beehive, using it to consolidate its identity. When persecution drove the followers of the church west, they took their bees with them to what is now Salt Lake City, Utah. *The Book of Mormon* features "Deseret," meaning "honeybee," and in the *Deseret Times* of 1881 the skep (see page 37) was described as a "significant representation of the industry, harmony, order, and frugality of the people, and the sweet results of their toil, union and intelligent cooperation." The state of Utah is now known as the Beehive State.

Freemasons, who had as their founding principles thrift and industry, orderliness and stability, used the beehive as a symbol in official documents and drawings.

In the nineteenth century, pioneers of the Cooperative Movement in Rochdale in Lancashire, England, used representations of the wheatsheaf (a bound bundle of harvested wheat stalks) and the beehive in their official seals, because "One ear of wheat cannot stand up on its own, with others in the form of a wheatsheaf it can; and one bee cannot survive alone, in a beehive with others, it thrives."

Meanwhile, across the Atlantic, the Church of Jesus Christ of Latter-day Saints or Mormonism sought to combine the ideals of the new America—a free nation, determined to forge ahead

ABOVE: *On a grave of an entrepreneur in Ohio, a beehive stands as a symbol of productivity.*

OPPOSITE: *Freemason George Washington, with the beehive symbol of Freemasonry visible in the bottom right-hand corner.*

THE SYMBOLISM OF HONEY

Honey has been said to have special qualities since ancient times. The earliest mention of bees and honey in sacred writings appears in the ancient Indian Hindu text the *Rig-Veda*, compiled around 1500–1000 BC. Both honey and the alcoholic drink that we know as mead feature frequently, each emanating from the gods and being offered up to them (see pages 13, 25, and 43). In classical mythology, the gods consume nectar and ambrosia, both associated with honey. Honey also represents richness and plenty: the Promised Land of the Israelites is "flowing with milk and honey" (Exodus 33:3). Its healing qualities have long been recognized too: according to Islamic tradition, "honey is a remedy for every illness." Before the cultivation of sugar cane, only God's righteousness was "sweeter . . . than honey, and the honeycomb" (Psalm 19:10). And those gifted with great articulateness were regarded as anointed with honey (see pages 14, 16, and 24) or, as they still are today, "honey-tongued."

The infant Jupiter (Zeus) is nurtured by the goat Amalthea, as depicted by Nicolas Poussin, c. 1638.

BEES IN MODERN LITERATURE

A. A. Milne's Winnie-the-Pooh must be the best-known honey-lover in fiction, going to enormous lengths to obtain honey and getting into numerous scrapes as a result. In the eponymous first book (published in 1926), he tries his paw at honey-hunting in a tall tree and gets chased by angry bees as he dangles from the toy balloon, which he uses to reach the nest (a scene famously depicted by illustrator E. H. Shepard).

❋

Sylvia Plath's series of five bee poems from October 1962 are based on a bee colony she acquired after her separation from her husband Ted Hughes. They are: "The Bee Meeting," "The Arrival of the Bee Box," "Stings," "The Swarm," and "Wintering."

❋

In "Royal Jelly," a short story by Roald Dahl, Albert—a beekeeper by trade—discovers that queen bees are continuously fed royal jelly throughout their larval life; it's what makes them queens. Albert decides it could help his daughter grow too . . .

❋

In *The Secret Life of Bees* (2002) by Sue Monk Kidd (also made into a film), Lily Owens, the troubled young heroine, runs away from her abusive father and seeks refuge with three sisters whose lives revolve around beekeeping and producing honey. Learning about bees and beekeeping takes Lily on a personal journey, helping her to face the world with confidence.

Honey in ritual

Birth

The Judaean prophet Isaiah, foretelling the birth of Christ, said that a virgin would conceive a son, who would be called Immanuel, and that he would eat butter and honey so that he might know to refuse evil and choose good (Isaiah 7:14). Presenting a newborn with honey is therefore meant to symbolize a connection with heaven, and is believed to ward off evil. Milk and honey featured as part of the Christian Eucharist as late as the seventh century AD.

In Hinduism, the ancient Laws of Manu stipulated a birth rite for a male child that involved sacred prayers, touching the child with gold, and feeding him honey and butter. This belief persists even today among Hindus in parts of India.

A priest offers honey in the Opening of the Mouth ceremony (see page 25).

Marriage

Over the centuries, honey and honey products have featured in many marriage ceremonies.

In Viking ceremonies, the married couple drank mead and ate honey-flavored cakes for their first month together, while an ancient Egyptian marriage contract tells us that a groom bound himself to his bride with twelve jars of honey each year. An old Hindu wedding ceremony involves anointing the bride's forehead, eyelids, ears, mouth, and genitals with honey. On kissing her for the first time, her husband then declares, "This is honey, the speech of my tongue is honey, the honey of the bee is dwelling in my mouth and in my teeth dwells peace."

In Africa, there are similar customs. In Gallaland, a country that bordered Abyssinia, a bridegroom had to bring a fair quantity of honey to his intended bride before the wedding. If the amount was unsatisfactory, the bride and her family rejected him as a future husband. In Morocco, the guests are given honey before the wedding ceremonies and the groom feasts on it afterwards, apparently because of its powerful aphrodisiac effect.

In Europe, too, honey has an important place among the wedding festivities. When a Polish bride reached her home after the ceremonies, she was led three times around the fireplace, her feet were washed, and when she entered the bridal chamber she was blindfolded and honey was rubbed on her lips. The Bulgarians offer a special soup to the bridal couple, called *okrap*, which is made from wine and honey. The wedding cake baked with honey is broken over the head of the bridegroom and some honey is rubbed on his face. During Swedish wedding festivities honey was liberally used. According to ancient records, in 1500 when the daughter of a wealthy Swede, named Krogenose, was married, half a ton of honey was consumed.

In modern Greece, some of the ancient customs remain. The groom's mother waits at the door of his house with a jar of honey. The bride must partake of this honey when she arrives so that the words of her lips may become sweet as honey. The remaining contents of the jar are smeared on the lintel of the door so strife never enters the home. In Rhodes, when the groom arrives in his new home, he dips his finger into a cup of honey and traces a cross on the door.

In Brittany, Westphalia, and Lincolnshire the betrothals are announced to the bees themselves, and the hives are decorated with red or white ribbons; part of the wedding cakes are placed before them and the new couples must introduce them-

BEES PROTECT ZEUS

A prophecy in ancient Greece stated that Kronos, who ruled during Greece's mythological Golden Age, would be usurped by one of his own children. Kronos attempted to prevent this by swallowing five of his children at birth. However, when Zeus was born, his mother, Rhea, gave Kronos a stone swaddled in cloth, which he swallowed in the belief that it was his newborn son. Rhea then hid baby Zeus under armed guard in a cave on Mount Dicte in Crete. In some versions of the myth, when the baby cried, the guards clashed their shields and spears to attract swarms of bees—these bees attacked intruders and provided honey for the newborn. In other versions, nymphs or bee-maidens fed the child honey. Zeus later fulfilled the prophecy Kronos had feared by tricking his father into disgorging his siblings and leading them in revolt against Kronos. Zeus banished his father and became the supreme god in the Greek pantheon.

OPPOSITE: *Skeps in a medieval apiary.*

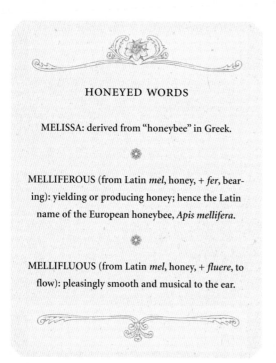

HONEYED WORDS

MELISSA: derived from "honeybee" in Greek.

❋

MELLIFEROUS (from Latin *mel*, honey, + *fer*, bearing): yielding or producing honey; hence the Latin name of the European honeybee, *Apis mellifera*.

❋

MELLIFLUOUS (from Latin *mel*, honey, + *fluere*, to flow): pleasingly smooth and musical to the ear.

selves to the bees to ensure that their married life is lucky. In Hungary the bride baked honey cake during a full moon and gave it to the groom to secure his love. After the First World War, the town of Kecskemét decided that every newly married couple should receive a beehive and a swarm of bees as a wedding present from the municipality to encourage apiculture.

While honey is both a pure and an irresistible pleasure, it is also symbolic of the idea that both the carnal and the chaste can and should coexist within marriage.

Death
Honey is a superb preservative: it is hygroscopic (absorbs water) and antimicrobial so, when it comes into contact with animal tissue, it slowly releases hydrogen peroxide which, in turn, inhibits bacterial growth.

The ancient world appreciated the preservative qualities of honey. Embalming with honey dates back to the second century BC in the ancient city

of Akkad (the remains of which are thought to lie beneath Baghdad in Iraq). Bodies were smeared with beeswax and buried in honey.

A grisly anecdote demonstrating this appears in *The Mummy* by the British Museum's celebrated Egyptologist E. A. Wallis Budge (1925):

> Once when he [medieval Egyptian Abd-Allatif-al-Baghadi] and several others were occupied in exploring the graves and seeking for treasures near the pyramids, they came across a sealed jar, and having opened it and found that it contained honey, they began to eat it. Someone in the party remarked that a hair in the honey turned round one of the

The Egyptians made use of wax for the mummification process.

fingers of the man who was dipping his bread in it, and as they drew it out the body of a small child appeared with all its limbs complete and in a good state of preservation; it was well dressed and had upon it numerous ornaments.

A cart carrying hives, 1555.

Immortality

One of honey's most unusual qualities is that it never entirely spoils, which may be why it was regarded as an immortal food and regenerative substance. The Greeks believed that ambrosia and nectar, the food and drink of the gods, contributed to their immortality. In one version of the Greek myth, Thetis anoints her infant son Achilles with ambrosia before placing him in fire to burn away his mortality.

For the ancient Egyptians, funerary rituals that featured honey were a way of preparing the dead for the life to come. As described in the *Book of the Dead* in the British Museum, the vital Opening of the Mouth ritual took place after mummification (see image page 21). A priest would use a special implement to open the mouth of the mummy, allowing the senses to be revitalized and bodily functions restored so that a dead person could enjoy the food and wine with which they had been supplied for the afterlife. Honey was offered at various points throughout the ceremony, and one incantation refers to the dead person going about as a bee, which is thought to have been a reference to the person's ka or life force.

LOVE AND HONEY

Honeyed words are associated with seduction, but love is bittersweet: honey can turn to poison, love to hate (the stinging honeybee is associated with the gods of love Eros and Kama—see page 13).

"Soft adorings from their loves receive / Upon the honey'd middle of the night"
John Keats, "The Eve of St. Agnes"

"And most of all would I flee from the cruel madness of love— / The honey of poison-flowers and all the measureless ill"
Alfred Tennyson, "Maud"

The origin of the word "honeymoon" is sometimes disputed. What is clear is that, while today the word has a positive meaning, it was probably originally a reference to the inevitable waning of love like a phase of the moon.

Honey-hunting and Early Beekeeping

HUNTING FOR WILD HONEY

Bees' nests are usually tucked away in virtually inaccessible cavities or hidden in trees. They are tricky to find and even trickier to access, so, in addition to the threat of being stung, collecting honey for the hunter–gatherer societies of the Stone Age was a precarious task. But sustenance meant survival, and honey-hunters were—and remain—brave and determined.

The earliest records of honey-gathering come from rock art of the Mesolithic era (Middle Stone Age) in southern Africa, Asia, Australia, and Europe. The oldest-known, in the Cave of the Spider (la Cueva de la Araña) near Valencia in eastern Spain, depicts a human perched perilously high on a rope ladder. To the left is the nest and in the figure's right hand is a basket for the honeycomb plunder (see page 12). In another cave, at Barranc Fondo, also in Valencia, a scene shows the honey-hunters working together as a group: four people on a ladder attempt to reach honey stored in a high crevice, while their accomplices wait below for the honeycombs to drop.

The basic methods used by these prehistoric honey-hunters are still practiced in pockets of the world today. The Bedouin in the Syrian desert, the Veddhas of Sri Lanka, and the Gurung people of Nepal, among others, all employ this technique. For many, such as the Indian Kattunaikkar tribe, honey-hunting provides a valuable source of income.

A biannual honey harvest also still takes place in the foothills of the Himalayas in central Nepal, demonstrating that various taboos and rituals associated with honey-hunting remain, including sexual abstinence and appeasement of the gods. Here, the honey harvest is preceded by offerings of fruit, flowers, and rice as well as prayer.

Only then will a honey-hunter commence his ancestral descent of the rugged cliff face to plunder the nests of one of the world's largest species of honeybee, *Apis laboriosa*.

Beekeeping is said to be the second-oldest profession.

Finding the honey

Early honey-gatherers devised many skills in tracking bees, such as looking out for their droppings on leaves. They also tagged them to make them easier to track—with red ochre, a blade of grass, or an ox hair. Columella (AD 4–70) in *De re rustica* describes a method of trapping bees inside a reed filled with honey, then releasing them one by one to ascertain the direction of the nest. It is certainly more pleasant than the eighteenth-century Eastern European method of soaking a bee in human urine to track its nest.

In North America in the eighteenth century, honey-hunters would sometimes attract a bee to a plate of flour. The flour would stick to the insect's hairy body and make it visible. Then they would follow the bee back to the nest for the honey. This is called "coursing" and it is still practiced in some areas of the world.

The splendidly named bird *Indicator indicator*, or Greater Honeyguide, shares a close relationship with humans in finding honey. While the bird is expert at tracking down bees' nests, it is not so

OPPOSITE: *Viking god Odin persuades the giant Baugi to drill into his brother Suttung's underground chamber where the mead of the poets is hidden.*

LEFT: *The Greater Honeyguide, with its white tail feathers.*

A FEW FACTS ABOUT BEES

There are 25,000 known species of bee.

Each type and species of honeybee performs its waggle dance (see pages 112–13) in a slightly different way.

The queen bee is wholly unable to care for herself. She has attendant bees that follow, feed, and groom her, and carry away her waste.

A honeybee dies after it stings, unlike a wasp.

The queen bee does have a stinger but it has smaller barbs and, unlike with other bees, may be extracted after a sting without killing her. Though her sting is most often reserved for killing other queens within the hive, this does mean that she is capable of stinging many times.

Male drone bees die soon after the act of mating.

Queen bees often make a "piping" noise—variously described as sounding like the quacking or tooting noise of a toy trumpet. It can be clearly heard outside the hive and has been said to be either a G# or an A in pitch.

expert at breaking into the nest for food. Humans, on the other hand, can open the nest, but require assistance in locating it first. Honey-hunters in tropical Africa discovered that this could benefit both parties. The chattering sounds and showy flashes of white on the Greater Honeyguide's tail feathers as it flies from perch to perch toward the honey attract the hunters and lead them toward their goal, and they will often reward the little bird with leftovers from the broken honeycomb.

EARLY BEEKEEPING

While honey-hunting still persists in some societies, honey is now largely obtained from kept bees. This beekeeping developed hand in hand with the growth of agriculture.

At the end of the Ice Age (around 10,000 years ago), small groups of humans moved seasonally to obtain food. By around 8000 BC, however, some of these hunter–gatherer groups had started to settle year-round in favorable sites around lakes and rivers, most markedly in the Fertile Crescent (an area extending from the eastern Mediterranean

Ancient hives in Tel Rehov, Israel, discovered in 2007 (see page 34).

shores to the Persian Gulf, which today would cross Israel, Lebanon, Jordan, Syria, Iraq, Kuwait, Turkey, and Iran). As agriculture developed, augmented over time by irrigation techniques, terracing, the cultivation of fruit, and domestication of livestock, so grew the foraging opportunities for wild bees. Bee numbers increased and, as bee colonies outgrew their nests, they sent out swarms to establish new nest sites.

Agricultural beekeeping may, therefore, have been triggered by accident, as swarms nested in woven baskets or baked-mud water pots left lying around outside in these early settlements. It is likely that human settlers would have been quick to capitalize on this—deliberately providing nest sites for bees close to home was a far more reliable and much less precarious way to obtain honey than hunting for them. (See also page 36.)

Ancient Egypt

The earliest form of organized beekeeping was in ancient Egypt. The annual flooding of the Nile delta plain created an area of rich silt that was ideal for agricultural development. Crops and wild flowers, such as chamomile, clover, wild celery, cornflowers, flax, and roses, flourished beside the river, the nectar of which would have been ideal for bees.

The bee was adopted as a royal symbol in pre-dynastic Egypt, and by the first dynasty (3100–2890 BC) there was already an official "sealer of the honey," suggesting that hive-beekeeping existed by then.

The earliest direct evidence of an organized honey industry in Egypt comes from a remarkably clear and detailed stone bas-relief found in 1900 at Abu Gurab in Lower Egypt in the Temple of the Sun, which dates to around 2400 BC. One of the four scenes that comprise the relief clearly shows nine horizontally placed pipe-shaped hives, probably made from Nile mud or clay. A man is depicted taking honey from the hives. In his hand he holds what is probably a block of dried cow dung, which was used to smoke the bees (see pages 173–75). The three other scenes show men filling jars with honey, pressing honey and sealing the jars, most likely with the owner's seal.

In 1978, Dr. Eva Crane—a beekeeper, physicist and founder of the Bee Research Association (later

ABOVE: *The Egyptian hieroglyph for "beekeeper."*
BELOW: *Relief from the Temple of the Sun of Neuserre in Abu Gurab, Lower Egypt.*

the International Bee Research Association), who was once described as "the queen bee among bee experts"—discovered hives in the Nile valley just like those depicted in ancient Egyptian drawings. The horizontally stacked cylindrical hives were made from interwoven twigs and reeds, which had been covered in sun-dried mud or clay; the gaps between them were filled with mud to increase stability. At each end was a mud disc. At the front was a flight hole for the bees and, at the back, the disc was removable so the beekeeper could pacify the bees with smoke before harvesting the combs.

The design of these hives meant that beekeepers would have been able to divide a colony of bees by removing a few combs containing brood (see page 103) and a queen from one cylinder, and placing them inside another, empty cylinder. This would

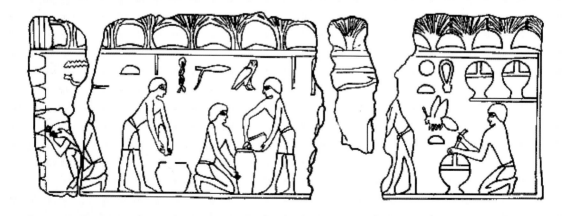

ABOVE: *A bee hieroglyph was used to denote the King of Lower Egypt. This depiction is carved in the columns of the Temple of Kalabcha, Egypt.*

BELOW RIGHT: *A Mayan representation of a honeybee.*

Ancient Maya

The Yucatán peninsula in Mesoamerica, dividing parts of modern Mexico, Guatemala, and Belize, was the home of the ancient Maya roughly 1,800 years ago. A highly developed culture, the Maya had their own language and hieroglyphic writing, and built spectacular temple pyramids, the remains of which still stand beneath the swampy green canopy of the Yucatán jungle.

Spanish explorers such as Bartolomé de las Casas, who discovered the Yucatán in the sixteenth century, were astonished to come across the hives of "tame" bees. Mayan beekeepers had learned how to divide existing hives to increase the number of hives and amount of honey produced, while taking care not to over-harvest (harvesting was done only twice a year as part of a religious ceremony). The visiting Spaniards found large apiaries (1,000–2,000 hives of native stingless bees) and a well-developed trade in beehives and honey. They were brought many gourds full of honey, which they reported to be "white and excellent" and not at all inferior to that of their native Castile.

have allowed a bee colony to expand while at the same time discouraging it from breaking away in a swarm and nesting elsewhere.

By around 1500 BC, beekeeping in Egypt was organized on a vast scale. There were state granaries, production lines, and accompanying administration: scribes noted down all the jars full of honey presented by peasants as payments, and an official was responsible for receiving honey and other valuable products as offerings at the temple. A colorful wall painting on the tomb of prime minister Rekhmire in Luxor, which dates to around 1450 BC, shows that honey vessels were stamped according to quality and color—"stph" denoting pure honey, light in color and "deshert" indicating a reddish-colored honey from the desert.

The hives that de las Casas found were made from hollowed-out logs in the shape of drums, and were beautifully carved with detailed figures and ornaments. Each hive also featured the sculpted sign of the owner.

The bees entered and left by a hole along the middle of the hive, and stone discs recently found in Belize are believed to have been the end stoppers. These discs, dating from somewhere between 300 BC and AD 300, are the oldest-known beekeeping artifacts in the New World.

A page of a Mayan codex. Many have been found to show representations of bees and hives.

The ancient East

Frequent references to sacred honey and mead in the Hindu *Rig-Veda* suggest apiculture may have been established in India in early times, though hunting for wild honey remains a lucrative business to this day.

Historians and archaeologists suspect that hive-beekeeping may actually have begun in the wake of the campaigns of Alexander the Great. These led him to cross the Hindu Kush (the mountains between Afghanistan and Pakistan) in 327 BC and establish Macedonian rule in the upper Indus basin (Persia). Certainly, the hives used in traditional beekeeping, as observed by Eva Crane in Kashmir in the 1980s, are remarkably similar to the horizontal hives discovered in ancient Greece, suggesting that beekeeping know-how was transported over long distances.

China's native honeybee, *Apis cerana*, is smaller than the Western honeybee, *A. mellifera*; it forages in a smaller area and produces less honey. Nevertheless, by the time of the East-Han dynasty (AD 25–150), beekeeping was clearly not only well established but highly profitable. Indeed, it is listed in a book from the period called *How to Acquire Wealth* by advisor and businessman Fan-Li. According to the *Kai Si*, a story written by Huang Pu Me in the AD 200s, more than 300 men went to learn beekeeping at Hanyang (in the modern-day Shaanxi province) and, by the Chin dynasty (AD 265–420), a government proclamation was issued to encourage productivity.

Ancient Rome

Although hives in Roman Italy were made from plant material that has not survived, writing from the period gives evidence of the existence of beekeeping in ancient Rome.

Roman writers on beekeeping almost certainly absorbed some of their knowledge from the Mediterranean regions—Italy, Sicily, Sardinia, Corsica, much of north Africa (including Carthage), most of Spain, Macedonia (a historical

ARISTOTLE THE BEEKEEPER

In Book IX of the *Historia Animalium* (written between 344 and 342 BC), Aristotle offers a detailed picture of beekeeping in ancient Greece. He includes notes about bee-masters removing the combs from the hives and taking care to leave enough food behind for the bees to survive the winter. And he describes how, when honey ran short, the beekeepers supplied the bees with figs and sweet-tasting articles of food. Aristotle states that a hive could yield anything from a meager 5 pints (3 liters) of honey to an exceptionally prosperous 18 pints (10 liters). He also describes a monotonous and peculiar sound that can be heard a few days before a swarm issues from a hive, and how a beekeeper should sprinkle a hive with sweet wine to attract the bees.

region of the Balkan peninsula) and Phoenicia, and the area west of Asia Minor, the region linking the Mediterranean Basin with the Great Asian hinterland—where hive-beekeeping had been practiced for at least 2,000 years. Practical descriptions are given of hiving swarms, how to dispose of unwanted queens, when and how to feed bees, how to protect bees from their enemies as well as how to keep the hive clean. There is also information about different types of honey.

The Roman scholar and papal envoy Varro (116–27 BC) described a range of hives, from those made from wood and bark (the best) to round wicker constructions covered with cow dung inside and out and ones made from earthenware (the worst because they provided the least insulation).

He also mentioned rectangular hives measuring 3 ft by 1 ft (100 cm by 30 cm) made from fennel stalks. These were known as ferula hives and origi-nated in Sicily. Being narrower at the middle, these hives, according to Varro, were meant to mirror the shape of a bee's body. The hives had holes on either side to allow the bees to enter and exit, and lids at either end so that beekeepers could remove the combs. The hives were kept on wall ledges in rows of two or three, with spaces between the rows.

Beekeeping was a profitable business in the Roman world. Two brothers by the name of Veianius inherited half an acre of land in Spain, created a garden, planted it with thyme, Cytisus (broom), and Apiastrum (mock parsley), and established an apiary, which brought them 10,000 sesterces a year (approximately $20,000 today).

The Roman writer who most conjures up the smallholder idyll of beekeeping, however, is Virgil. Book IV of his treatise on small-scale farming, *Georgics*, which appeared around 30 BC, contains an epic poem on bees and beekeeping. Virgil describes in elegant detail the practicalities of keeping bees and procuring stocks (via "bugonia"—see pages 14 and 53); he discusses the roles of the different bees in the hive and he describes his preference for hives of cork and willow with small entrances and cracks stopped up to save the bees labor. He even warns against what can be offensive to bees, such as the smell of yew—which, being poisonous, would produce poisonous honey—loud noises, and even the odor of burnt crabs!

BIBLICAL BEEKEEPING

In 2007 archaeologists, scientists, historians, and bee-keepers alike were thrilled by the discovery of 30 ancient hives during an excavation at a site in Tel Rehov in Israel (see page 29). Amazingly intact, and stacked in rows of three, the hives are similar in style to those shown on the Egyptian Temple of the Sun bas-relief (on page 30). They are made of straw and unbaked mud, cylindrical in shape, and measure 32 in (80 cm) in length and 16 in (40 cm) in diameter. One end of each cylinder is closed and has a small hole for the entry and exit of the bees. The opposite end is covered with a clay lid that can be removed to extract the honeycombs. These hives have been dated to the ninth or tenth century BC, to the time of King Solomon, and prove that, while there is no mention of beekeeping in Hebrew scripture, it was well established in the Holy Land during biblical times. Experts estimate that the hives would have produced around half a ton of honey each year—such a valuable yield may explain why the hives were situated in the heart of the largest city in the area.

BEEKEEPING: THE MIDDLE AGES AND BEYOND

Keeping bees in trees

Before the advent of purpose-built beehives, honey-gatherers in medieval Europe sought out mature trees with resident wild honeybees nesting in their cavities. Once they found a nest, the bee-keepers protected and nurtured it, often by cutting a small flight entrance into the tree for the bees to continue to go in and out naturally, and then fitting a wooden door to protect the nest from bad weather and predators. Straw, secured by ropes around the trunk, helped insulate the nest in winter. If the nest was high up, the gatherers improved their own access to it by cutting footholds out of the trunk of the tree, or by affixing rope ladders.

Alternatively, honey-gatherers created the right conditions to attract swarms of bees to start nests by lopping the tops off growing trees to reduce their height and to thicken their trunks, and then carved artificial cavities in the trunks in an attempt to create a desirable habitat for the bees.

Whichever solution was adopted, throughout spring and summer the early beekeepers left the bees alone to get on with the job of making honey. The reward for their patience was the honey harvest in early autumn.

Throughout Russia, Poland, Germany, and the Baltic region, tree beekeeping developed wherever the climate was warm enough and the trees large enough. This often led to the creation of "bee forests," usually owned by the aristocracy or the Church. A bee forest could contain 100–500 tree cavities, with up to 20 cavities occupied by honeybees at any one time—the number limited by the available forage in the grassy clearings.

Keeping bees in hives

Bee forests, though they played an important role in the local economy, had their downside. The bee nests were widely spaced and attending to them was time-consuming. One solution was to cut down the part of the tree that contained the nest and transport the resulting "log hive" closer to home, where it would sit either on the ground or on a slab of stone or wood next to other hives, creating a woodland apiary. These hives were often carved or painted and, in Poland and eastern Germany particularly, the carvings could be very sophisticated. The largest log hives were often carved into human forms, with a back access for the beekeeper and an entrance in the chest or navel for the bees. Smaller logs might be carved into bears or other native animals. Some of these carvings survive and examples can be found in the Jan Dzierzon Museum in Kluczbork, Poland, where there is a permanent exhibition related to the history of apiculture and to the beekeeper and priest Jan Dzierzon, who designed the first successful movable-frame hive (see page 61).

In England, the earliest hives were of pliable willow or hazel wands, woven around a circle of stakes that joined at the top. They were coated, or "cloomed," with a mixture of wet cow or oxen dung and ashes or gravelly soil in the spaces between the woven stems; this mixture set hard, sealed the hive, and became waterproof.

In wetter parts of Continental northern and western Europe, where willow was commonly available, wicker was used in hive-making. In drier, cereal-growing and grassland areas, straw, reeds, and sedges were twisted into ropes before being coiled in ever-increasing circles and stitched with blackberry briars to make basket-weave hives or "skeps."

Wooden hives in Lithuania, similar to ones used by medieval eastern European beekeepers.

ABOVE: *A skep, still used in some areas today.*

BELOW RIGHT: *Driving bees from a skep to another skep held above it at an angle.*

BEEKEEPING IN AFRICA

Beekeeping has a long history not only in Egypt, but elsewhere in Africa. African bees are very productive but some races can also be defensive. Beekeeping in sub-Saharan Africa requires techniques different from keeping European bees in the rest of the world. Beekeeping is still being encouraged in Africa. It is easy and cheap to start, requires little land and is therefore ideal for small-scale, resource-poor farmers. In addition, it can provide valuable food and medicine for the family, and cash crops of honey and beeswax to take to the local market. Parts of Africa, such as Chad and Upper Volta (now Burkina Faso), still use tree stumps and clay pots of various sizes as hives as they have for centuries.

Skeps

Tribes of the Elbe region of Germany are believed to have been the first skep-makers (the word "skep" is derived from the Old Norse *skeppa*, meaning "basket" or "bushel"). They were originally made of a coiled rope of straw and were dome-shaped. An alternative, made elsewhere, was the conical wicker hive made in a similar way to making a basket. Sticks were often inserted near the top of the skep to provide a support from which the bees could build their combs. A little roof, or hackle, made from reeds in a wigwam shape to fit over the top of the skep, offered more protection from the weather.

Skeps evolved into two main types: the cloomed and hackled covered ones that usually sat on individual stands; and the uncloomed (uncovered) ones that were lined up in rows beneath a lean-to roof or penthouse. Honey was originally harvested from a skep by a method known as "sulfuring": the skep was placed over a pit containing burning sulfur, candles, or paper darts impregnated with sulfur. The poisoned, dead bees would then

be shaken from the combs before collection. By the mid-nineteenth century, the practice of "driving" was used instead: an empty skep was placed above the full one while the beekeeper drummed rhythmically on the lower one; the bees crawled into the upper skep, leaving the combs behind for harvesting.

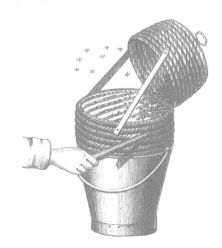

The skep has never lost its role in beekeeping; it is an airy and pleasant environment for the bees. To hive a swarm from a tree branch (see below), many modern beekeepers hold a skep beneath the swarm and pull the branch down sharply, so that the whole swarm drops into the skep in a ball. The swarm is then transferred to a hive.

Before modern beehives, bee boles and shelters were practical ways of keeping the wind and rain away from the skep and the bees inside.

Bee boles

In Britain, France, and parts of Belgium, sturdy stone constructions were sometimes specially made to house skeps. These "bee boles" were often set in a south- or southeast-facing wall, either part of the house walls or those of walled orchards from where bees could pollinate the fruit trees.

OPPOSITE: *A Dutch beekeeping scene, c. 1565.*

There are bee boles from the Middle Ages as well as some modern ones, but construction was at its peak in the eighteenth and nineteenth centuries, while there were still numerous large country estates.

An unnamed apiarist of the seventeenth or eighteenth century, quoted by Malcolm Fraser in *History of Beekeeping in Britain* (1958), advised:

> Let the hives be set as near the dwelling place as conveniently can be, or to rooms most occupied, for the reddier [sic] discovery of rising swarms, or to be apprised of accidents. Besides, the bees habituated to the sight of the family will become less ferocious and more tractable.

Bee boles generally feature a row of separate, often arched, open niches, about 24 in (60 cm) high, 20 in (50 cm) wide, and 18 in (45 cm) deep; each niche contained a yellow-straw skep with a front

TANGING A SWARM

Bees in skeps swarmed as soon as the nest became overcrowded. This meant that skep-beekeepers became practiced at catching—or hiving—a swarm. To encourage a swarm to settle nearby, the beekeeper banged together metal objects, such as pots and pans, to make a loud noise. Once the swarm settled, the bees were encouraged to take up residence in an empty skep. This ancient tradition—known as tanging and still practiced today— also alerted the neighborhood that bees were swarming (as in Jan van der Straat's etching from c. 1580).

Bee boles at the Lost Gardens of Heligan, Cornwall, England.

entrance. There were, however, many different designs, including square boles, ones with stone lips at the front on which the bees could land, and others with wooden doors. Sometimes there were two

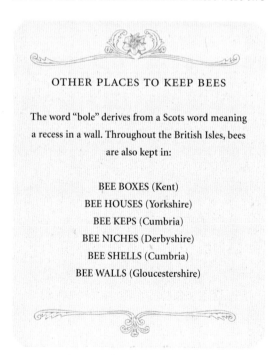

OTHER PLACES TO KEEP BEES

The word "bole" derives from a Scots word meaning a recess in a wall. Throughout the British Isles, bees are also kept in:

BEE BOXES (Kent)

BEE HOUSES (Yorkshire)

BEE KEPS (Cumbria)

BEE NICHES (Derbyshire)

BEE SHELLS (Cumbria)

BEE WALLS (Gloucestershire)

or even three rows of niches, one above the other, and occasionally just a single bole.

Many bee boles and shelters can still be seen in ancient buildings, ecclesiastical remains, and country houses or old farms; there are also examples of skeps and hives used as architectural ornamentation and in coats of arms carved in stone. There are more than 1,400 known sites of bee boles in England, Ireland, Scotland, and Wales (these can be viewed online at the International Bee Research Association: http://ibra.beeboles.org.uk). Many are accessible to the public, including a spectacular one with two rows of boles, some with doors, in the Lost Gardens of Heligan near Mevagissey in Cornwall. An incredibly ornate, free-standing stone example was discovered in Nailsworth, Gloucestershire; it has now been moved to St. Mary's churchyard in Hartpury. A bee shelter, not a traditional bole, it was built in the mid-nineteenth century for an ornamental garden and has spaces for 28 skeps in two rows, and further accommodation in larger, arched recesses in the base.

BEEKEEPING AND THE LAW

Once beekeeping became part of the economy, it was usually incorporated into law. The Hittite empire in central Anatolia (in modern-day Turkey and Syria) is the source of the first-known legislation surrounding bees and beekeeping. At Bogazkoy, the site of the Hittite capital of Hattusa (in north-central Turkey), a number of clay tablets from around 1500 BC are inscribed in cuneiform (an ancient form of writing). Two paragraphs not only reveal the penalties for stealing bees or hives—a forfeit of five shekels of silver for stealing a swarm (equating a swarm's value to that of a sheep)—but also mention earlier, harsher penalties.

The Romans, however, created the laws that were—and still are—a cornerstone of rulings on animals in many countries. According to Roman civil law, bees were wild animals. Hive bees, however, belonged to the hive owner. If hive bees swarmed, the owner had the right to collect them from a neighbor's land but, once they were out of the owner's sight, they were deemed to have reverted to a wild state and the owner had no claim to them.

An Old English document, *Rectitudines Singularum Personarum*, dating from AD 1000, records the rights and obligations of workers and tenants on an English estate before the Norman Conquest in 1066. Each worker is listed in order of rank; the beekeeper or "beo-ceorl" is ranked second lowest, next to the swineherd. Despite this, the beo-ceorls were authorized to collect honey both from wild nests and from man-made hives, and heavy punishments were inflicted on anyone who stole from these hives.

The *Domesday Book* of 1086 was a survey of existing settlements, stock, and land. It contains many references to the collection of wild honey; the law during William I's reign (1066–87) began to state exactly who had rights to this wild honey and what sanctions there were against stealing.

Two centuries later, and two years after it was first written, the *Magna Carta* (one of the most important legal documents in the history of England) triggered supplementary legislation, including the Charter of the Forest. This charter, sealed in 1217 by Henry III, repealed the death penalty for taking royal game; it also allowed people to take the natural produce of the forest, which

A skep in a single bee bole.

included honey and beeswax—and it remained in law for over 750 years, until 1971. The charter gave greater security to the people not just in allowing access to food sources, but financially: at that time, many landowners were monasteries and it was common to pay rent either partly or wholly in beeswax, which the monks needed for votive candles (see page 48). Towns often also combined coin and honey in their taxes to the king.

The Ancient Laws of Ireland, known as the *Senchus Mor*, were codified in the fifth century, but are thought to date back to before AD 250. Not only does the document frequently mention beehives, it also describes what is to be fed to children: the offspring of the poor could expect a "stirabout" of oatmeal and buttermilk or water; the offspring of aristocratic chieftains had barley meal with new milk and butter; sons (and daughters) of kings, however, might receive wheaten bread with new milk and honey—an indication of the high value placed upon honey.

Keepers of bees were responsible for damage to fruit and flowers caused by their bees but, as it was clearly impossible to police, each beekeeper had to pay a fixed share of honey to any neighbors living within the range that a bee could fly and allegedly cause damage—this was estimated to be the range of a crowing cock or a church bell.

Modern laws concerning bees and beekeeping have been concerned both with combating the spread of disease among bees (see page 74) and protecting neighbors in urban environments.

Page from a manuscript, showing bees flying in and out of a beehive and foraging, 1200.

It was traditional in northern Europe (and the custom spread to the New World with settlers—see pages 64–69) to tell the bees if their beekeeper was dead. If the bees were not asked to stay with their new master or mistress, it was believed they would die or abscond. One method was to tap each hive quietly with a large metal key three times, then tell the bees. The custom is described in this extract from John Greenleaf Whittier's poem "Telling the Bees" (1858) in which the beehives are "ranged in the sun" by the brook:

Before them, under the garden wall,
Forward and back
Went drearily singing the chore-girl small,
Draping each hive with a shred of black.
Trembling I listened, the summer sun
Had the chill of snow;
For I knew she was telling the bees of one
Gone on the journey we all must go!

Fragment of the Ebers papyrus, New Kingdom, c. 1550 BC, which contained numerous prescriptions using honey for medicinal use.

USES OF HONEY

Though honey is now mainly used as a foodstuff—a sweetener in its own right—ancient cultures used it for healing, preserving, and embalming (see page 24). It was also used as taxes or tributes from subject or conquered peoples or given as an offering to the gods. It is little wonder, therefore, that beekeeping was such an efficiently run endeavor and that a guaranteed supply of such a valuable commodity was a sign of wealth and power.

In ancient Egypt, there is evidence of special honey cakes made for children in the 1400s BC, baked in the shape of a person, rather like the gingerbread men of today. There were also triangular "shat-cakes" made with honey and date flour.

By 594 BC beekeeping in Athens had reached such a scale that Solon passed a law stipulating that stocks of bees had to be placed 300 ft (100 m) apart. Attic honey, from the thyme-covered slopes of Mount Hymettus, was reputed to be the best, a feast's "crowning dish." Cheese-cakes, steeped in rich honey, along with honeycombs and honey wine or mead, made Athens a gastronomic center of the ancient world.

According to the Ebers papyrus from 1550 BC (one of the oldest Egyptian medical documents), a blend of honey, ground alabaster, natron powder (sodium bicarbonate), and salt could be used to beautify the body. Honey was also used in anti-wrinkle cream, but then so was turpentine!

When the Spanish conquered the Mayan people in 1542, they demanded tributes. Out of 173 towns and villages in 1549, 163 paid some in beeswax and 157 gave some in honey; the total payments included around 3 tons of honey and 277 tons of wax.

BEE PRODUCE

Honey is the only natural food that, as long as it is kept airtight, does not spoil.

❈

After honey has been extracted, the comb can be mixed with water and boiled down into firm yellow cakes, from which the color disappears if exposed to the air for long enough. Once thoroughly bleached, it is again melted down into cakes, becoming known as the "white wax of commerce." Before oil lamps, the white wax of commerce was used in the manufacture of candles (see page 292), burning better and without the smoke or odor of tallow candles.

❈

Bee venom can be used to treat rheumatic diseases, especially arthritis and multiple sclerosis (MS). It can be applied directly or by intramuscular injections.

Honey as a healer

The first record we have of the use of honey in a prescription appears on a Sumerian clay tablet, found in the Euphrates valley (in present-day Iraq); the tablet has been dated to around 2100–2000 BC. The Chinese also seem to have used honey medicinally as far back as 2000 BC. It was the Egyptians, however, who seemed to have embraced honey most widely—it features as an external application in no fewer than 147 prescriptions in the Ebers papyrus (see page 43) and in 102 prescriptions for internal use; it was taken

to cure, among others, respiratory conditions such as phlegm and asthma. And the Kahun papyrus, which lists remedies for gynecological complaints and dates to around 1900 BC, describes a contraceptive paste made of honey, crocodile feces, and saltpeter.

Honey was also invaluable in wound care for the ancient Egyptians, Assyrians, Chinese, Greeks, and Romans. The Ayurvedic surgeon Susruta (who lived around 1400 BC) treated wounds with a paste of honey, butter, barley, and herbs, while the Greek physician Hippocrates (469–399 BC), regarded as the father of modern medicine, cited its use for sores, carbuncles, and ulcers, and also as part of a simple dietary remedy to tackle pain, thirst, and fevers.

From early times, honey was regarded as good for health, a "giver of life."

During World War I, when supplies of conventional antiseptic ran out, the Germans used honey as a substitute. To this day, a licensed form of honey is still used in wound care in hospitals and to store human corneas for transplant operations.

Mead

Mead, or honey wine, is brewed from honey and water and fermented with yeast. It has been around since ancient times, mainly in Europe, Africa, and Asia, although its origins are lost in prehistory.

The earliest archaeological evidence for mead production comes from around 7000 BC: pottery vessels containing a mixture of mead, rice, and fruits along with organic compounds of fermentation were found in Northern China. The first written description is in the *Rig-Veda*, while Aristotle mentions it in *Meteorologica* and Pliny the Elder in his *Naturalis Historia*, in which he also differentiates wine sweetened with honey from mead.

Mead was a common historical drink and, even as its production declined (because of heavy taxation

EARLY BIOLOGICAL WARFARE

Xenophon and his Greek army retreated ill from Persia in 399 BC. While none of the men died, the effects apparently ranged from a feeling of drunkenness, to vomiting and purging, and madness that lasted three days. There were reports that this was the result of toxic honey being used against them (see pages 137–38 for more on bad honey).

A later use of toxic honey is recorded by the Greek geographer and historian Strabo in his *Geographica*. The honey had been used to ambush Pompei's soldiers in the Third Mithridatic War in 65 BC the inhabitants of Pontus placed the maddening honey on the soldiers' route, the soldiers helped themselves, were rendered senseless and were easily overcome, costing Pompei over 1,000 men.

The great Norse god Odin was said to have a magic goat that filled a pitcher every day with the mead of poetry, the most precious drink in the universe, which would turn anyone who drank it into a poet or scholar (see page 27).

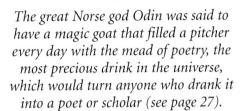

and regulation on alcoholic drinks), some monasteries kept up the old traditions of mead-making as a by-product of beekeeping, especially in areas where grapes could not be grown.

The honey collected by the Anglo-Saxon beoceorls (see page 280) was mostly used for mead; indeed, the mead hall was at the heart of Anglo-Saxon communities and features in the epic poem *Beowulf* (AD 700–1000), set in the fifth century. In northern climes such as Scandinavia, the mead hall was owned by the lord, who allowed its use by his warriors as a place of rest, feasting, and refreshment after battle; in 2004–5, a mid-sixth-century Viking hall measuring 160 ft (50 m) in length was excavated southwest of Lerje in Denmark.

Virgil writing the last of the four volumes of his Georgics, *which he devoted to the subject of bees.*

TYPES OF MEAD

BALCHÉ/PITARILLA: mead fermented with the bark of the balché tree

BRACKET/BRAGGET/BRAGOT: beer mixed with honey or mead

CAPSICUMEL: mead fermented with chili peppers

CHOUCHENN: mead produced in Brittany, sometimes with fresh sea water

CYSER: fermented apple and honey

CZWORNIAK/DWOJNIAK/POLTORAK/TROJNI-AK: types of Polish mead with a varying ratio of honey to water

HYDROMEL: weak mead

MELOMEL: mead with added fruit

METHEGLIN/METHEGLEN: mead with spices or herbs

MULSUM/OMPHACOMEL/PYMENT: grape wine sweetened with honey

RHODOMEL: mead fermented with roses and honey

TEJ: a mead made by Ethiopians, to which powdered leaves and bark are added

USES OF BEESWAX

Beeswax, the substance secreted by bees for forming the honeycomb, is a material that is easily shaped, and this quality has been exploited throughout the ages. For thousands of years, it has been used for modeling. The ancient Greeks and Romans made models for creating molds into which molten metal was poured for making jewelry and sculptures. Known as lost-wax casting—the wax was covered in plaster and then melted to create a mold—it is a technique still employed to this day (see page 290).

Although beeswax had been used in many cultures for thousands of years, it was not until the eighteenth century that anyone had a clear idea of how bees produced it (see page 57).

Wax models have also been used for less salubrious purposes. Egyptian magicians made waxen images of men or animals, usually to inflict harm on them. The earliest example of this sorcery comes from the Westcar papyrus, which tells of Aba-aner, an influential priest who lived around 2830 BC. Aba-aner's wife had a lover, so Aba-aner created a model of a crocodile, which apparently came to life, and then disappeared from sight with the hapless lover clamped in its jaws.

Wax itself could be used as a substitute for papyrus, if a permanent record wasn't required, or the message needed to be erasable. The Romans adopted the tradition of using wooden tablets covered with a layer of beeswax in which they wrote their message using a stylus. The recipient then rubbed out the message on the tablet and wrote a reply on the smoothed sheet of wax.

Though the wax the Maya presented to the Spaniards was of a lesser quality than honeybee wax, so much wax was needed for European candles that, by the mid-1600s, it was still one of the chief exports from the Yucatán.

With no lasting inscription, the system became popular for transmitting secret love messages. While wax tablets did not persist beyond Roman times, beeswax maintained its link with letter-writing. Mixed with turpentine and later colored with vermilion, it was used from the Middle Ages for sealing letters and envelopes, helping to ensure that a letter was read by no one but its intended recipient.

Wax has been used as a sealing agent in other contexts. The Egyptians used it in shipbuilding and there is evidence that it was also employed for this purpose in Bronze Age Britain. In 1992 a subterranean pumping station was being built at Dover on the southeast coast of England; archae-

The wax tablet and stylus were popular among Romans for writing love notes.

OTHER USES OF BEESWAX

AS A FILLER: to fill the screw holes and the seams
between the slates in pool tables

IN MUSICAL INSTRUMENTS: blended with pine
rosin by accordion makers to fix reed plates to the
structure inside the instrument. It is also the traditional material from which to make didgeridoo
mouthpieces and the fret for a type of boat lute in
the Phillipines, the kutiyapi.

FOR FOOD: as a coating to protect food as it ages,
particularly for cheese. (Some cheese makers have
replaced wax with plastic, but many feel that this
gives off unpleasant flavors and therefore still use
beeswax.) It is also a food additive, E901.

AS A LUBRICANT: in guns that use cap and ball,
and cartridge-firing firearms that use black powder
(gunpowder)

ologists uncovered a long piece of wood in a water-
logged hole. The excavated site revealed the
remains of a prehistoric wooden boat (dating from
around 1500 BC), the timbers of which were lashed
together and sealed with beeswax to make them
watertight. Longbows were also covered with a
water-resistant coating—of wax, resin, and fine
tallow—to help them last longer.

Church candles

Beeswax was used for making candles as early as
3000 BC, by the ancient Egyptians and by the
Minoans in Crete. During the Middle Ages, there
was great demand for candle wax for the vast cathe-
drals of Catholic Europe. Many European monasteries and nunneries even kept their own bees for
the wax they produced. Wax could be made from

rendered domestic animal fat, but the result was
common tallow that, when burnt, sputtered and
smelt rancid. As the symbol of God, therefore, candles for churches had to be made from beeswax.

Often money was left in the wills of the rich and
powerful to pay for wax candles to be burned in
perpetuity for the welfare of their souls. And, as
Christianity spread and grew stronger, so did beekeeping. There were even monks whose designated
job it was to care for the bees, collect and process

MADAME TUSSAUD

The world's best-known sculptor in wax is Marie
Tussaud (1761–1850). Tussaud, then Marie
Grosholtz, learned her craft in Paris, where her
mother was housekeeper to a doctor who made
demonstration human figures in wax for his stu-
dents. In 1789, at the start of the French Revolution,
she and the doctor made models of severed heads
straight from the guillotine: the head was encased in
plaster; after it had set, the plaster was removed and
used as a mold to be filled with beeswax. When the
wax solidified, the plaster cast was chipped away to
reveal precise features, wrinkles and all; paint, hair,
and eyelashes were added as the final delicate touch
to a macabre process.

In 1802, by then married to Francis Tussaud,
Madame Tussaud and her waxworks entourage went
to England to exhibit her work, together with the
original guillotine blade and other relics; after her
death in 1884, the collection settled in Marylebone
Road in London, where it remains despite fire
(1925), earthquake (1931), and bombing (1940), and
is constantly updated. There is even a waxwork of
Madame Tussaud herself (aged 81).

Before the advent of commercial beekeeping in America, many farmers and villagers kept a few colonies of bees in hives to supply their own needs and those of their friends, relatives, and neighbors.

the wax and make candles and tapers.

Then came the Reformation and the dissolution of the monasteries in the 1530s. Hundreds of bee colonies were lost or died out. The austerity of the reformed, Protestant churches had little need for beeswax candles and demand fell dramatically.

With the return of the Catholic Queen Mary in 1553, the old regime was reinstated; a giant 300 lb (140 kg) beeswax paschal (Easter) candle was placed at the shrine of Edward the Confessor in Westminster Abbey, London, in 1558 by the Worshipful Company of Wax Chandlers (see box).

The Catholic restoration in England was short-lived (Mary died in 1558 and her Protestant sister Elizabeth I took the throne); the Catholic Church elsewhere, however, kept up the tradition of using pure beeswax candles right up until the twentieth century, when this requirement was finally waived by the Pope.

THE WAX CHANDLERS

In ancient and medieval times, when blocks of wax were used to pay tributes and fines in England, a wax industry grew up. In 1358, the Worshipful Company of Wax Chandlers was established as a Livery Company of London. The company received its first Royal Charter in 1484, from Richard III, and was granted arms in 1485. Though the company is no longer a trade association of wax candle-makers (it's a charitable institution), the medieval structure and management of the company remains in place today: there is a Master Wax Chandler, Upper and Renter Wardens, and a Court of Assistants. The company is also a patron of the National Honey Show and the British Beekeepers' Association (BBKA), and involved in the presentation of annual awards and prizes for beekeepers who achieve the highest marks in the BBKA examinations.

Wax Chandler.

Scientific Advances in Beekeeping

THE WRITERS OF ANTIQUITY who kept bees made detailed observations of the bee communities they tended. During his lifetime (384–322 BC), Aristotle spent an extraordinary amount of time researching the anatomy and habits of honeybees. It is said that he made an observation hive in order to watch his bees working and reproducing, but they covered the outside in wax, obscuring his view. Nevertheless, Aristotle managed to provide plenty of accurate observations of the life and activities of bees.

For several centuries after the beginning of the Christian era, however, the influence of Christian philosophers and clergy can be seen in texts about natural history and there is a decline in objective observation. Originating in Alexandria (AD 140–410), the *Physiologus* or "Naturalist" was written in Greek and was probably the work of several authors. In stark contrast to the work of Aristotle and Pliny, which included a wealth of knowledge on beekeeping, it consisted of descriptions of animals, birds, and mythical creatures, together with the moral and symbolic qualities of each. The creatures were used to represent Christian teaching and most descriptions were invented to fit the moral comment, rather than to record any real details based upon observational science. The *Physiologus* was translated into European and Middle Eastern languages and revised and copied over the next few centuries into a form known as the bestiary, one of the most widely read book forms in medieval times. It was only in the twelfth century that the bee was added to the lists of creatures.

The invention of the printing press by the German printer Johannes Gutenberg in around 1440 made books far more widely available and gave the European middle classes access to the written word that had, until then, only been available to the aristocracy and clergy. Though still in Greek and Latin, among the first volumes printed were Aristotle's collected works, which, together with Virgil's *Georgics* (Book IV of which was devoted to beekeeping—see pages 33–34 and 53), Pliny the Elder's *Historia Naturalis* (AD 77–9) and several encyclopedias of agriculture from the late 1500s, give an insight into pre-Christian-era views, including those on bees and beekeeping.

BEEKEEPING KNOWLEDGE BEGINS TO EXPAND

There is little evidence of an increase in the knowledge of bees and beekeeping between the eleventh and fifteenth centuries. The first original beekeeping manual to be written in English was Edmund Southerne's *A Treatise Concerning the Right Use and Ordering of Bees*. Published in 1593, Southerne's book was based on his own observations as a bee owner and he recognized that (male) drones had a role to play in the hive. Although he could not identify their purpose, he felt that they should not be destroyed.

Surviving estate records and account books of farmers and landowners in Britain between 1600 and 1800 expand our knowledge of beekeeping in the UK. In Yorkshire, in the north of England, the records of Henry Best show the hiving of swarms using a white sheet, still common beekeeping practice today. A swarm is captured in a skep or a bucket (see page 37) and is then tipped onto a white sheet spread out at the hive entrance; the bees readily stream up the sheet into the hive in orderly lines. The dark bees are easily visible

OPPOSITE: *Beekeepers preparing to take a swarm, from an eighteenth-century edition of Virgil's* Georgics.

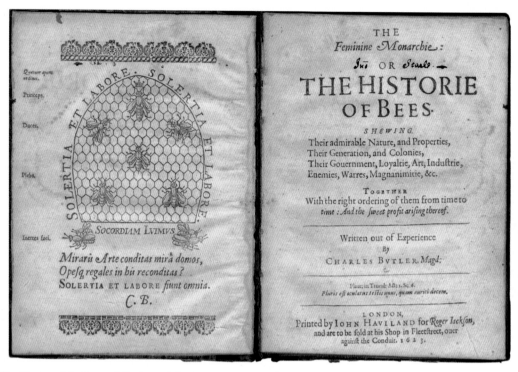

The title page of the Reverend Charles Butler's The Feminine Monarchie, *1623.*

against the whiteness of the sheet, enabling the beekeeper to identify the queen and check on her health. In his records, Henry Best also recommends taking the honey as a suitable job for the household maids. The maids would wring the honey out of the combs into a tub; the empty comb and loose hive debris were then put in another tub full of clean water, from which mead would be made.

In the second half of the sixteenth century, a number of influential books on agriculture and natural history were published in Europe, among them one by German physician Georg Pictorius of Villingen, *Pantopolion* (1563), which was partly a celebration of the natural world and includes a treatise on beekeeping. *Pantopolion* was the first of these texts to be translated into English (by Thomas Hyll of London in 1568), as *The arte of gardening*, and the section that dealt with beekeeping was republished six years later under the title *A profitable instruction of the perfite ordering of*

bees. And, in 1655, *The Reformed Common-Wealth of Bees*, believed to be the first government publication to deal with beekeeping, appeared in England. It was compiled by Samuel Hartlib (c.1600–62), an English government official responsible for, among other things, the dissemination of information to the general public.

After the dissolution of the monasteries and destruction of the monastic apiaries, less wax and honey had been available; mead became expensive, too—out of the price range of common people. At the same time, the Atlantic trade in sugar cane, and the increasing availability and cheap price of sugar, meant that sugar rapidly began to take over from honey. While, in 1410, honey sold for the US equivalent of $1.83 per lb, with sugar at $.16 per lb, by 1530 honey prices had risen to $2.58, but the price of sugar had plummeted to $.05.

Samuel Hartlib and his friends, including the members of the Council of Trade, were keen to

BUGONIA

Unlikely as it sounds, for thousands of years bees were believed to be born out of the carcass of an ox—referred to as "bugonia," from the Greek for "ox birth." The belief came from seeing small, amber-brown winged creatures flying out of a carcass in ancient Egypt; it was presumed from this that bees had miraculously generated from the rotting flesh. The fourth book of Virgil's *Georgics*, dedicated to the art of beekeeping, refers to this belief and unflinchingly tells readers how to create a swarm of bees. Roughly paraphrased, his instructions are:

First, place a bullock in a confined space, block its nostrils and mouth. Next, bludgeon it to death without breaking the skin. Leave the bruised body in a closed room and, once the bones have softened and fermented, you will have a swarm of bees.

Blocking the nostrils and mouth was important to prevent the animal's soul leaving the body—the soul was destined to enter the bees—and the best time to beget bees was said to be when the sun entered the constellation of Taurus, the bull.

This miraculous misconception of spontaneous generation—not just of bees, but of snakes, mice, and maggots—was not actually disproved until 1668 by Italian physician Francisco Redi, and we now know that these creatures seen rising from ox carcasses were not bees, but similar-looking hover-flies (see page 82), which, unlike honeybees, lay their eggs on rotting matter.

encourage beekeeping and once more produce cheaper and more plentiful honey as a substitute for sugar, in food and in brewing. Though Hartlib diligently assembled a great deal of well-researched information about bees and their habits, he was not a beekeeper and it was a badly organized publication. It also, unfortunately, starts with an account of Mr. Carew of Cornwall, who was alleged to have produced bees from calves' carcasses (see box).

PIONEERING BEEKEEPERS AND SCIENTISTS

The Reverend Charles Butler

By the end of the sixteenth century a new breed of writer had emerged, that of the parson–naturalist. These men had both the time and the interest to look more closely at their natural surroundings and write about them for the education of others. One such writer was the Reverend Dr. Charles Butler (1559–1647), a scholar and a musician, as well as a doctor of divinity and author of several musical works. Having tried to rear silkworms commercially without success, Butler turned his attention to bees and his 1623 essay *The Feminine Monarchie or a Treatise Concerning Bees, and the Due Ordering of Them* earned him the title "the father of English beekeeping." His work was the first of its kind, in that it was written by a practicing beekeeper who took a directly observational approach, based totally on his own, detailed studies and practical experience. It was easy to read, well laid out, and had a user-friendly index and margin references—it was considered a masterpiece.

Laying out the physical differences and separate functions of queens, workers, and drones, Butler described the drones as:

grosse hive-bees without stings, quick to feed, slow to work, for whosoever they brave it with their round velvet caps, full paunches and load [sic] roar, yet they are but idle companions, living by the sweat of others' brows . . . for albeit he be not seene to engender with the honey-bee, either abroad as other insects doe or within the hive, yet without doubt he is the male-bee by whose masculine virtue bees are secretly conceived and bred.

Given that even most modern beekeepers have never seen the impregnation of virgin queen bees during their flight through the drone congregation, this was an incredible deduction.

———— ❋ ————

Reverend Charles Butler attempted to express through musical notes the sounds made by bees in the hive.

———— ❋ ————

Charles Butler's careful and intelligent observations enabled him to provide beekeepers with practical information on the bee year and how to manage their bees through the seasons. He knew when swarms were most likely, when and what forage was available, the best time to take the honey harvest, and how to care for the colonies in winter. He advised that a wet spring followed by a dry summer gave a good harvest, and he wrote of the danger to bees from wasps and bees from other colonies attempting to rob the honey stored in the comb. His advice was to use narrow entrances to the hives or, in the case of very docile or weak colonies, keep the hives closed until 10 a.m. or 11 a.m., thus helping the colony to defend its hive. He also gave detailed advice on difficult operations, such as finding and hiving swarms and the equipment required, as well as how to drive bees from a skep using the smoke from burning hay.

When a colony is about to swarm, certain workers, known as scout bees, will fly out to search for a suitable new home. On their return to the hive or swarm, these bees will enact a range of bee dance

Some of the Reverend Charles Butler's bee music.

(see pages 112–13) to indicate the distance and direction of the potential new home. Butler identified the scout bees, or "spies," as he called them, and also noted that, shortly after the dances took place, the colony would leave the hive, though he did not deduce that the dance was itself a guide (it would be another 300 years before this link was made—see page 112).

There are a host of ways in which *The Feminine Monarchie* helped to establish a new approach to the art of beekeeping, and some of Butler's observations are still of value today. For example, every modern beekeeper knows that their bees dislike certain body odors, such as perfume, sweat, or strong breath (particularly caused by garlic, onions, or alcohol); this is a basic yet important and practical point, and Butler was the first to identify it.

Robert Hooke, Jan Swammerdam, and the microscope

Historians tend to disagree about who invented the microscope, probably because this great leap in technology was achieved over a period of time. However, Dutchman Anthony van Leeuwenhoek (1632–1723) is generally credited with making the first really workable microscope and he, in turn, had been inspired by the work of the influential

English chemist and physicist Robert Hooke (1635–1703) and the Dutch naturalist Jan Swammerdam (1637–80).

Hooke is believed by many to be the greatest experimental scientist of the seventeenth century. Among his outstanding achievements were beautiful drawings that depicted what he saw through his compound microscope. In 1665 he published *Micrographia*, illustrated with a series of 83 meticulously engraved plates, including the anatomy of a flea, the sting of a bee, and cork cells. The book was a bestseller, described by contemporary Samuel Pepys in his famous *Diary* as "the most ingenious book that I ever read."

❖

Swammerdam's observations laid the foundations of our modern understanding of insect physiology and development.

❖

Jan Swammerdam took microscopy a stage further than just viewing. Using a simple microscope with only one lens that he had made himself, he carried out meticulous dissections under natural light. After detailed studies lasting almost seven years, his results were eventually published (posthumously) in 1737–8, and in English in 1759 as *The Book of Nature*.

Jan Swammerdam's interest in natural history had been piqued by his father's collection of curiosities, which contained many insects. Among his insect dissections, Jan carried out many on bees, using minute scissors, and knives so small that they had to be sharpened under the microscope; he then carefully drew his findings. One of his most famous illustrations shows the dissected ovaries of the queen bee, accompanied by a 1,000-word description; his discovery of the ovaries was the final proof that the "king" bee was female.

Swammerdam also dissected and wrote the first accurate description of the mouth-parts and sting

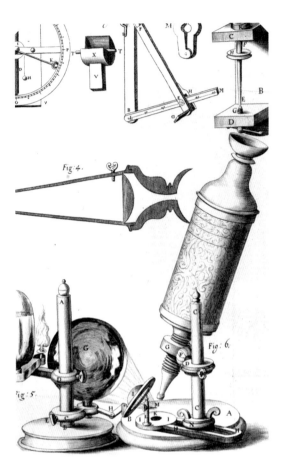

Figure from Micrographia *showing Robert Hooke's microscope.*

of the honeybee, as well as the bee brain, with the optic nerve projecting from the two compound eyes. He was the first to describe the anatomical differences of queen, worker, and male larvae, and their development. He also observed the way in which worker bees cluster around the queen on the comb, and he deduced that she must give off a strong scent (what we now know as the honeybee sex pheromone secretion, the so-called "queen substance," which maintains the queen's dominance in the colony). He demonstrated this by carrying her on a stick some distance from the hive and noting how the workers followed behind.

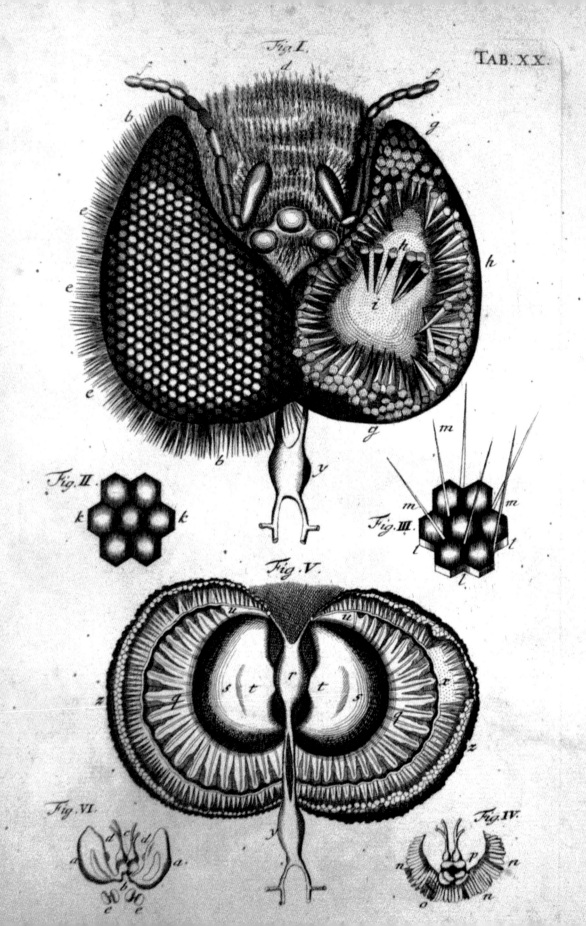

TAB. XX.

Fig. I.

Fig. II.

Fig. III.

Fig. V.

Fig. VI.

Fig. IV.

Swammerdam noted that colonies of all species of social insects consist of three distinct forms—sterile females (in honeybees these are called workers), one fertile female or queen (or sometimes several), and some males—and he tried to explain the existence of this type of social organization. He observed that the "common" bees (workers) did all the work of cleaning, feeding, and rearing the young (although he made the mistake of believing that workers had no ovaries). He also emphasized the importance of having only one queen in each colony and described the battles that arose if there were more than one.

His work represents a huge change in the understanding of insect anatomy and social behavior, and many natural-history illustrators maintain that his drawings cannot be improved upon.

The Royal Society and the learned establishments

In 1660 King Charles II became patron of the newly formed scientific academy, the Royal Society. Other European monarchs were also beginning to support similar learned scientific establishments, such as the Accademia dei Lincei (literally "Academy of the Lynx-eyed," to highlight the sharp observational skills needed in science) in Rome (1603) and the Académie Royale des Sciences in Paris (1666). Scientific research from all over Europe began to be published, making new information more available, and works about bees and beekeeping finally found a much wider audience. The eighteenth century also saw continuing improvement in international communication, helped by the translation of works and treatises into one language, English, thereby establishing English as the main language of modern science. With the wider dissemination of scientific works in general, and bee publications in particular, there was a surge of interest in the sci-

THE DISCOVERY OF WAX SECRETION

The origin of wax had proved particularly elusive for scientists. In around 340 BC, Aristotle in his *Historia Animalium* stated that bees collected wax from olive trees. (What he had actually seen was bees collecting the almost-white pollen of the olive flower.) Since that time no one had contradicted his findings, although wax scales had in fact been observed on the undersides of bees by a number of bee scientists.

However, in a paper he read to the Royal Society on February 23, 1792, eminent Scottish physiologist and surgeon John Hunter (1728–93)—whose sound anatomical knowledge established surgery as a scientific profession—described how he had observed bees in a glass hive producing wax from glands on the underside of their abdomens. He showed how wax was formed as an external secretion of oil.

ence of beekeeping among the educated classes and gentlemen-farmers.

François Huber's new observations

The achievements of Swiss natural historian François Huber (1750–1831), a man who began losing his sight from his late teens yet became a brilliant naturalist and researcher, are extraordinary. Despite becoming completely blind, he laid the foundations of an understanding of the honeybee life cycle, something that had eluded previous naturalists.

OPPOSITE: *Jan Swammerdam's drawing of the compound eye of the bee.*

Huber's wife Maria read all the relevant literature out loud to him and acted as his record-taker while his manservant/assistant François Burnens, though a man with little formal education, carried out all Huber's experiments.

In 1792, Huber published many of his findings in *Nouvelles observations sur les abeilles*, translated in 1806 as *New Observations on the Natural History of Bees*. A great work, its subject matter ranged from bee antennae to hive construction; it explains numerous previously misunderstood or mis-interpreted aspects of apiculture, such as how bees reproduce. After studying bees in the glass observation hives designed by the French naturalist René-Antoine Ferchault de Réaumur (1683–1757; see page 63), Huber became certain that any bees living in such circumstances would not behave naturally. So he invented his own hive, known as the "leaf hive," that consisted of a series of 12 small wooden boxes each 12 in (30 cm) square and 12½ in (32 cm)

ABOVE: *François Huber, who laid the foundations for our understanding of the honeybee life cycle.*

BELOW: *François Huber's leaf hive.*

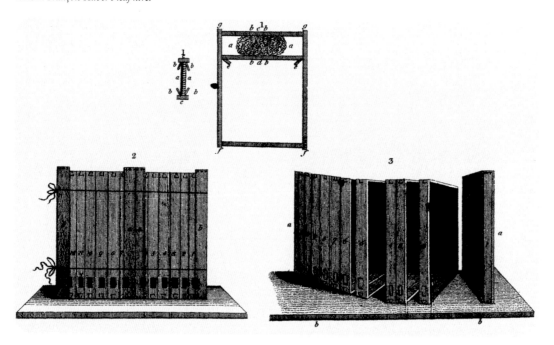

thick, hinged together along one vertical side, so that any adjacent two boxes could be parted like the pages of a book (see opposite). A piece of comb was then affixed to the upper part of each frame, bees were introduced, and the whole thing was closed up until the bees had secured each comb onto a frame and then built the comb into the frame's lower part. Once the bees were settled, it was easy to open each pair of leaves in turn and inspect both surfaces.

In the twentieth century, it was found that, unknown to Huber, many of his findings had been preceded some twenty years earlier, in 1771, by Slovenia's pioneer apiculturist and celebrated beekeeper Anton Janscha (1734–73).

Using these leaf hives, Huber and Burnens carried out a series of elegant experiments. They paid particular attention to the queen's ability to reproduce, and eventually managed to disprove Jan Swammerdam's theory that insemination was the result of the strong-smelling aura, *aura feminalis*, that emanated from the drones and penetrated the body of the queen. In one instance, they removed the drones from the hive into an adjacent box with holes in it that allowed the aura to escape, but the queen, known to be a virgin, remained so; no eggs were laid. Another approach involved leaving the drones in the hive but pre-venting the queen from leaving; still there were no eggs. Eventually Huber and Burnens came to the conclusion that the mating process had to take place outside the hive. Following hours and hours of observation, Burnens noticed a virgin queen leaving the hive; on her return she bore a thick whitish substance and the severed sexual organs of a drone in her lower body. Two days later she began laying eggs. Unfortunately—and not for want of trying—he never observed the actual mating.

INNOVATIONS IN BEEHIVE DESIGN

As knowledge about beekeeping spread, so did the search for more sophisticated forms of beehive. While skeps were still widely used, there was continual experimentation with different types of straw for skep-making and with different types of wooden hive. The chief aim was to create a hive that prevented the keeper from having to kill the bees in order to harvest the honey.

Thomas Wildman's multi-layered skep

One experimenter, the Englishman Thomas Wildman (1734–81), in his book *A Treatise on the Management of Bees* (1768), criticized the prevailing custom for destroying bees:

> Were we to kill a hen for her egg, the cow for her milk, or the sheep for the fleece it bears, everyone would instantly see how much we should act contrary to our own interests; and yet this is practised every year, in our inhuman and impolitic slaughter of the bees.

In the same book, Wildman introduced an innovative new system for keeping bees. He used flat-topped, tub-shaped straw skeps with five wooden boards nailed across the top of each. He started with a strong swarm and put it into a single skep; when the bees had filled it with honeycomb, he placed a second one beneath it and so on until the queen was lying in the bottom of four skeps and the upper ones were filled with honey. At this point, he suggested, the top skep could be removed and banged on its side to flush out any bees; after this, the full honeycomb could be pulled out and the empty skep placed back under the others. Wildman was tapping into what bees do naturally—building combs steadily downwards, as they would in the wild, and storing honey in the upper parts of the hive.

Wildman and his nephew Daniel were actually better known outside beekeeping circles for their

Exhibitions, such as this "bee tent," also helped raise interest in, and knowledge of, the honeybee.

showmanship. Thomas, in particular, became popular in the 1760s for his demonstrations of "mastery over bees" in which he encouraged the bees to cluster on his arms, chest, head, and chin in an extraordinary bee beard. His circus act also involved getting the bees to follow him around the ring, while he rode ahead on horseback; for his finale the swarm would land on his shoulder. In order to do all this, Wildman had to mimic the conditions of a natural swarm: he fed the bees sugar syrup, after which he cupped the queen in his hands. The bees, calmed by the syrup—just as if they had eaten their own honey in preparation for swarming—then followed the queen's pheromone scent. Wherever Wildman and the queen went, the bees followed.

Daniel was a showman like his uncle and also wrote a book: *A Complete Guide for the Management of Bees Throughout the Year* (1780). The crowds loved the Wildmans' shows, which helped to make beekeeping fashionable among the leisure classes. Sales of fancy new hives—including expensive mahogany models with glass partitions that Wildman sold in his bee supplies shop in Holborn, London, were boosted.

Despite his critics, Thomas Wildman was a skilled beekeeper who was able to hire out his beekeeping services and take care of others' apiaries.

Movable-frame hives

The Langstroth hive

Previously, bar hives, similar to Wildman's (see pages 59–60), and box hives, in which bees attached their hive combs directly to the roof and walls, had been used. But neither allowed for easy collection and reuse. It is probable that the first example of a movable-frame hive was designed by Englishman Major William Augustus Munn, the author of *A Description of the bar-and-frame hive*, published in 1844. However, it is the American-born Reverend Lorenzo L. Langstroth (1810–95) who is now credited with the invention of the fully functioning movable-frame hive. It was he who recognized the concept of "bee space," the vital component in his and all subsequent hive design.

Langstroth did not take up beekeeping until he was nearly forty. He used bar hives, but became frustrated when he found the covers of his box hives became stuck down with propolis (the resinous substance bees use to seal up holes and gaps in hives or rock cavities). He cut rabbet out of the boxes, which enabled him to drop the hive bars down to about $^3/_8$ in (9 mm) below the cover. Having solved this problem, he then realized that, by manipulating the amount of space in the interior of the hive itself, he could improve its overall design.

His new hive consisted of a box with two thin strips of wood, or runners, attached inside the box opposite each other at the top, and from which the frames were hung. The critical aspect of his design was the space between the edges of the frames and the walls and floor of the box—an opening wide enough for a bee to pass through and hence termed the "bee space." Langstroth initially allowed a space of $^1/_2$ in (12.5 mm). (Subsequently, it has been discovered that, if the space is greater than $^3/_8$ in or 9 mm, the bees will build bridges of wax—brace comb—joining the frames with the insides of the box, thereby making it impossible for the beekeeper to lift out individual frames of honey-filled comb without breaking the comb and spilling the honey. The approximate limits of the bee space are between $^1/_4$ and $^3/_8$ in (6 and 9 mm)—the variations reflect the range of sizes of different races of bees; see Part Two: Understanding the Honeybee.)

In January 1852, Langstroth applied for a patent for the design, and in 1853 he gave up his teaching job and made a set of frames, fitting them with a small piece of comb from which the bees could build more comb. In his book *The Hive and the Honey-bee*, published in 1853, he says that, after discovering the bee space, he felt like Archimedes and wanted to run down the street shouting, "Eureka!"

Langstroth built over 100 movable-frame hives and quickly sold them all. Today 75 percent of all the hives sold throughout the world are based on Langstroth's hive design.

Further innovations

In 1863 Frenchman Charles Dadant (1817–1902) and his family emigrated to America and settled in Illinois. He had been a beekeeper in France, and he believed that the brood box Langstroth had designed was too small. So he designed a larger one, the Dadant hive, which became popular in parts of Europe and Russia. He also established a factory, the first of its kind, to manufacture beekeeping tools.

A variation of Langstroth's hive was also developed by T. W. Woodbury of Exeter in England. It had movable frames, but incorporated a layer of straw for insulation in the sides of the hive boxes and the cover.

All these designs over a half-century period made up a breakthrough in inventions and techniques that made life easier for beekeepers, allowed bees to live through the honey harvest, and hugely increased the honey yield from each hive.

In his book *Langstroth on the Honey-bee* (1860), the "father of American beekeeping" acknowledges his debt to François Huber:

The use of the Huber hive had satisfied me, that with proper precautions the combs might be removed without enraging the bees, and that these insects were capable of being tamed to a surprising degree. Without knowledge of these facts, I should have regarded a hive permitting the removal of the combs, as quite too dangerous for practical use.

observations via glass hives in a garden next to the Paris Observatory. Meanwhile, at around the same time, the Anglo-Irish chemist and physicist Robert Boyle (1627–91) apparently had in his study "a Transparent Hive, whence there was a free passage into a neighbouring garden."

From the late seventeenth century onwards European scientists were looking for more effective ways of observing bee behavior, and glass hives became popular in a range of experimental designs that included square or octagonal boxes, bell jars, and types of skeps that included little glass windows. Observation hives also became a popular "must have" among the aristocracy, resulting in the emergence of many unusual types. Studying bees was on the increase, and it was not just specialists who were getting involved.

Observation hives

Observation hives have a long history. Aristotle reputedly made one of glass (see page 50). In his *Historia Naturalis*, Pliny the Elder mentions a form of transparent hive, in which parts of the walls are believed to have been made from thin pieces of horn or mica inserted into the sides or top of woven baskets or bark hives. From watching the bees through the see-through sections, Pliny was able to describe bee development in the cell.

The earliest-known hive incorporating a flat pane of glass is described in *Elysium Britannicum* by the English diarist and writer John Evelyn (1620–1706). This hive was octagonal and made of wood, with a window of glass measuring 4 by 6 in (10 by 15 cm) set into the panel opposite the door, and was shown to Evelyn in 1654 by Dr. Wilkins at Wadham College, Oxford.

Later, in about 1680, Giacomo Maraldi (1665–1729), an Italian-French astronomer and mathematician, also wrote an account of his bee

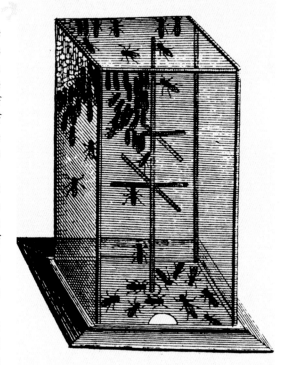

Réamur's glass hive (1740) for observations and experiments.

IMPORTANT BEEKEEPING INVENTIONS AND NEW EQUIPMENT IN THE LATER NINETEENTH CENTURY

JAN DZIERZON (Poland) 1848: Inventor of the first movable-frame side-opening hive.

LORENZO L. LANGSTROTH (USA) 1852: Movable-frame hive with frames that didn't touch each other, enabling the frames to be easily removed.

JAN MEHRING (Holland) 1857: Manufacture of wax foundation (fitted to the removable frames inside a hive to encourage bees to build their honeycomb) by pressing pure beeswax between metal rollers.

ABBÉ COLLIN (France) 1865: Perforated queen excluders that confined the queen to the brood chamber (lower hive box) and enabled the worker bees to enter the supers (upper hive boxes).

FRANCESCA DE HRUSCHKA (Italy) 1865: Development of a tangential centrifugal honey extractor.

A. GRIM (USA) 1870: Practice of large-scale transportation of queens from Europe to the USA.

FREDERIC WEISS (USA) 1873: The roller foundation mill.

T. W. COWAN (UK) 1875: Development of a self-reversing, radial centrifugal honey extractor.

MOSES QUINBY (USA) 1875: Development of a smoker with improved bellows.

E. C. PORTER (USA) 1891: Effective bee escape to enable bees to be cleared from honey supers (upper-hive boxes) before extraction.

E. B. WEED (USA) 1892: Perfected production of beeswax sheets using long rollers.

In his *Mémoires pour servir à l'histoire des insectes* (1734–42), French entomologist René Antoine Ferchault de Réaumur (1683–1757) discussed the use of hives that held several parallel combs, as well as simple square hives with glass sides, in which cross-sticks could be inserted to support the combs. He included pictures and descriptions of hives—tiered boxes with large glass panels protected by hinged wooden doors.

Particularly useful for observation was the vertical glass box containing a single frame of honeycomb and removable shutters on either side, which were necessary to reduce heat loss. The box also had to be big enough for the bees to move easily around but small enough to prevent them from covering the glass sides with wax as if these were neighboring frames. In his *Traité . . . sur les abeilles* (1810), Charles Romain Féburier (1764–1800?) describes the required dimensions for such a hive: 18 to 24 in (45 to 60 cm) high and 12 to 18 in (30 to 45 cm) wide, with a crossbar of about 2 in (50 mm) by 1½ in (34 mm) along the top, from which the bees would hang their comb.

It is rumored that Reverend Langstroth based the design for his hive on a discarded champagne case.

Modern observation hives differ little from the simpler, early designs: a wooden frame holding two vertical glass sides, each protected with a door that can be opened for observation, and an access, usually via a tube to the outside, through which bees can come and go.

The Global Spread of the European Honeybee

AROUND 150 MILLION years ago, as a result of continental drift or tectonic plate movement, the plates that formed what are now North America, the Pacific Islands, Central and South America, and Australasia became detached from the rest of the landmass. This happened before the evolution of honeybees, so the European honeybee, *Apis mellifera*, and its

In the seventeenth century, many English ships attempted to cross to the New World. Some cargos included European honeybees.

subspecies were never native in those areas of the world. It was only from the sixteenth century onwards, during the period of colonization of other continents by the nations of Western Europe, that the honeybee was established in the New World.

TRANSPORTING THE EUROPEAN HONEYBEE TO THE NEW WORLD

North America

By the seventeenth century, the race was on to develop and exploit the Americas. In 1606, the Virginia Company of London was charged with the provision of supplies and ships for a voyage to Virginia. In 1609, a fleet of their ships carrying settlers, their cattle, apple trees, and bees attempted to cross the Atlantic. They failed when a hurricane sank one ship and sent the flagship more than 600 miles off course. Records of later crossings also show the Virginia Company transporting European honeybees across the Atlantic. For example, on December 5, 1622, the council of the Virginia Company in London wrote to the governor in Virginia to advise that ships had been dispatched carrying "divers sorte of seed, and fruit trees, as also pidgeons, connies [rabbits], Peacock maistives [mastiffs], and beehives." For anyone, let alone northern European honeybees, a journey of some two months across the Atlantic by sailing ship was potentially hazardous.

However, some crossings were successful, and Europeans from a number of countries made their homes in the New World. On arrival, they found that the Native Americans had cultivated farmland and forest, using slash-and-burn techniques to clear the ground. The burned remains of these large trees proved excellent for swarming for the newly arrived European honeybees, and it helped them to disperse quickly across the newly settled lands.

Established in England in the mid-eighteenth century, the Shakers were accomplished fruit growers and recognized the importance of bees in ensuring a good crop. As settlers in the New World, the Shakers took their bees with them and each homestead had its own round bee house, often two stories high and containing many hives. The philosophy of a fulfilling life of work on the land was expressed in a hymn, claimed by Shaker Anna White to have been given to her in a vision:

Like the little busy bee,
I'll gather sweets continually
From the life-giving lovely flowers,
Which beautify Zion's bowers.
No idle drone within her hive,
Will ever prosper, ever thrive.
The seeds of industry I'll sow,
That I may reap wherever I go.

As European immigration continued, many Protestant religious sects, originally from England and Germany, moved south to West Virginia (1622), North Carolina (1697), and Kentucky (1760), and west from Pennsylvania into Ohio (1754), Indiana (1793), and Illinois (1800). With them went their bees. In addition, feral bees that had escaped from the hives spread to the northern

Up until 2010, it was illegal to keep beehives in New York City. Chicago, however, has a city-owned beehive on the roof of City Hall.

and central states and to Mississippi by 1770 and Florida by 1773. When Spain ceded Florida to the English and the Spanish residents moved to Cuba, their bees went with them.

By 1776 European honeybees had reached eastern Canada, but it wasn't until 1858 that records show them arriving in western Canada. Similarly, it was 1856 before honeybees reached the west American coast, taken by John Harbison of Pennsylvania to an area now called Harbison Canyon in California.

It took another 70 years for honeybees to reach the furthest outposts of America—beekeeper J. N. McCain finally managed to over-winter seven colonies in Alaska in 1927.

THE BEES THAT SAVED AMERICA

Out walking one day during the American War of Independence (1775–82), a young Quaker girl, Charity Crabtree, is said to have come across a wounded soldier, who asked her to ride his horse to George Washington to advise him that the English Lieutenant General Lord Cornwallis and his army were planning an imminent attack.

Charity galloped off, closely followed by British Redcoats. To prevent the enemy troops closing in on her, Charity threw down the bee skeps that she was holding, and her bees attacked the Redcoats. Washington is supposed to have subsequently remarked: "It was the cackling geese that saved Rome, but it is the bees that saved America."

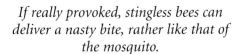

If really provoked, stingless bees can deliver a nasty bite, rather like that of the mosquito.

Central and South America

In Central America, particularly the Yucatán peninsula, the native stingless bee, *Melipona beecheii*, had been kept in traditional hives for more than 2,000 years (see page 31), though most of the South American peoples remained hunter–gatherers, obtaining their honey from wild bees.

European honeybees reached Mexico from Europe, possibly as early as the 1500s (there were later additional introductions from Texas and California). However, European bees didn't arrive in Central America until the nineteenth century, being introduced into Costa Rica and Guatemala in 1830, and into El Salvador in 1855. The date of their arrival in Nicaragua is less certain, probably 1850–1900, and they only reached Belize in 1957 (brought by veterinarian John Robbins) and Panama in 1960. Beekeeping became concentrated in the upland areas of Costa Rica, Guatemala, and Honduras, where the majority of people lived and where the bees thrived because it was cooler.

The earliest accounts of honeybees in South America are based more on deduction than record. Historical events suggest that the first European bees were brought by Spanish adventurers between 1520 and 1540, although alternative possibilities are that a chance swarm came across the Atlantic on board ship or as an introduction from Cuba. More definite introductions include Chile in 1840 (with German settlers), from where they spread to Paraguay around 1900. Feral bees then traveled to the border of Argentina.

OPPOSITE: *A political cartoon targeting discrimination: bees, normally the symbol of hard work, are shown representing the enemy of industry.*

SWAT THEM ALL!

Australian aborigines chase bees on a bee hunt.

Australia

In March 1810, the *Anne* arrived in Sydney harbor, along with Yorkshire-born Reverend Samuel Marsden (1765–1838), 197 convicts, Maori chief Ruatara, and several hives of Marsden's honeybees. The bees could have been from England but they could also have been acquired from a port en route. And, though they arrived safely, the bees did not survive for long.

The next recorded attempt to import bees was made in 1822 when Captain Wallace brought twelve hives with him from England on the convict ship *Isabella*. This time, the bees survived.

Australia remains the only continent free of the devastating Varroa destructor mite (see page 240).

AUSTRALIA'S NATIVE BEES

Australia has more than 1,500 species of native stingless bees, of which many are solitary (see page 81). Others are social bees of the genus Trigona, small and dark-colored, often measuring as little as ¼ in (4 mm) from the head to the base of the abdomen. These bees produce wax and a thin honey; the Aboriginal peoples call them sugar-bag bees, and collect their honey for bush food.

In most countries, only men harvest the wild honey but in Australia the women do this. Because bush fires have destroyed many of the trees, native bee nests can be found in low bushes or even in disused termite nests.

Once in Australia, honeybees spread by swarming; the first wild nests were noted in 1838. Their spread may have been limited, however, by the lack of water, low rainfall, and frequent bush fires that resulted in limited bee forage. By 1846 hived honeybees had also been taken to the south, then to Queensland (1851) and Western Australia (1881). But it was not until the 1930s that they became more widespread across the continent.

New Zealand

The first successful establishment of European bees in New Zealand is believed to have been by Mary Bumby who, traveling with her missionary brother, took two baskets of black bees from England via Sydney to the Wesleyan Mission Station at Mangungu, Hokianga Harbor, on March 19, 1839.

Three years later, the Reverend William Charles Cotton (1813–79) took more colonies from England to New Zealand. These apparently survived the crossing. In *My Bee Book* (1842), Cotton describes his plans:

> I will not get Bees from India—nor Bees from South America—nor Bees from New Holland, but carry them direct from England, sixteen thousands miles over the sea. How is this to be done?—By putting them to sleep, by keeping them at a low temperature, by burying them, and keeping them dry . . . At all events, I will try: I will take care that my Bees shall send word to their relations in England how they fare . . . I shall be fully satisfied if they arrive safely at New Zealand in five months, or even a little more . . .

Using as his example how North American salmon were shipped to Calcutta—when the dead fish were well packed in ice, well insulated, and did not deteriorate—Cotton created a system by which to accomplish the same with his live bees.

His hives were packed in a hogshead barrel, which was divided across the middle at its widest part and lined with felt. Three skeps containing winter bees (bees produced in the autumn, which do not forage during the cold months) were wrapped in canvas and packed into the upper half, surrounded by dry cinders and provided with a tube through the top for ventilation. The lower half was filled with ice from Scottish lochs, and the water was drawn off as it melted.

TRANSPORTING THE EUROPEAN HONEYBEE WITHIN EUROPE

The northern forests of Siberia's Altai mountains were suitable for honeybees, and honey-hunting was widely practiced, but in 1776 descendants of the Old Believers, a branch of the Russian Orthodox Church who had been banished to Siberia after a religious schism in 1685, asked for some hive bees to be sent to them so they could produce honey.

Unfortunately, there were no survivors from the 30 hives that came to them from the Urals. A further consignment of 24 hives was dispatched in 1784, this time from Kiev. These hives were carefully packed in felt and straw and loaded on to sleighs for the four-month journey. These bees died soon after arrival. A third dispatch finally survived.

In 1802, under Alexander II, the government offered parcels of land to the people of Siberia on condition that they kept bees for at least five years. Beekeeping not only flourished but also steadily spread—from Siberia, across Russia, reaching the Ukraine in 1904 and Mongolia by 1959.

MIGRATORY BEEKEEPING

The seasonal movement of stock to fresh pastures or warmer sites is as old as mankind's domestication of animals. Bees are no exception, and transporting them further afield has the advantage of giving them access to plants with different flowering seasons, thus ensuring continuity of forage throughout the summer months.

The earliest record of migratory beekeeping dates back to Lower Egypt around 250 BC (see page 30). Hives were carried either on the beekeeper's back or in hammocks slung between two poles, with a bearer at either end. For longer distances, pack animals, such as mules and camels, were put into service, often attached to carts and wagons. More recently, in all parts of the world,

Pack animals are used to move hives if suitable tracks exist.

mechanized transport and improved road systems have allowed hives of bees to be moved over much greater distances, quickly, reliably, and in larger numbers.

In the second half of the twentieth century, significant changes in farming practices and the plowing of previously wild areas and grassland have resulted in the loss of native flora in many countries. Bees have therefore become dependent upon a range of agricultural crops for their main source of forage: the beekeeper gets the honey; the grower gets the crops pollinated, which is essential if they are to produce fruits and seeds; and the bees have a few weeks when nectar and pollen are just outside the hive door and easy to collect. The bees are then moved from field to

Carrying a hive up into the mountains in Greece.

field as the different crops come into flower, meaning that migratory beekeeping forms the basis of modern commercial beekeeping—in the UK, France, Spain, the USA, Australia, South Africa, and Argentina—as well as being used to some extent elsewhere.

Examples of migratory beekeeping

Europe

From early on, European beekeepers have moved their bees to where the best flowering was to extend the honey-producing season. Early spring in the UK brings the first fruit blossoms of pears, plums, and apples. In arable areas, hives will be moved on to oil-seed rape (canola) as soon as the first flowers appear, followed by field beans and borage in limited local areas. Large-scale growers of raspberries and strawberries need migratory bees—if soft fruit is not well pollinated, it will grow misshapen and unsaleable. Hives can also be taken into greenhouses for specialist crops, such as eggplants.

The final crop of the season—and a risky one—is heather. In a good year, it is worth the effort as heather honey is much prized and can command a higher price than other UK honeys. It is less worthwhile when the weather is wet and cold: the bees cannot fly and, when they can leave the hive, they may actually come home poorer, in terms of honey, than before they went to the heather site.

In France there is a similar sequence in Provence. Here, the season starts with almonds and rosemary in February and March, followed by apples, apricots, and peaches in April and May. And, from the end of June to the beginning of August, there is lavender, which is grown for its perfumed oil in northwest Provence. Here, there are both resident and migratory beekeepers, who cover the hillsides and the edges of the lavender fields with rows of hives.

Central Asia

In Uzbekistan, Tajikistan, and the southern parts of Kazakhstan in Central Asia, *Apis mellifera carnica*, a subspecies of the European honeybee, is used in migratory beekeeping as it can tolerate temperature extremes. Some beekeepers run between 150 and 300 beehives, which are kept permanently in wagons designed for migratory beekeeping. The wagons are also home to the beekeepers and their families, and carry facilities for the extraction and processing of the honey.

Depending on the climate of the country, the foraging season will start in either April or May, when the nectar-rich flowers start to bloom. The wagons will move to the best wildflower places, such as the Kyzyl Kum desert (in parts of Uzbekistan and south Kazakhstan), the mountains to the north of Samarkand (eastern Uzbekistan), and the Fergana valley (stretching across eastern Uzbekistan, Kyrgyzstan, and Tajikistan). The countries of Central Asia are also famous for their cotton growing, and cotton is a significant forage crop for the migratory bees, providing around 50 percent of the honey crop over a long season from July to September.

Migratory bee wagons in Kyrgyzstan.

America

In 1884, Nephi Ephraim Miller, a young man from Utah, traded a bag of oats for a colony of bees. Before long he gave up his job to become a full-time beekeeper and sought a way of making money from beeswax—if he could rid the wax of its impurities. In 1907, Miller went to California to improve his beekeeping skills. Immediately after his arrival, he realized that beekeeping in the balmy climate of a California orange grove in winter could be much more lucrative than in cold Utah, if only he could find a way to get his bees there. The Union Pacific Railroad proved to be the solution; it could transport Miller's bees west for the winter and back to Utah for the summer crops. This was the beginning of commercial beekeeping, on a large scale, in the United States.

For the rest of the twentieth century, migratory beekeeping remained the most profitable way to make money from the industry in the USA, though road rather than rail has become the usual way to transport hives.

The orchards in the USA are immense and the fruit growers of America now take the role of honeybees as pollinators very seriously. They recognize the bees' crucial role in maximizing the crop, and they are willing to pay large fees to beekeepers. The figures are staggering: every spring up to 40 billion bees are trucked across America to the almond orchards of California, in the first leg of a journey that will take them on to Florida for citrus fruit, north for apples and cherries, and east to Maine for the blueberries for which the United States is famed.

Beekeeping becomes an industry: a nineteenth-century honey farm in America.

"Telling the Bees" *by Charles Napier Hemy (1841–1917).*

THREATS TO THE HONEYBEE

The movement of bees and honey around the world has brought many advantages, with improved strains of bees, an exchange of genetic material, and valuable trading opportunities. But it has also brought dangers.

Threats to native wildlife

In the early days of the exportation of the European honeybee to new lands by travelers and settlers, the main threat from the bees was to the native populations of flora and fauna. And this isn't just a historical problem. Australia, with its unique wildlife, has been particularly affected in this regard, although measures are finally being put in place to combat this. In the late 1990s, there

was concern that the introduced honeybees were competing for forage with native nectar-collecting insects (especially stingless bees) and honey-eating birds, of which there are more than a hundred species in Australia. There was fear, too, of competition for nesting holes in trees with other hole-nesters: native possums, bats, and birds, including owls, kingfishers, tree creepers, and the iconic kookaburra.

There is now legislation against importing honeybees into certain areas of the country. There has also been a proposal to remove wild honeybee nests from the national parks and other wildlife conservation areas altogether, which would entail a honeybee eradication program in designated areas.

Disease

Moving bees around the world inevitably carries a greater threat of disease. A perfect example of this is the spread of the bee mite *Varroa destructor* (see page 240). Russian soldiers, who took their bee-hives with them wherever they were posted, returned home from the east of Russia to the west of the country, carrying colonies containing the mites. Once the mites were across the natural mountain barrier, they were able to embark upon their unstoppable progress westwards through Europe, reaching as far as the UK in the early 1990s.

These diseases show how poor bee husbandry and the wrong sort of hive are exacerbated by moving bees around, and illustrate the dangers faced by the global bee industry and population.

American Foulbrood

American foulbrood (AFB) (known as American to differentiate it from European foulbrood, or EFB) is a contagious, bacterial disease of the bee brood (egg, larvae, and pupae). It can strike anywhere, but was introduced early on into New Zealand via imported bee stocks. In the 1890s, beekeeper Isaac Hopkins campaigned for legislation to attempt to control the spread of AFB in particular and bee diseases in general. He became apiarist to the New Zealand government and his campaigning led to the introduction of the Apiaries Act, passed in 1906, which was the first of its kind in the world. The act made it illegal to keep bees in fixed-frame hives—which make it difficult to examine the combs, to identify AFB's presence and prevent its spread—and put in place measures to control the disease. This legislation was supported by the timely invention of Langstroth's movable-frame hive (see page 61), which forever changed the nature of beekeeping by aiding the early identification of and prevention or elimination of AFB and other bee diseases.

Isle of Wight disease

In 1906, beekeepers in the Isle of Wight, about 5 miles off the south coast of the English mainland, observed that their bees were crawling on the ground around their hives and dying so fast that whole colonies were wiped out at the height of summer. The disease recurred at least three times from 1906 to 1919.

The April weather of 1906 was wonderful: hot enough to draw crowds of visitors to British beaches. It was followed by a devastating May with killing frosts and chilling temperatures as low as 23°F (5°C), even in London. The flowers of the fruit orchards were critically damaged, as were those of early crops and wild plants. Honeybees were not able to leave their hives and, because the hives were full of young, spring bees, conditions became over-crowded and stressful for bee communities.

A period of confinement in spring may cause dysentery and possibly paralysis virus in bees. Their inability to leave the hive because of bad weather means that they are forced to defecate on the comb, and this can spread the gut parasite *Nosema apis*, if it is present. Also, continuous close proximity in a tightly packed colony helps to spread *Acarapis woodii*, the tracheal mite, to which young bees are especially susceptible.

In that early part of the twentieth century there were around one million hives of honeybees in Britain, four times more than there are just a hundred years later. Beekeeping practices of the time were less rigorous: hives were often kept close together in the apiaries, so that bees, returning from foraging, could return to the wrong hive by mistake, carrying disease with them; beekeepers commonly—and using recommended practice—exchanged combs and bees among colonies, thus spreading any diseases present. Many beekeepers had homemade recipes for disease control, but

OPPOSITE: *A Russian Tartar beekeeper of the eighteenth century.*

these, and the commercial remedies available at the time, were more likely to kill rather than cure.

Investigations in 1919 revealed the existence of the tracheal mite in colonies on the island and this was identified at the time as the main cause of Isle

---

Isle of Wight disease had wiped out most of the honeybees on the island by 1907; it then spread to the mainland.

---

of Wight disease. But the bees' crawling behavior strongly points to infection with Chronic Bee Paralysis Virus (CBPV). It is also now known that Nosema is associated with a number of viruses (see page 149). So, while bees were being attacked by many of their natural enemies between 1906 and 1919, they were also suffering the effects of bad weather and bad husbandry practices, making the situation much worse.

The stress of migration

Migratory beekeeping, though successful, can be very stressful for bees, taking its toll on bee health and making colonies more susceptible to disease. This applies even more in the twenty-first century than historically as it forms the basis of today's commercial beekeeping, and is particularly true of the USA, where transportation over long distances is normal.

Hard-working bees need all the nectar they can get to maintain high energy levels, and the life of a worker bee during the foraging season is very short, around 4–6 weeks. Every exhausted, dying bee that has been pollinating the crops and does not return to the hive has to be replaced by a young, new forager, so the energy and protein requirements of the colony to support itself are high. As a result, much of the honey and pollen may be consumed for the colony's survival and there may be little honey to harvest at the end of the season.

MAINTAINING A HISTORIC RELATIONSHIP TODAY

The global spread of the European honeybee has brought huge benefits to commercial beekeeping, increasing honey production throughout the world, with concomitant crop pollination underpinning the agricultural economy. The downside is that this, plus bee migration on a large scale, means that disease is likely to spread over a much wider area than previously.

Disease is a major threat to bees today. In the USA, for example, the mysterious condition known as Colony Collapse Disorder (CCD—see page 246), destroyed 30 percent of commercial honeybees in a single year. On the other hand, we now have at our disposal a far more sophisticated, science-based knowledge of bees and beekeeping, acquired over the last few centuries, plus a better understanding of the causes of disease. Disease has threatened bees in the past and the threat of disease in the future can never be eradicated; it will remain an ongoing battle, just as it is with human disease. With care and persistence it is hoped that this current threat will be overcome and the historic relationship that has persisted between bees and humans for millennia will endure.

OPPOSITE: *A lavender field in France, used by both local and migratory beekeepers for forage for their honeybees.*

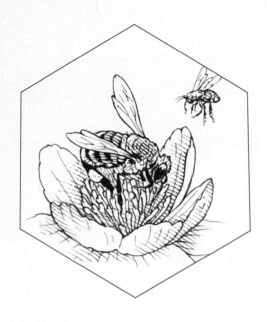

PART TWO

UNDERSTANDING THE HONEYBEE

A Special Kind of Insect

Bee Origins

A Selection of Honeybee Races Across the World

Occupants of the Hive

From Egg to Adult

Social Life in the Hive

Bee Communication

Swarming

Winter Survival

Man's Best Friend

Honey, Wax, and Propolis

When Bees Sting

Bee Habitats

A Bee-Friendly Garden

Threats to Bees

A Special Kind of Insect

ESPITE MANY THOUSANDS OF YEARS of concerted exploitation by humans, there is still a widespread and deep-seated, but completely mistaken, belief that it is the black and yellow striped and furry bumblebees that make honey in the hive. An increasingly urban human society is losing touch with the rural staple of flowery meadows and buzzing insects, and is regularly being bombarded with misleading images in advertising, films, and television.

Honeybees are actually genuinely nondescript, mid-brown, medium-sized insects, but nature further clouds the issue by having evolved a whole series of lookalike insects. Some, like wasps, are also armed; others, such as hover flies, benefit from mimicking dangerous, stinging creatures, when in reality they are unarmed and entirely harmless.

Bees, wasps, and ants are part of one of the largest and most important groups of insects: the order Hymenoptera, meaning "membrane-winged," from the Greek words *hymen* (membrane) and *pteron* (wing). More than 100,000 different hymenopteron species are known, with the possibility of many thousands more in the deep tropics.

This large group of insects is extremely diverse. It includes sawflies (whose toothed egg-laying tube is used to saw slits in plant tissues), gall wasps (which make strangely shaped growths on leaves and stems in which their larvae feed), and ichneumons (which parasitize caterpillars and other insects by laying their eggs inside the living host). Within the Hymenoptera, bees, wasps, and ants form the division Aculeata (meaning "pointed"). They are popularly lumped together as "stinging" insects.

Ants, being mainly small, wingless, and subterranean, are relatively easily recognized, but bees and wasps are often confused. There is one very straightforward difference.

Although wasps visit flowers for nectar, and fallen fruit for sweet juices, they are primarily carnivores, killing and eating small insect prey. More importantly, it is solely this prey that they feed to their larvae.

Bees are purely herbivorous, feeding only on pollen, nectar, sap, and other plant secretions; thus, their larvae are entirely vegetarian. Unfortunately it can be difficult to identify this behavior correctly in individuals busy on foraging trips.

TOP: *Wasps are primarily carnivores.*

BOTTOM: *A honeybee feeding; bees are purely herbivorous.*

As an indication of just how closely related bees and wasps are, there is no very easily observed structural distinction between them. A few generalities can be suggested (see page 83), but they broadly overlap in size, shape, color, and pattern.

A common mistake is to confuse honeybees and social wasps. Social wasps (hornets, *Vespa* species, and yellow jackets, *Vespula* and *Dolichovespula* species) live in large colonies underground or in attics and trees, but the bright yellow, orange, or brown and black markings of these insects are usually strikingly different from the more somber orange, browns, and black of honeybees.

Generally wasps are not hairy; if they have a few bristles, these appear as simple spines under the microscope. However, all bees have hair, some thicker than others. Honeybees can appear shiny and smooth in some lights, but close examination shows that they, too, have hairs in bands across the abdomen, and sticking out from the thorax.

The key difference is that no matter how many or how few hairs a bee possesses, under a lens they appear feathery, branched with tiny plumes sprouting along the shaft, unlike the wasps' sleek tubular bristles. These plumes are thought to have evolved because pollen sticks to them better than to simple unbranched hairs.

There are at least 25,000 known species of bee. They range in size from the tiny tropical stingless bees (*Trigona* and *Melipona* species), which are only about 1/16 in (1½ mm) long, to giant carpenter bees (*Xylocopa* species), which have bodies up to about 1½ in (40 mm) in length. At about ½ in (12-18 mm) long, honeybees lie in the middle of this huge range.

---- ✳ ----

Honeybees are brightly colored to warn potential predators (and honey thieves) that they have a stinger to defend themselves.

---- ✳ ----

TOP: *Bumblebees are densely furry all over.*
BOTTOM: *Honeybees have hairs in bands across their abdomens and sticking out from the thorax.*

Honeybees should not be mistaken for their much furrier relatives, the bumblebees. It is with the vast number of solitary bees that most doubt lies when identifying a honeybee. Solitary bees, however, have not evolved the complex colony behavior of honeybees. Instead an individual female provides a small nest with just a few rough cells in which she lays her eggs.

Most of the 25,000 known types of bee fall into this broad category, and among them are plenty of medium-sized brown ones, which look rather like honeybees. These include, among others, the lawn or mining bees (*Andrena* species, which nest in the soil), leaf-cutter bees (*Megachile* species, which line their nests with cut leaf fragments), sweat bees (*Halictus* species, which sometimes pester the perspiring entomologist for salts and liquid),

Leaf-cutter bee.

Carpenter bee.

and carpenter bees (*Xylocop*a species, which bore into wood to make their nests).

The distinction between these and many other thousands of bee species is a complex and trying process and continues to baffle entomologists around the world. In the end, the best way of identifying a honeybee is simply from its general appearance by becoming familiar with its image.

In addition, there is another identification problem that frequently confounds the novice naturalist—drone flies. Named for their large-eyed resemblance to male honeybees (which have much larger eyes than the queens and workers—see pages 83 and 97), drone flies are hover flies, *Eristalis tenax*. These and several other closely related species tend to breed in flooded tree rot-holes (where branches have torn off from the trunk), ditches, liquid sewage, and putrescent carrion. They are almost exactly the same size as honeybees and are marked with a similar variety of brown and orange patterns.

They mimic the honeybee so closely that it is very common to find photographs in books, magazines, and on the Internet that show them incorrectly labeled as bees. It is also this elementary confusion that gave rise to the myth of the ox-born bee, which survived for many thousands of years (see pages 14 and 53). The main distinction between bees and hoverflies is that bees have two pairs of wings while hoverflies have only one pair, although this can be extremely difficult to see in insects buzzing around a flower.

Honeybees are not the only stinging insects to be mimicked by hoverflies. This diverse group also has species that mimic hornets, social wasps, solitary wasps, and bumblebees. Again, spotting the difference is often a question of becoming acquainted with the different insects and their images, but there are a few simple guidelines that will work most of the time. The identification table opposite is a quick guide to those bee-like creatures likely to be seen visiting flowers (so the queen honeybee is not included).

IDENTIFYING THE HONEYBEE

	HONEYBEE	BUMBLEBEE	HORNET/WASP	HOVERFLY
WINGS	Four: front wings large, back wings small, but held together by microscopic hooks to make a single aerofoil, usually held parallel along the back and overlapping when at rest			Only two, often held out at an angle when at rest
ANTENNAE	Obvious, stout and relatively long			Insignificant, short, knob-like
MOUTHPARTS	Jaws present but often obscured by hairs; tongue relatively long		Large powerful jaws; tongue almost hidden	No jaws; elbowed tongue with soft blob-like tip
EYES	Oval or kidney-shaped; relatively small, except drone honeybees			Large, often huge in males (head appears as one big eye)
COLOR	Brown and black with orange bands on the abdomen	Black, white, red, orange or brown, in large bands	Black or brown with yellow or orange bands	Variously colored to mimic all bees and wasps
BODY	Smooth, but with some hairs on the thorax and bands on the abdomen	Densely furry all over	Almost hairless	Usually almost hairless but some bumblebee mimics are densely furry
BACK LEGS	Prominent combs of hairs forming a pollen basket, often obscured by the mass of harvested pollen (absent in drones)	Similar pollen basket, but this is missing in males and "cuckoo" species (queens lay eggs in the nests of other bumblebees)	No pollen basket	No pollen basket
STING	Present in workers (not drones), and quick to use if hive is threatened	Present in queens and workers (not males), but rarely used unless the bee is picked up	Present in queens and workers (not males), and quick to use if nest is approached	Not present, harmless
BEHAVIOR	Busy, focused foraging usually in large numbers at many different groups of relatively open flowers	Busy, focused foraging, usually in low numbers, at more tubular flowers with deep corollas	Visit some flowers, but mainly hunting for their insect prey	Visit many different flowers; skilled at hovering, especially territorial males

Bee Origins

Despite the huge diversity and numbers of insects buzzing about today, their fossil record is sparse. Being near the bottom of the food chain means that most insects end up being eaten by larger creatures, so their bodies are not available for the processes of fossilization. Those that do make it into the soil don't have the hard bones of vertebrates, and usually rot away to nothing.

The oldest known fossil bee, the tiny, wasp-like *Melittosphex burmensis*, was discovered in 2006, cocooned in amber in a mine in northern Burma. At less than ⅛ in (2.95 mm) and 100 million years old (34–45 million years older than any other known bee fossil), this bee confirms the consensus that bees first appeared some time in the Cretaceous period (about 145–65 million years ago). The fossil's "wasp" features (the lack of any obvious pollen-carrying structure and narrow legs) and the feathery hairs that characterize bees today seem to support the generally held view that modern bees and wasps all evolved from more waspish ancestors during this period.

The Cretaceous era was an explosive time for biological diversity: the first mammals appeared as well as the first true flowering plants. The appearance of flowers on the landscape is one of the seminal changes in the biological history of the world, and the appearance of flower-visiting insects—bees in particular—is intricately linked with this change. For over 100 million years, flowers and bees have evolved together.

It is not clear precisely how bees evolved—the details seem to be lost in the shrouded mists of prehistory—but, by comparing the shapes and forms of modern insects, and their behaviors, life histories, and ecologies, as well as the latest techniques of DNA analysis, a picture is gradually emerging.

The evolution of bees from wasp-like ancestors involved a transition from predatory hunting insects to those feeding exclusively on pollen and nectar (and feeding it to their larvae). The turning point coincided, not surprisingly, with the availability of flowers.

Bees inherited several very important features from their wasp-like predecessors: a constricted narrow waist, a sting, and an all-pervasive nesting instinct. The narrow waist allows the abdomen great flexibility: it can be pointed in almost any direction, including forwards underneath the insect, between its front legs. Originally this was useful for stinging insect prey held in the wasp's claws, but, in bees, it is important for depositing eggs deep into the cells of the honeycomb. The sting was once used for killing—or sometimes merely paralyzing—its prey, but has now become a defensive weapon against marauding honey and brood thieves (see pages 144–45). The nesting instinct has developed in bees (in honeybees in particular), making them arguably the most complex of all insect societies.

EVOLUTION OF SOCIAL BEES

A honeybee colony is potentially immortal. As individual bees die, they are quickly replaced—workers week by week and month by month, and queens every few years—but the colony can go on indefinitely. This happens in honeybees and their tiny close relatives in the tropics, the stingless bees, but we can trace an evolutionary path to this point from simpler nesting strategies.

All bees and wasps make nests. In most cases, these are simple tunnels in the soil or burrows in dead wood or inside hollow plant stems. The tunnel is divided into cells, each with an egg laid inside it and each with a stock of food for the larvae—insect remains for the wasps and a mixture of pollen and nectar for the bees. Each nest is constructed solely by a lone female working on her own; other nests may be constructed nearby, by other females, but this is simply because they have chosen the same suitable nest site, not because they are working together. Such congregations are often likened to bee (or wasp) villages.

In the vast majority of bee and wasp species the mother dies and never gets to meet her offspring. The larvae feed on the food laid up for them, transform from maggots to pupae to adults, and emerge in the world to start the process all over again.

A few simple tunnel-nesting bees have taken a small step on the road to sociability. In these species, the founding female is later joined by her daughters, who excavate some extra cells in the same nest for her to lay eggs in. This is the beginning of a worker caste, who forage for food and build the nest to assist their "mother."

In the next level of truly social insects—the bumblebees and yellow-jacket wasps—the daughters become very numerous: a few hundred and many thousand respectively. (The males, as in all

A wasp nest.

Honeybees are social insects; they work together in a highly structured social order.

---

bees and wasps, do nothing for the nest; their sole purpose is to impregnate the next generation of females.) The worker females are much smaller than the mother queens and sometimes have distinct colors or patterns. More importantly, they behave very differently. The workers spend all their time foraging for food and building, expanding, providing for, and protecting the nest. The founding queen rarely or never leaves the nest; she now has

a single task—laying the eggs. The workers are incapable of mating and, although they are physically capable of laying eggs, they rarely do so, and their resulting offspring can only be males.

From the smallest tunnel-nesting colony of mother and daughter bees, with perhaps a few dozen roughly shaped cells in the earth, to the largest and most complex beach ball–sized paper carton nest of a social wasp (yellow jacket), with over 20,000 hexagonal cells in eight to ten brood combs, all these social colonies share one important aspect in their varied social lives. At the end of their single season, every one of these colonies dies. The nest is abandoned and becomes derelict. The workers disperse, straggle, and fade to death.

The males enjoy a brief mating spree, but they have equally short lives. Only the mated queens

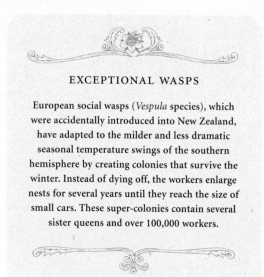

EXCEPTIONAL WASPS

European social wasps (*Vespula* species), which were accidentally introduced into New Zealand, have adapted to the milder and less dramatic seasonal temperature swings of the southern hemisphere by creating colonies that survive the winter. Instead of dying off, the workers enlarge nests for several years until they reach the size of small cars. These super-colonies contain several sister queens and over 100,000 workers.

live on. They hibernate through winter and, if they survive, they must start new colonies from scratch, beginning the whole process all over again.

Only in the honeybees (and the tropical stingless bees) are the colonies perennial. The secrets of this long life are two particular strategies they have evolved: honey storage away from the brood combs (containing eggs, larvae, and pupae) and colony creation by swarming.

HONEY STORAGE AND SWARMING

The storing of honey against lean times—of cold, wet, or when flowers are unavailable during a summer drought—allows large numbers of honeybee workers to survive where, in other bees (and wasps), only the sexual reproductives can survive. This storage of honey means that they can continuously heat or air-condition the nest—warming it by huddling and shivering in winter and cooling it by water evaporation during the heat of summer (see pages 111 and 118). Without the energy from

the stored honey to power this energy-expensive activity, the bees could not keep the nest habitable the whole year round.

Colony creation by swarming, involving the fission of a large colony into two smaller units, means the bees that move out to found a new nest have a strong head start over those species that must begin at the beginning and start the whole process again with a single female building up a colony from her vulnerable embryo nest of just a handful of cells.

Although "honey" of some sort is produced by most species of bee, it is usually made in only small quantities and is rather watery and little changed from the original nectar, or it is mixed with pollen to make a stiff cake. Even in the largest of bumblebee nests, for example, there is precious little and it is difficult to separate the honey from the wax cells, which are either constructed haphazardly or contain the eggs and larvae for which the honey is collected. It is these two behaviors—honey storage and swarming—that have brought honeybees to the close scrutiny of humans hungry for the sweetness stored in the comb.

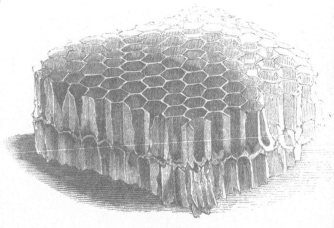

Honeycomb and its hexagonal cells.

A SELECTION OF HONEYBEE RACES
ACROSS THE WORLD

Honeybees belong to the bee genus *Apis*, which occurs naturally in Europe, Asia, and Africa. There has long been debate among entomologists (and beekeepers) on just exactly how many different species there are; over the years some have claimed as few as four; others as many as eleven. There are certainly at least four very distinct types.

DWARF HONEYBEE
Apis florea and A. andreniformis

Diminutive, 1/4 in (7 mm) long. Occurs through parts of Iran, China, and India, through southeast Asia into Indonesia and the Philippines. The nests are small and simple (see page 92). The upper part of the comb containing the honey store is broadened and expanded out into a small horizontal platform from which bees leave and come back to the nest. Scout bees land here and perform a communication dance (see pages 112–13) to alert the others to new flower forage sources in the area. Below this is a band of pollen storage cells and beneath these are the main brood cells.

Only a few fluid ounces of honey is ever stored in the comb of the dwarf honeybee, but this is still enough for humans to harvest. In some parts of rural Asia wild nests are cut down and taken back to the village, where they are secured to the branch of a small tree. The honey cells are cut down and squeezed out and, after a few weeks, the bees will have repaired the comb and restocked it so the honey can be harvested again. This is a very unreliable method as the bees are unlikely to remain beyond the second raid on their cells.

GIANT HONEYBEE
Apis dorsata

Occurs over a similar range to the dwarf honeybee, from southern Afghanistan, India, and Sri Lanka to southern China, Indonesia, and parts of the Philippines. (In northern India, Nepal, and Tibet, giant honeybees are now classified as a separate species, *Apis laboriosa*.) Unlike dwarf honeybees (see page 87), the nest of the giant honeybee is a single, large, exposed comb (see page 92).

Comb organization of the giant honeybees is similar to that of dwarf honeybees, with honey cells at the top, then pollen, and brood cells below. Toward the bottom of the comb is the "mouth," the active area where the bees take off and land, and where scouts return to perform their food discovery dances to recruit more foragers. Unlike the honeybees in enclosed domestic hives, giant honeybees need to be able to see the sun clearly to communicate the direction of new forage sources.

By necessity, the large combs are extremely strong and often contain 50 lb (20 kg) of honey. The large bees are noted for their defensive behavior—they will pursue a raider for more than 300 ft (100 m)—but the size of the nests makes them tempting targets for honey hunters; the honey is highly prized and commands a premium in local markets.

In some places, the bees are migratory, arriving to coincide with the end of the rainy season and the blooming of certain species of nectar-producing flowers. Bee hunters wait until the height of honey production before climbing the trees and collecting the honey. Left untouched, most colonies seem to migrate away as the rains return and their honey stores become depleted.

ORIENTAL HONEYBEE
Apis cerana

Also known as the Eastern honeybee. Wide natural distribution across most of Asia, from India through southern Russia to China, Japan, Korea, southeast Asia, the Malay Peninsula, the Philippines, and Indonesia. Its wide range has led to the evolution of several distinct geographic races, including *A. cerana indica* in India and Burma, *A. cerana japonica* in Japan, *A. cerana javanica* in southeast Asia, and *A. cerana koschevnikovi* in Indonesia. Over many years these have been variously argued to warrant full and distinct species status or have been demoted to subspecies.

The more northerly temperate and subtropical varieties store more honey (a response to cooler winter weather when foraging is more limited). The tropical varieties store less honey, but are more mobile—swarming, absconding, and migrating more easily in response to wet and dry season changes rather than just temperature.

Oriental honeybee nests consist of several combs hanging together in parallel plates (see page 92). The multiple honeycombs are always separated by a uniform gap, the bee space (see page 155), allowing workers to gain access to all cells in all parts of the nest, even in the deep interior.

A CLOSE RELATIVE OF THE ORIENTAL HONEYBEE
Apis nigrocinta

Found in several Philippine and Indonesian islands and often considered to be a distinct species. Tends to be larger than *Apis cerana*, and can be distinguished by its yellowish lower part of the face. Like *Apis cerana*, it builds its nests in dark cavities (see page 92).

WESTERN HONEYBEE
Apis mellifera

The most geographically widespread bee on Earth (see page 64). It has been domesticated, or at least exploited, by humans for so long that its origins are now unclear. Probably arose in Africa and then spread into Europe (recent studies of honeybee DNA have suggested that there may even have been three different prehistoric invasions of Europe, from Africa), and later, with colonial expansion, to North America (1620s) and Australia (1820s).

Like the Oriental honeybee, it nests in darkness and has multiple combs (see page 92). And, like its Oriental cousin, this nesting behavior made it the honeybee of choice when humans realized that they could keep these bees in artificial nests.

Exactly when Western honeybees were first "domesticated" is not recored, but it is thought that they arrived naturally in Europe, from Africa, about 10,000 years ago. The ancient Egyptians were accomplished beekeepers 4,500 years ago and the later Greek and Roman civilizations understood many of the complexities of honeybee colony life (see pages 30–33). Although records of beekeeping in China go back 3,000 years, it is the history of European colonial expansion and trade in the last 600 years that has taken the Western honeybee well beyond its original geographic range (see pages 64–71).

Geographic isolation is the main evolutionary driver of race and subsequent species formation; mountain ranges, seas, and even rivers prevent cross-breeding. Throughout its native distribution, the Western honeybee had started to evolve different characteristics in different parts of the world, until beekeepers started to breed selectively for increased honey production, for less defensive and more docile behaviour, for better nest hygiene, and for reduced swarming behavior. Although there are no strains of bee that can be described as truly "domesticated," more recently, pest and disease resistance have become highly desirable traits to breed into honeybees.

There are now very many Western honeybee races and forms throughout the world, but a few of the more important ones are listed opposite.

CARNIOLAN HONEYBEE
A. m. carnica

ITALIAN HONEYBEE
A. m. ligustica

Dusky brown-gray with more muted orange bands. Originally from Austria and the Balkan regions, now transported around the world and the second most popular race after *A. m. ligustica*. Gentle, disease-resistant, and well able to defend itself against hive pests. High yields of honey.

Distinctive yellow- or orange-banded abdomen. Originally from Italy and Sicily, this was the species of choice imported to the UK after Isle of Wight disease devastations (see page 74) and is now the most widely distributed race in the world, particularly to the UK, USA, Australia and New Zealand. Less hardy in cooler regions, but a gentle, good forager and disease-resistant.

DARK EUROPEAN OR GERMAN BLACK HONEYBEE
A. m. mellifera

Hardy, adapted to cooler northern climate. Found in western Europe, including Britain, north and west of the Alps.

CAPE HONEYBEE
A. m. capensis

Southern form of African honeybee, from the Cape Peninsula in southwestern South Africa, but now introduced to other parts of southern Africa. Unique among honeybees because workers are able to lay fertile female eggs (see page 85). Introduction into African honeybee range has meant that queens enter *scutellata* nests undetected and unchallenged, and cuckoo-parasitize the colonies, by laying their own eggs in the cells. Resulting workers continue laying eggs and eventually the colony collapses.

AFRICAN HONEYBEE
A. m. scutellata

Native to southern and central Africa. Similar to *A. m. ligustica* in appearance. High yields of honey, very disease- and pest-resistant, but also very defensive. Accidental release in Brazil in 1957 led to cross-breeding with more docile imported European races and gave rise to Africanized bees, also rather sensationally called "killer" bees because of their ferocious defense of the colony.

EGYPTIAN HONEYBEE
A. m. lamarcki

Small dark bee with yellow abdominal bands. Native to the Nile Valley, but some genetic similarity to bees introduced into California. Defensive and low-yielding, but with good housekeeping. Probably the race used by Ancient Egyptians; immortalized in carved reliefs from temple remains.

BEES' NESTS

Each of the species of honeybee also has its own type of nest. The nests of the dwarf honeybee are small and simple, comprising a single exposed comb hanging from a branch or just inside a cave, usually 3 to 26 ft (1 to 8 m) from the ground. They include the horizontal platform from where the scout bees perform their communication dance. Although the nest is exposed, the bees tend to hide it by using a small branch cloaked with dense foliage. To prevent ant raids, both ends of the branch, either side of the nest base, are coated with a sticky band of propolis (plant gum) ¾ to 1½ in (2 to 4 cm) wide. The comb is protected by a thick layer of bees, three or four deep (usually about 75 percent of the colony's workers). If the nest is threatened, this living coat shimmers as the bees synchronously waggle their abdomens from side to side and buzz loudly. If the intruder remains, the bees take off and attack, or they may abandon the nest, forming a new nest nearby.

The giant honeybee's nest couldn't look more different. These nests are made up of a single exposed tombstone-sized comb (3 ft or 1 m across and up to 7 ft or 2 m high), hanging from a cave roof or cliff-face overhang. *Apis dorsata* is more of a forest species and often makes its nests high up in mature *Dipterocarpus* trees, 80 ft (25 m) or so above the ground. Certain favored trees can have 10 to 20 nests hanging from the branches and sometimes many more; "bee trees" with over 100 nests have been found.

Rather than the single comb of the giant honeybee, the Oriental honeybee nest consists of several combs hanging together in parallel plates, separated by a uniform bee space. These are usually found in dark, hollow trees and caves. Similarly, the Western honeybee builds its nest in darkness, in hollow trees and logs, caves, rock cavities, and other natural voids. The nest also consists of multiple combs, hanging parallel and separated by a regular and uniform bee space of the same dimensions as the space between Oriental honeybee combs.

It is this behavior of nesting in dark places that humans have been able to exploit for millennia by providing suitable dark artificial nesting voids such as terracotta pots, wooden boxes, baskets, and shaped logs in which to domesticate wild colonies of honeybees (see page 29). Periodic removal of some of the combs allows the colony to continue even as the honey is harvested.

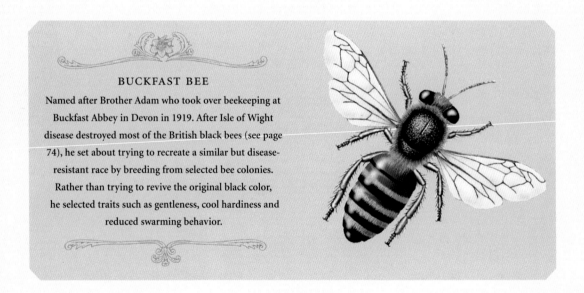

BUCKFAST BEE

Named after Brother Adam who took over beekeeping at Buckfast Abbey in Devon in 1919. After Isle of Wight disease destroyed most of the British black bees (see page 74), he set about trying to recreate a similar but disease-resistant race by breeding from selected bee colonies. Rather than trying to revive the original black color, he selected traits such as gentleness, cool hardiness and reduced swarming behavior.

A dwarf honeybee nest.

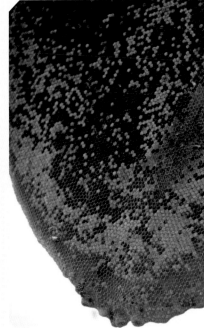

A giant honeybee nest.

A Western honeybee nest in a hollow tree.

An Oriental honeybee nest.

Occupants of the Hive

A T THE height of summer, a vibrant, healthy hive may contain about 20,000–50,000 individual honeybees. Almost all of these will be workers (sometimes called "neuters" in older books and essays on the subject). Workers are infertile females whose entire lives are spent foraging for the colony, creating the wax combs, loading cells with honey and pollen, protecting the hive from honey-stealing marauders, and nursing the brood of larvae through to adulthood.

Scattered among the workers will be a few hundred males (drones), who appear to do nothing except sip honey—until they are eventually ejected from the hive by their sisters.

At the very heart of the colony is one fertilized reproductive female, the queen. She spends her entire life laying eggs into the wax brood cells, while being fed by a constantly rotating court of workers.

With many insects, distinguishing males from females is often possible only under a microscope, or at least a hand lens. But male and female honeybees (both queens and workers) can be distinguished with the naked eye. First, however, it is important to understand the bodily structure of the different types of honeybee, starting with the most abundant, the ubiquitous worker.

THE WORKER

A worker honeybee is, at first sight, a typical insect. Its body is roughly ½ to ¾ in (12 to 15 mm) long, relatively stout, and almost cylindrical. Mostly dark brown or black, the races around the world are variously decorated with abdominal bands of chestnut, orange, or yellow. There are four wings, but the hind pair is much smaller than the front pair and difficult to make out in the living, moving insect. The wings are clear and trans-parent, marked with a tracery of dark veins, which give structural support to their flimsy construction. Fore and hind wings are attached: a series of curved hooks along the trailing edge of the front wing (called the hamuli) grasp a reinforced fold along the forward edge of the hind wings. The two wings are thus joined into a single, and more effective, aerofoil for stronger, faster, and more energy-efficient flight.

The head is rather broad, with two lozenge-shaped compound eyes composed of about 6,000 hexagonal facets called ommatidia. Each ommatidium functions as an individual light receptor. Light is focused by the transparent lens covering onto the elongated cone of light-sensitive pigments (called the rhabdom) and nerves inside. The image collected by a bee's compound eye is

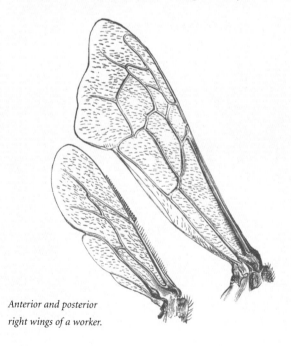

Anterior and posterior right wings of a worker.

relatively coarse compared to mammalian or avian visual systems—like a mosaic of small overlapping sub-images. Nevertheless, sight is one of the bee's most powerful senses, allowing it to recognize flowers and to construct a visual map of the immediate vicinity of the hive.

One of the most important aspects of bee sight is the ability to detect the direction of light polarization, since it allows them to determine precisely where the sun is in the sky above, even on overcast days (see page 113). It does this using the "pol" parts of its eyes, the marginal two or three rows of slightly larger facets toward the top edge of the eyes.

On top of the head are three small shining domes arranged in a triangle. These are the ocelli, sometimes also called the "simple" eyes. Their function is not yet fully understood, but the focus of the lens and arrangement of light receptors inside suggest they measure differences in overall light intensity instead of gathering usable images. One explanation is that they monitor the contrast between light (sky) and dark (the Earth) at the horizon, allowing level flight when navigating to forage zones.

Attached to the front of the bee face are two antennae. These comprise a long basal segment (the scape), an articulated elbow segment (the pedicle), and a whip-like series of 10 segments that make up the flagellum. The antennae are covered with sensory hairs, pits, and small plates called sensillae; apart from the eyes, these are the bee's most important sense organs.

In the dark of the hive, the antennae are sensitive organs of touch, and are constantly brushed over neighboring workers or the queen. Some of the hairs on the antennae are touch sensors, with nerves at the base of the hairs reacting to mechanical movement of the bristles as each comes into contact with another solid object.

It was once thought that bees were deaf. It was also thought that they could not hear airborne vibrations. However, a special sense detector called

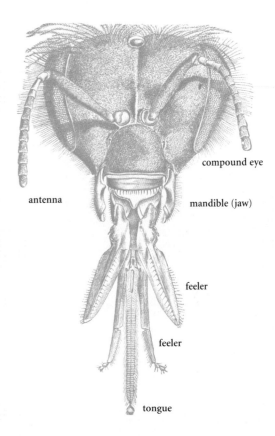

antenna · compound eye · mandible (jaw) · feeler · feeler · tongue

The head of a worker honeybee.

Johnston's organ has recently been found in the pedicle. Sound vibrations, especially those made by bees using the communication dances in the hive (see pages 112–13), cause the whip-like flagellum segments to vibrate and the movements at the base of the flagellum are detected by Johnston's organ.

In the hive, and out in the open, the antennae are also the prime organs of smell. The sensillae, which cover the antennal segments, come in many shapes and forms, including long and short hair-like structures, flat plates, holes, and bristle-lined pits. These are covered all over with microscopic pores, which allow the slow movement of air to the interior, where chemoreceptors react to the molecules in airborne scents. These include perfumes released by flowers, pheromones (chemical secretions) released by the bees themselves in the hive, and alarm pheromones released during stinging (see page 130). The sensors on the antennae also react to temperature, humidity, and carbon dioxide (particularly in the breath of mammalian enemies intent on raiding the nest).

Below the antennae are the mouthparts. The most prominent are the two moderately large triangular jaws (mandibles) that meet in a smooth scissor-like action along the midline and are used in handling objects, manipulating pollen, and to shape the wax to build cells.

Beneath these is the proboscis made up of various small segments, which are fused or held together to make a long tubular sucking tongue (the glossal lobes), feelers (palpi) to touch and manipulate the food, and a thick protective sheath (the stipes and galeae). When not in use, the proboscis is folded away underneath the head.

The most obvious uses of the mouthparts are for biting at flowers and pollen, sucking up nectar for regurgitation to other nest-mates, manipulating and shaping the wax during comb formation, moving debris and dead bees during housekeeping, biting and holding intruders in the nest, and licking the queen.

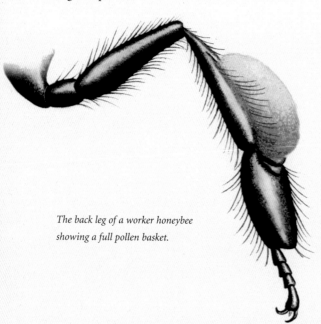

The back leg of a worker honeybee showing a full pollen basket.

The feathery hair on the bee's body is an ancestral adaptation for collecting pollen. In some "primitive" solitary bee species, pollen is brushed from the body when the bee returns to the small nest burrow; in leaf-cutter bees the underside of the abdomen is thickly covered with long hairs used to trap the pollen and transport it back. In honeybees (and bumblebees), pollen collection has become a sophisticated process and the bee has specially shaped back legs to carry large loads.

Pollen collection begins when the bee becomes covered with the grains as it visits flowers for nectar. As well as drinking this plant juice, a honeybee may bite the anthers, the pollen-producing parts of the flower, to release their contents. The bee then uses its legs, which are covered with stiff bristles, to comb the pollen from the head and thorax and transfer it to the back legs.

Workers do not mate and cannot produce fertile eggs, but occasionally lay infertile eggs, which give rise to drones.

These back legs are armed with a special series of nine comb-like rakes of stiff bristles (the pollen brush) which, as the bee rubs its legs together, force the pollen into the joint between the lower leg (tibia) and the swollen first segment of the foot (basitarsus). This hinged joint between the bee's leg segments is called the pollen press; as the joint articulates, it compresses the pollen from a loose but sticky dust into a hard cake, which passes on to the broad, smooth, concave outer face of the tibia. Here it is contained by a fringe of long, curved hairs called the corbicula or pollen basket. The pollen basket is particularly well developed in honeybees (and bumblebees), and the brightly colored pollen loads are often clearly visible as the bees return to the nest. The pollen

OCCUPANTS OF THE HIVE AT A GLANCE

THE WORKER

Lighter in weight than a queen or drone, she has pollen-collecting apparatus on the legs. There is a ridge on the underside of the abdomen, indicating the wax glands used to build the comb. The worker's sting is barbed (and the worker dies as she tries to withdraw the sting after use).

THE QUEEN

A queen's body is elongated, with a pointed abdomen, and her legs are long and stout, with no pollen-collecting sacs. Her sting is curved and less barbed than the worker's (it can be used and withdrawn without endangering her life). She has a shorter tongue than a worker.

THE DRONE

The drone honeybee is heavier and stockier than the queen and workers, with a blunt abdomen and large compound eyes that meet at the top of the head. Drones have no pollen sacs or wax glands, their tongues are very short, and they have no sting.

from different flowers is differently colored, so the forage targets of the bees can often be identified by eye.

At the tip of the worker's tail is the bee's defensive sting. This has been inherited from its wasp-like ancestors, which used the sting as a weapon of attack—to kill or paralyze prey before feeding it to the larvae. In worker honeybees the sting is only used to defend the colony. The sting is modified from part of the egg-laying apparatus, so only females possess one. See pages 128–30 for more on honeybee stinging behavior.

Although invisible to the naked eye, worker honeybees possess two other very important internal organs: the honey sac (sometimes called the honey stomach) and the wax glands (see page 124).

THE QUEEN

The queen is narrower and longer than the workers (up to ¾ in (18-19 mm)). Her head is smaller and so are her eyes (only 3,500 facets), which are also more densely hairy (she does not need to measure the arc of the sun or identify the correct flower to visit). Her tongue is shorter (she never drinks nectar from a flower, but is fed by her attendants). Her jaws are similar in size to those of the worker, but, instead of meeting at a straight edge, they are toothed at the tip and have a deep cusp in the middle (this may be linked to her well-developed mandibular glands, which secrete the all-important queen pheromone—see page 114). Her thorax is narrower (she makes no long foraging flights), but her abdomen is longer (it is full of eggs). Her legs are stouter than a worker's, and much less bristly; the outer face of the rear tibia is convex not concave, and there is no fringe of hairs to make a pollen basket (the queen will never collect pollen from any flower). She lacks the ridge on the underside of the abdomen of a worker, which suggests she has no wax glands (she takes no part

in building the combs), and her sting is less barbed than a worker's (she takes no part in defense of the colony).

THE DRONE

The drone (male) is nearly ¾ in (15-17 mm) long, heavier, more robust, and more densely pubescent all over than any female. Like the queen, a drone never takes part in foraging, nest-building, or defense. His sole aim is to follow a virgin queen on her "nuptial" flight and try to mate with her. His head capsule is not much larger than that of a worker, but most of it is taken up by huge eyes; with over 10,000 facets each, they meet and touch each other at the top of the head, displacing the ocelli forward onto the front of the face. The antennae are stouter and longer and, as with the males of almost all bees and wasps, there are 11 segments to the flagellum where females have only 10. His jaws are small and covered in dense hairs and, like the queen, his tongue is short. His thorax is broad and stout, his abdomen is broad, the tip is blunt, and there is no sting. His legs are almost naked of hairs; the front four are very slender and the hind legs lack any pollen-collecting apparatus.

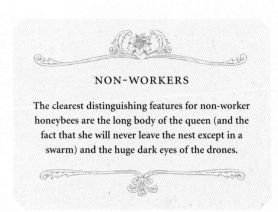

NON-WORKERS

The clearest distinguishing features for non-worker honeybees are the long body of the queen (and the fact that she will never leave the nest except in a swarm) and the huge dark eyes of the drones.

From Egg to Adult

IN A THRIVING colony during early summer, the queen is able to lay about 1,500 eggs each day. That is more than her own body weight in eggs. Over the course of a year these eggs are likely to produce 200,000 workers.

Normally a queen presides over a colony for two or three years, but there are records of some living for four or five years, or even longer. Whatever the span of her life, the queen is able to keep laying eggs just on the store of sperm she collected on her brief mating flight, during which she may have mated with a single drone or, more likely, several. She ekes out the sperm store over the years, and almost all of the eggs she lays are fertilized by the sperm that she keeps in reserve (see page 116).

SEX DETERMINATION
Honeybees, like all bees, wasps, ants, and a few other insects, were originally thought to use a mechanism of sex determination called haplodiploidy. This suggested that the gender of a honeybee was determined by the total number of chromosomes—16 in the male (the haploid, or single chromosome condition) and 32 in the female (the diploid, or double chromosome condition). During egg production a halving of chromosome number, from 32 to 16, was thought to occur in the queen's ovaries, but this was not matched by a halving in the sperm, which contained the same 16 chromosomes as in every other cell in the drone's body. Whenever the queen laid an egg she could choose whether or not to fertilize it with stored sperm. An egg fertilized by a sperm would contain 16 maternal and 16 paternal chromosomes, giving a female total of 32; would it become a female, either a worker or a queen. An unfertilized egg would contain only the 16 maternal chromosomes and would therefore become a drone male.

However, sex determination in honeybees is now known to be determined by a single gene called the "complementary sex determiner" (csd) gene.

A queen (marked with the yellow spot) is laying eggs while the workers are feeding larvae.

LAYING EGGS, HATCHING, AND FEEDING BROOD

After mating, the queen's ovaries expand and enlarge, stimulated by special nutrient-rich food she receives from the workers. Within three or four days she begins laying eggs.

When an egg hatches on the third or fourth day, the attendant workers immediately start to visit the tiny larvae in their open cells and feed them. A worker larva receives over 140 small meals in total, provided by the brood-attending bees, also called nurse bees. After just five days of feeding, worker and queen larvae will have finished growing; drones, however, take one extra day.

Thus, by day nine after laying, the larva is replete and the workers seal the cell with a wax cap. This is a more complex procedure than at first appears: the bees make over 100 visits and the construction takes at least 6 hours.

Inside the closed cell, the larva transforms into a pupa and the metamorphosis into an adult occurs. By day 16 after the egg was laid, a queen emerges from her pupal cell. A worker usually emerges by day 21 and a drone by day 24. The pre-

ABOVE: *Comb cells showing an egg, a larva and a pupa. A nurse bee is attending the larva.*

cise timings for the development of bees through larvahood to adulthood are dependent on local conditions and it is remarkable, therefore, that these times generally correspond between the Western honeybee (*Apis mellifera*), and the southern form of the Oriental honeybee (*Apis cerana indica*), despite the fact that the Western lives mainly in temperate regions and the Oriental in the much hotter tropics. This similarity in timing occurs because honeybees are extremely good at maintaining a constant temperature—about 95°F (35°C)—in a core area of the hive, even in the deepest cold of winter or in the tropical heat (see pages 109 and 118).

The differences that do occur in emergence time are mainly a combination of differences in the final body size of the adult insect (drones being larger and heavier than workers) and the different

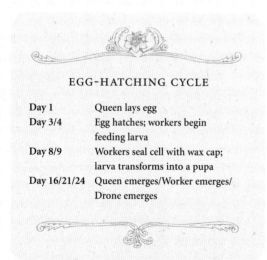

EGG-HATCHING CYCLE

Day 1	Queen lays egg
Day 3/4	Egg hatches; workers begin feeding larva
Day 8/9	Workers seal cell with wax cap; larva transforms into a pupa
Day 16/21/24	Queen emerges/Worker emerges/ Drone emerges

OPPPOSITE: *The emergence of a bee.*

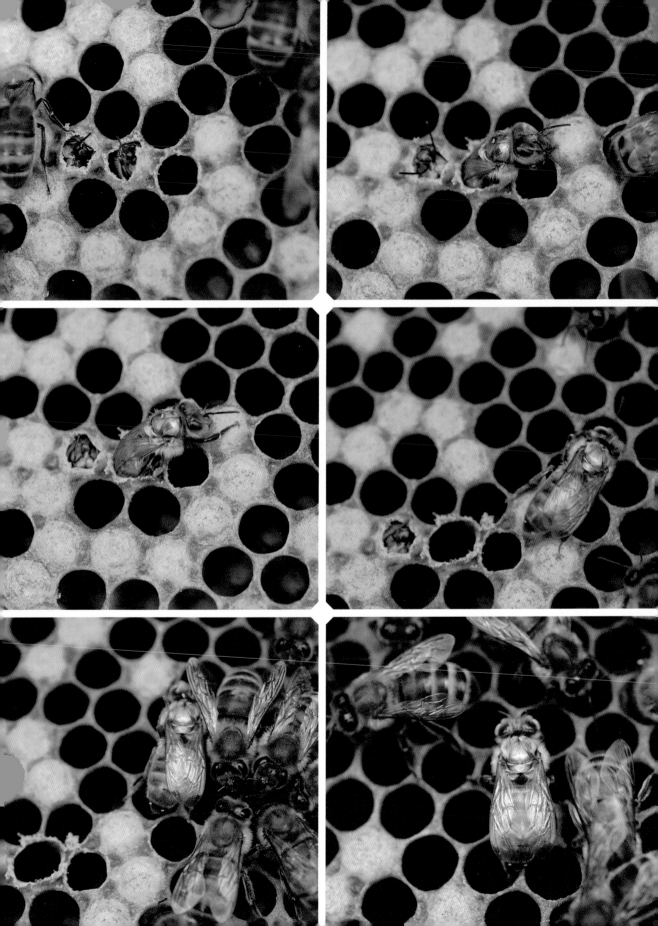

Cells containing larvae. The capped cells contain developing pupae and food. The yellow stores are pollen.

nutrient value of the food they receive. This is particularly important for female larvae.

For many years it was not known how the same type of fertilized egg could give rise to the different body types of both worker and queen. During the first days of its life (usually until day three or four), the larva destined to become a worker receives exactly the same sort of food as the one marked out to be a queen: rather than straightforwardly regurgitating honey and pollen to the larva, the nurse bees secrete a protein-rich substance from the hypopharyngeal and mandibular glands inside the throat. The protein in this secretion is mainly derived from the large amounts of pollen that the bees eat.

Later, the food given to worker larvae becomes diluted with honey and pollen while queen larvae continue to be fed on the hypopharyngeal and mandibular secretion. So much is provided for them that it builds up into a milky white food store in the especially large queen cells in which they are growing.

It was the discovery of this—the nutritional difference between worker and queen larvae—that gave rise to the name "royal jelly," but, since all

Cold weather can dramatically slow down development during larval or pupal life.

bees receive at least some of it for a short time, it is also sometimes called "brood food."

It seems likely, too, that it is not simply the type of food available to a larva that will determine its caste, but also the quantity. During the first days after hatching, all larvae receive unlimited quantities of food; after that, worker larvae are fed regular, but small, quantities. Only queen larvae continue to receive a superabundance of food.

The decision whether to feed larvae in the cells is still not fully understood, but the type of cell in which the larva is growing is of great importance. During the normal early-season growth of the colony, almost all the brood cells are for rearing workers. But, as the honey flow increases, a few larger cells are created, usually at the edges of the brood combs; these are for drones.

During summer and autumn drones occur in the nest in low numbers: a few score in small nests and a few hundred in large colonies. Their presence

LIFE SPANS

WORKERS live on average between four and six weeks during the active season, but those reared late in the season may survive until spring, as they do little during the cold months except feed and keep warm.

The average life span of a QUEEN is one to three years.

A DRONE dies immediately after mating with the queen, his genitals left in her. Drones are prevalent in colonies during the spring and summer months. As autumn approaches, they are evicted from the hive, and soon die of starvation and exposure.

Worker and drone (larger) cells.

Queen cells are larger than worker and drone cells and hang perpendicular to the comb. This cell has been opened to show the queen pupa.

is insurance in case the colony reaches the point at which it must quickly rear new queens and swarm to create new colonies, as the drones will be required to mate with the virgin queen before she can start egg-laying.

While a fully functional and healthy queen is in residence, the colony continues as normal. But, if the queen is removed or dies, or even becomes unhealthy, the workers in the colony detect a change within hours. When the colony reaches a certain critical (large) size (see page 114), the worker bees' behavior will also change. They will immediately start to construct large queen cells: spacious and abnormally shaped cells from which new queens will be reared. If the queen is removed suddenly, before a fertilized and therefore female egg can be laid into a queen cell, the workers will enlarge a (previously made) worker cell that already contains a larva still inside the three-day brood food nutritional window, and feed it so that it becomes a new queen.

A GENETIC PUZZLE

When Charles Darwin (1809–1882) first proposed his theory of evolution by natural selection in 1858 (later expanded in his book *On the Origin of Species* in 1859), he questioned whether honeybees might be the unraveling of his ideas.

Natural selection worked when an individual organism (plant or animal) outperformed its competitors by having some very slight advantage of size, shape, color, or behavior. This individual organism was therefore more successful at passing its genes on to its offspring, who would inherit that same slight advantage. Individuals without the same advantage would suffer against their competitors; they would underperform and be less successful at passing on their genes to any offspring, who themselves would also be at a disadvantage and would eventually die out.

How was it then that a system had evolved in honeybees (and bumblebees, wasps, and ants)

Charles Darwin devoted part of On the Origin of Species *to the subject of honeybees.*

where the workers constructed, protected, and cared for the nest and yet were sterile, had no offspring, and therefore lacked the means to pass on their "helping" genes to a future generation? Further, worker honeybees also commit themselves to death when protecting the hive by stinging an attacker. Passing on a suicide gene is a biological contradiction: if you are dead, you cannot pass on anything.

According to natural selection, therefore, bees should not have survived this long. This has long puzzled biologists, but two newer theories have helped to explain why caste-based societies, such as that of the honeybee, have outperformed their competitors and are still around today.

Group selection

Group selection suggests that a colony that benefits from such helpful behavior in its workers will fare better than a colony in which all females seek to achieve genetic immortality by competitively laying their own eggs. Thus the caste-based society of worker honeybees more successfully breeds

other caste-based societies. In effect, rather than natural selection working on each individual bee, it works at the level of the colony or group— group selection.

While this partially seems to explain how social forms might continue once they have evolved, however, it cannot demonstrate how they arose in the first place.

Kin selection

In 1964 the English evolutionary biologist William D. Hamilton (1936–2000) suggested that the haplodiploid sex determination system in bees (see pages 99–100) might have contributed to the evolution of a worker caste. Worker bees inherit 100 percent of their father's genes (all 16 of the drone's chromosomes), but they inherit just 50 percent of their mother's (16 of the queen's 32 chromosomes). Thus, workers share, on average, 75 percent of each other's genes. This is more than they would share with their own offspring (50 percent) if they were able to reproduce.

In terms of successfully passing their genes on to future individuals, therefore, it pays the workers to forsake their own reproductive potential (even giving their own lives to protect the colony) in order to maximize the rearing of their mother's later offspring—the eventual new queens that the colony will produce and that are still the workers' sisters.

One of Hamilton's evolutionary colleagues, John Maynard Smith (1920–2004), named this model "kin selection," reflecting, as it did, the close kinship between the workers.

Another suggestion, one that is also compatible with kin selection, is the concept of "maternal manipulation." It is suggested that in this process, a reproductive female (the queen) reduces the physical ability of some offspring (the workers) by manipulating them through her pheromones (chemical secretions), but in so doing she also optimizes the success of a select few of her other offspring—those new fertile queens.

A honeybee swarm.

True social nesting (mothers and daughters living and working together in a colony) appears to have evolved in insects on at least 12 different occasions; 11 of these are in the Hymenoptera (the twelfth is in Isoptera, the termites). While haplodiploidy is universal among the order Hymenoptera, however, many subsections in the group have never evolved any social behavior.

Biologists have still not solved the contradictions of colony evolution. Their ideas are based on theories, mathematical models, implication, and extrapolation, but there are still large gaps in experiment and observation, and a full understanding of honeybee behavior is lacking.

Social Life in the Hive

I T HAS LONG BEEN RECOGNIZED THAT, in order for a colony of 50,000 bees to construct, maintain, and protect such an elaborate nest, and to ensure the future of new offspring colonies, there must be a division of labor between the numerous workers. There are many tasks to perform, including scouting for new forage sources, collecting nectar and pollen, collecting water, and, back at the hive, receiving food from returning workers, feeding some to the queen, the brood, or other workers, and storing some in the honey cells. Bees in the hive build new comb, clear out debris—including dead bees and unwanted drones—challenge incomers at the nest entrance, and defend the colony against honey and brood thieves. They also fan the combs to cool them. Eventually there is the task of swarming with a queen to create a new nest.

This division of labor was once thought to be innate and fixed, and dependent on the age and condition of each individual bee, with each insect moving through the various tasks as its life progressed inexorably from hatching to death. However, the career path of a worker turns out to be less fixed than was once thought; experiments have shown that bees of all ages can perform any task, and their longevity is linked to nutritional intake rather than energy expenditure.

There is, nevertheless, a typical progression, through which most bees pass.

ABOVE: *Workers aged around four or five days begin to feed the brood.*

OPPOSITE: *A worker enters the hive with pollen while the guards protect the entrance.*

THE LIFE STAGES OF A WORKER BEE

Birth to early flight

Having hauled herself out from her pupal cell, a newly hatched adult bee spends a few hours grooming until she is clean and fluffy. During the first few days of adulthood she stays in the nest, doing little other than soliciting food from passing workers by sticking out her tongue at them. Eventually she will get a response from one, who opens its mandibles, lowers its tongue slightly and regurgitates a droplet from its honey sac. The young bee inserts her proboscis into this droplet and drinks it down. These feedings will be her only source of nutrition for several days.

After about three days, the new bee starts to feed herself from the honey cells, and to eat pollen from the pollen stores. During this period she spends much of her time in apparent idleness on

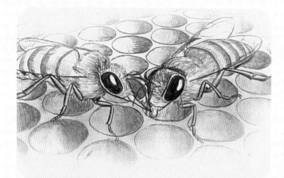

TOP, LEFT AND RIGHT: *A worker feeds a young bee.*

BOTTOM: *A worker removes a dead larva.*

Workers aged around 10 or 12 days begin to build new comb.

the comb. In fact, all bees spend more time resting than working, but even then they are performing the useful task of combining their body heat, which keeps the combs at the optimum temperature for brood-rearing 88–95°F (31–35°C). If the worker bee does any work it may be "household duties," such as cleaning out previously used brood cells in readiness for new eggs to be laid in them.

At day four or five a worker may start to feed some of the brood. At first she can only feed the older larvae with regurgitated honey and pollen but, at about day six, her hypopharyngeal glands will have started to mature, converting the pollen she has eaten into the protein-rich brood food with which she can now start to feed the younger larvae.

At 10 to 12 days, the brood food glands are nearly exhausted, but the glands on the underside of her abdomen will have started to produce wax so she can begin comb construction and repair.

At about this time the worker will also start to leave the hive. On her maiden flight, the bee defecates for the first time. It may even be that the build-up of waste matter in her gut is actually the stimulus to fly as no healthy bee willingly defecates inside the nest.

Reconnaissance flights take place in warm weather toward the middle of the day or early afternoon, and allow the bee to familiarize herself with the immediate surroundings of the hive. They begin with hovering flights just outside the hive entrance but, as she gains experience of the area, she will fly farther afield and at a greater speed.

A guard bee repels an intruder trying to enter the hive.

Hive work

For the next week or two the bee still spends much of her time in the hive receiving the incoming forager bees. Instead of each forager penetrating deep into the interior of the hive and seeking out a suitable honey cell in which to deposit its newly acquired feed-load, the bees work a relay system. Incomers regurgitate nectar to the younger household bee; she takes the liquid into her own honey stomach and matures it (see pages 123–24). This maturation involves the removal of water, thickening the plant juice into sticky honey. She then places it into the honey storage combs. She also packs away the pollen loads, which are only roughly deposited by the foragers near the pollen storage cells, sometimes being little more than kicked off as they pass by.

During these first weeks in the hive the bee continues to clear up any debris, including dead bees, in the brood combs or those that have fallen under the nest. Eventually, after about three or four weeks, the young bee becomes ready to leave the nest and become a forager. She pays more and more attention to returning scout bees. These are the advance foragers, which, after finding a new source of nectar and pollen, perform the communication dances (see pages 112–13). It seems to take many hours, or several days, before a new recruit is able to follow the coded dance instructions correctly, but eventually she will join the flower visitors outside the hive.

Some bees also take on guard duties. Adopting a characteristic pose with the front legs held off the floor as if in readiness to pounce, guard bees stand at the entrance to the hive and challenge all incoming bees to prevent intruders from stealing the precious honey store. This duty will only be taken on by a small proportion of the bees, however; most move directly on to proper foraging duties.

Foraging

The primary forage target for a bee under normal operating conditions is nectar. This is the source of virtually all the bees' and the brood's carbohydrate, containing the three most important sugars: glucose, fructose, and sucrose. The bees' ability to store this important foodstuff as honey is one of the key factors in the success of honeybees around the globe.

---※---

A forager effectively works herself to death.

---※---

For protein, bees harvest pollen; they do this selectively, not just as a by-product of visiting flowers for nectar. In times of high pollen availability, they often deliberately discard pollen on their way back to the hive. When there is a greater need for pollen, for example when large numbers of larvae are being reared and the bees need pollen to create brood food, more is brought back to the colony.

As well as collecting nectar and pollen, foragers also visit pond- and stream-sides, puddles, ditches, dew, animal dung, and wet mud to return with water, which is used to dilute honey when it is being fed to the larvae, and to smear onto the combs to cool them by evaporation.

They visit trees to collect propolis, the sticky sap that exudes through wounds in the bark. The forager bees bring this substance back to the nest where it is used to seal up gaps to keep out predators and pests, and to improve the structural strength and integrity of the multiple combs (see page 126 for more about propolis).

During the height of the active summer period, a forager works so hard that her total life expectancy is rarely more than five or six weeks.

This is in contrast to bees that hatch in autumn, which can expect to live for five or six months, and whose life span is not just dictated by the frenetic activity of foraging. During the early adulthood of these autumn-hatched bees, when they gorge on pollen to produce brood food, they also store energy in fat reserves. Summer bees use up their fat reserves fast as there are plenty of larvae to feed; during autumn there are fewer larvae for the brood food so the bees keep hold of the vital fat energy reserves, which is what gives them their increased longevity.

LEFT AND RIGHT: *As well as nectar and pollen, a forager also collects water for use in the hive.*

Bee Communication

FORAGING is a highly organized operation: bees are able to communicate the presence, direction, distance, and nutritional strength of nectar and pollen sources to their nest-mates. It was long thought that the "dances" carried out by bees on the comb were in some way simply co-ordinating forage efforts; it was not until 1923 (later expanded on in 1946) that a better under-standing of forage bee recruitment developed.

It was the Austrian zoologist Karl von Frisch (1886–1982) who artificially manipulated forage sources and was then able to measure aspects of the bees' dances. He described two types: the round dance and the waggle (or figure-eight) dance. Both are carried out on the comb and immediately attract the attention of other bees, which huddle close to monitor the scout's move-ments in the darkness.

If a forage site is relatively close to the hive, less than say 250 ft (75 m), a returning scout signals to others on the comb by doing the round dance. The bee moves in one or more circles clockwise then counter-clockwise and so on. Scent cues brought back by the scout then stimulate forager recruits to leave the hive and look in the immediate vicin-ity for similar-smelling flowers.

The most remarkable dance, though, is the wag-gle. Some returning scouts perform a distinctive dance in a figure-eight pattern. A semicircle clock-wise is followed by a straight line in which the bee vigorously vibrates its abdomen from side to side, then another semicircle counter-clockwise and another straight-line waggle. This is repeated sev-eral times by the dancer, who is closely followed by several other bees.

What is most astonishing about this dance is that it not only alerts other bees to a new forage source, it also tells them in which direction to fly, how far to fly, and what to expect when they get there. The figure-eight dances are not always ori-ented in the same direction on the comb; the central straight-line waggle part of the dance gives the direction to the other bees. The angle between the waggle path and the vertical axis of the comb is the same as the angle between the forage source and the sun's position above the horizon (the azimuth). On leaving the hive, the bees fly at this angle from the sun to find the new forage source. In addition, the length of the cen-tral waggle path is an indicator of how far away the source may be. Bees are now thought to judge distance by measuring the movement of the envi-ronment past their eyes (optic flow) and to use this to compensate for wind speed, because fly-ing with the wind will take less time than flying into the wind. Waggles are correspondingly shorter or longer. Finally, the richness of the for-age source is indicated by the apparent fervor

The waggle dance. The angle x between the vertical and the central movement is the same as the angle between the sun and the forage source.

A worker with yellow pollen sacs executes the waggle dance to indicate the food source.

with which the scout vibrates its abdomen during the waggle.

Although von Frisch used glass-fronted observation hives (see pages 62–63) to measure and study the bees' dances, recruit bees "observe" a scout dancing on the comb in complete darkness, which means that they must sense the orientation, vigor, and size of the dance by non-visual means. The recent discovery of Johnston's organ in the antenna (see page 95) shows that they are hearing vibrations in the 200–300 Hz range made by the scouts' closed wings during the straight-line waggle. The communication is not just one-way; attendant bees signal to the dancer by pressing the thorax down onto the comb and vibrating their wing muscles at about 350 Hz. The dancer detects these vibrations through the comb using sensors in its legs. When a dancer "hears" the vibrations it may stop moving and regurgitate a sample of food to its audience.

EFFECTS OF THE WEATHER ON FORAGING

Warm sunny weather is ideal for bees to forage. They are unwilling to fly in heavy rain, as they soon become bedraggled, which affects their ability to fly properly. There is also evidence that rain dilutes the nectar in open flowers, making bees' efforts more fruitless. Bees can visually detect the sun's presence even behind clouds, by sensing the brightest portion of the sky. Even on more overcast days, they can still identify the sun's position from the wavelengths of the light reaching the eye. As sunlight is scattered by the upper atmosphere, short wavelengths are disrupted more than long wavelengths. Bees can detect even shorter length ultraviolet light, and the only part of the sky low in ultraviolet is occupied by the sun. In addition, if just a small area of blue sky is in sight, bees are able to detect the polarization of the sun's light: as the sun moves through the sky, the direction of vibration of the sun's light waves changes in a regular pattern visible to bees, but not to humans. The bees use these patterns to navigate, and adjust their dances to reflect the changes in pattern.

Swarming

ALTHOUGH BEEKEEPERS now aim to prevent or control swarming (see pages 198–208 for some methods) because it disrupts normal hive activity and interrupts the honey flow, it is the natural means by which new honeybee colonies are created. Before the advent of movable-frame hives (when skep use was widespread; see page 37), swarming bees were a welcome and valuable part of the beekeeper's year.

The basis of swarming is very simple. When the colony gets to a certain size and density (approximately 36 bees per cubic in of hive volume; about 2.3 bees per cubic cm) new queens are reared in special large queen cells and fed an enriched diet of brood food (royal jelly). Shortly before a new queen hatches, the old queen leaves the nest with about 10,000–20,000 workers (and maybe a few drones) and creates a new nest somewhere else. Only honeybees (and some stingless bees) swarm in this way; their behavior should not be confused with the "swarming" of ants or termites, which is really just a mass of males and females on a synchronized mating flight.

Worker honeybees prepare for swarming many days in advance of the event. The most obvious sign of this is the construction of queen cups, large broad cell bases usually created at the edges of the combs, which will be extended into the extra large cells where new queens will be reared. The signal for the workers to start this construction has long been thought to be some form of pheromone (scent) from the queen.

If a queen is removed from the nest or isolated from direct physical contact with the workers, they immediately start building emergency queen cells, expanded from worker cells.

A cloud of bees forming a swarm is a spectacular sight.

Swarms on their way to a new nest sometimes choose surprising resting places.

It is now known that the queen pheromone, a complex cocktail of (so far) 24 identified and many other yet-to-be-identified chemicals, is secreted from the queen's jaws. The two major components are 9-oxy-(E)-2-decenoic acid (9-ODA) and 9-hydroxy-(E)-2-decenoic acid (9-HDA). This chemical signal is picked up by the many attendant workers that constantly monitor, groom, feed, and exchange body fluids with the queen. With the frequent exchange of foods and fluids across all the members of the colony, this pheromone spreads through the entire workforce.

As the number of bees increases, the pheromone becomes more diluted across the colony, or exchange becomes hampered by the sheer numbers of workers on the comb, and there comes a point beyond which the workers receive less than their daily requirement of $\frac{1}{28,000,000}$ per 1 oz (0.001 mg). This is one of the main triggers of swarming behavior.

The queen also secretes chemicals from her feet, and these mark the comb as she walks over it. As the colony becomes too large for her to patrol all of it regularly, the absence of queen footprints is another trigger for the workers to start building queen cells.

A few days before a swarm leaves, scout bees (usually older bees that know the area well) seek out potential new nest sites. The swarm has two forms: a flying cloud of bees or a resting seething mass. Both are spectacular sights, once seen never forgotten. Observing a swarm in flight is difficult, but there are many anecdotal reports. The cloud of bees can fly many hundreds of yards. It starts off roughly spherical in shape and moves slowly, about ½ mph. Then it gradually speeds up to about 6 mph, and becomes flattened into an ovoid as it moves about 10 feet up in the air.

The scout bees will lead the swarm to the new site from inside and outside the mass of bees. They streak through the flying cloud of bees, heading in the direction of the new site. Other bees land at the chosen site and start releasing a special scent called the Nasonov pheromone from glands at the tip of their abdomens. The smell of this scent will guide the flying cloud to land. The bees may find a resting place on the way, but the scout bees will urge the swarm on to the final site.

On arrival at the nest site, the workers immediately set about making new honeycomb. They are able to do this because, before leaving the old nest, they gorge on honey, each bringing with them an average of about ⅟₈₀₀ per 1 oz (36 mg), roughly 40 percent of the bee's weight. It takes only a few days for new usable comb to be created and the new nest to be established.

MATING

Back in the old hive, meanwhile, the new queens start to emerge from their wax cells. Only one queen can live in a nest so, as soon as the first queen is groomed and dried, she seeks out all the other queen cells on the comb, tears open the cells with her jaws and stings the other queens, queen grubs, and pupae to death.

The new queen then leaves the nest on her "nuptial" flight. Contrary to the usual insect behavior of males seeking out virgin females, the fresh queen seeks out clusters of drones that have previously congregated at treetops and other elevated places, where they have released special attractant pheromones (chemical scents). Once the males see the queen, they pursue her in a mating "swarm," also called a drone comet because of its shape. The mating flight lasts about 30 minutes and may travel over several miles as one or more drones successfully inseminate the queen.

High-speed films have revealed that a mating drone approaches the queen from below (his greatest visual acuity is at the top of his head). The drone first grips the queen with his legs and inserts his endophallus, the penetrative part of the male genitalia, into her sting cavity. He then releases his grip and flips backwards. This sudden flexing body movement compresses the internal organs of his abdomen, forcing an explosive ejaculation of sperm into the female and the breaking of the endophallus with an audible snap. The act of mating is over in seconds and the castrated male falls to the ground and dies. The queen returns to the nest with the sperm stored and the broken endophallus still in place, protruding from her vagina—this is the mating sign, long known as an indication that the queen is no longer a virgin.

The drone mounts the queen during her mating flight.

The broken endophallus may have evolved to serve as a plug, to help the queen retain the large store of sperm that she will keep all her life in order to fertilize her eggs. It does not, however, prevent further drones from mating with the queen, since part of the endophallus's shape and form is adapted to help a subsequent male scoop out a previous mating sign (and hopefully some of the sperm too) and leave his own in its place.

Back at the hive, the workers remove the mating sign.

"A swarm of bees in May is worth a load of hay. A swarm of bees in June is worth a silver spoon."

Winter Survival

A hive in winter in France.

THE WESTERN HONEYBEE (*Apis mellifera*) is one of the honeybee species that is best adapted to a cool climate. This has allowed it to be domesticated far beyond its original subtropical Afro-Asian range. It now occurs throughout almost all of the temperate world, in particular northern Europe, Eurasia, Japan, North and South America, Australia, and New Zealand. It can survive the frequent daily and weekly swings of temperature experienced by wet and windy oceanic countries like Scandinavia, the UK, Ireland, and New Zealand, and also the huge seasonal temperature ranges of continental land masses such as Europe and North America, where long, hard snowy winters give way to searing summers.

A. mellifera has been able to cope with such a temperature range because of its habit of winterclustering during extreme cold weather. Like all insects, honeybees are poikilothermic (rather inaccurately sometimes called "cold blooded") because their internal body temperature is greatly influenced by the external ambient air temperature of their surroundings. Consequently their body temperature drops during winter (unlike homoeothermic "warm-blooded" mammals and birds that maintain a constant core temperature, usually to within 2°F/1°C all year round).

Below a body temperature of about 54°F (12°C) honeybees develop a chill coma; muscular paralysis sets in and breathing movements of the abdomen cease. Above 60°F (15°C) the bees start to become active and can move around, but they are unable to fly below about 82°F (28°C).

Throughout much of the honeybee's geographic range the winter temperatures stay well below these thresholds for prolonged periods, but individual bees are still able to make brief cleansing flights to defecate. This is because they are able to maintain temperatures higher than ambient.

In the depths of winter, no foraging or broodrearing takes place; instead, the bees that have survived the huge energy expenditure of summer activity (the longer-lived autumn bees, see page 111) cluster together with the queen on a few combs at the very center of the hive or nest. Here they form a more or less spherical mass, the winter

A winter cluster of bees.

cluster. Despite the almost complete lack of movement, there is still metabolic activity in the bees: by uncoupling their wings from the large thoracic flight muscles, the bees are able to shiver by rapidly contracting and relaxing the muscles without actually moving the wings. This metabolic activity generates heat to maintain the cluster as close as possible to an optimum 95°F (35°C) temperature; brood fail to thrive below 90°F (32°C) and die above 97°F (36°C).

During the middle of winter the actual temperature in the cluster may fall as low as 68°F (20°C). As the ambient temperature falls lower, the bees pack ever more tightly together, effectively lessening the surface area of the ball and reducing heat loss. If individual bees in the cluster become too hot they migrate outwards to the edges of the cluster to cool off.

As the temperature starts to rise in spring, the bees begin to move more actively and brood-rearing restarts. The first brood cells are near the center of the winter cluster and all produce workers. Drone-rearing cells tend to be near the edge of combs but these are not brought into brood use until later in the season, although, in exceptional cases, a large, strong colony will have expanded its brood rapidly to include comb edges so drones may be produced unusually early. However, if the workers rousing from winter dormancy encounter old drone cells near the middle of a comb, they will usually tear them down and replace them with worker cells.

Man's Best Friend

THE OBVIOUS MATERIAL BENEFITS of bee-keeping are the copious quantities of honey and wax produced by the bees, which have been harvested, valued, and eulogized about for millennia. The true worth of honeybees, however, both to the environment and to humankind's economy, is in the act of pollinating flowers.

Pollen is the plant world's equivalent of sperm and, in the same way, large quantities are produced to ensure that at least some eventually reaches the target egg. As in animals, the egg is fixed in the female part of the flower and the pollen must make the journey from the male flower (or male parts) to the female. Only then can fertilization take place and a new embryo grow in the developing seed.

Almost all grasses, and many trees, have wind-blown pollen. (This is the pollen that causes hay fever.) Tiny airborne pollen grains are cast into the winds until eventually some end up on the right female flowers.

The colony simultaneously exploits many different flower species, but an individual bee tends to forage on one particular type on each foray.

Flowering plants, on the other hand, have taken a different evolutionary approach. By investing in an elaborately colored and structured flower and the

The open flower of the apple tree allows easy access for the honeybee to forage.

provision of sweet nectar, these plants attract insects (and other creatures such as bats and birds) to act as go-betweens and, inadvertently, to transfer pollen from one plant to another as the nectar-feeder moves about.

This investment in flower and nectar production takes a toll on the plant's resources, so it also needs a mechanism to ensure that pollen from the male plant correctly makes the journey to a female of the same species. Although many insects are rightfully credited with pollination, honeybees (and bumblebees) offer a particular behavior that makes them the supreme pollinators: flower fidelity. When a honeybee is out foraging, it tends to visit only one type of flower on each foray. Not only does this greatly increase the chances of successful pollen transfer to the correct flower species, it greatly diminishes the wasteful loss of a flower's pollen to other flowers of different species.

The rise of bees from their wasp-like ancestors 100 million or so years ago is intricately linked to the arrival of flowering plants on the prehistoric scene. Although plant fossils have been dated as far back as 425 million years ago and most insect groups had appeared only about 250 million years

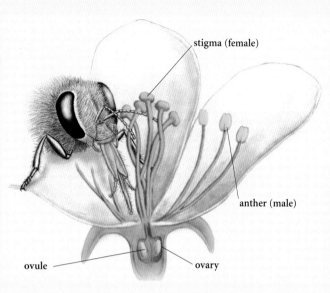

The forager's tongue reaches inside the flower to retrieve nectar. It collects pollen by brushing past the anthers, or by breaking open the anthers with its mandibles (jaws).

ago, the earliest known flowering plants date from just around 130 million years ago. Today, these angiosperms (flowering plants) are by far the most diverse and wide-ranging of plants, and arguably the dominant plant group on the planet. Though bees alone were not responsible for this explosion in diversity, they are a major part of the evolutionary history of flowering plants.

Today, the relationship between plants and their pollinators is still being unraveled. Some solitary bee species are linked to just a single plant species, or genus of closely related species, and they will forage only on this limited range of flowers.

Different bumblebee species have varying tongue lengths so they will visit different flowers with shorter or longer corollas (the tubular back of the flower leading down to the nectaries).

Honeybees have a relatively short tongue, only about ¼ in (6 mm) long, compared to some bumblebees with tongues over ½ in (10 mm) long; nevertheless honeybees are recorded visiting a

OPPOSITE: *Bees transfer pollen from the anther of one plant to the stigma of another.*

ALFALFA

Ironically, one very important animal feed crop, alfalfa, is poorly pollinated by honeybees even though they visit it to take nectar. On first visiting, a bee triggers an explosive release of pollen from the flower's anthers, which strike the underside of the bee. Honeybees seem to dislike this and soon learn to take nectar through a gap in the flower's petals without activating the pollen-release mechanism.

huge range of flowers. They are perhaps best known for pollinating the *Rosaceae*, a vast family of plants that includes, as well as ornamental and wild roses, some of the important human crops: almond, apple, apricot, blackberry, cherry, peach, pear, plum, quince, raspberry, and strawberry. These are "open access" flowers, where the petals are arranged in a broad cup around the pollen-giving and pollen-receiving organs.

These flowers are relatively poor nectar providers and bees have to visit many flowers to fill their honey stomachs before returning to the hive, but this shows how plants and bees have evolved in partnership over many millions of years. The bees are able to forage with ease; they do not have to push heavy convoluted petals out of the way (as in peas and beans) or face a barrage of stiff hairs (as in foxgloves). The plants benefit because a bee must visit many flowers to fill its nectar load, thus ensuring good cross-transfer of pollen.

The financial implications of honeybee pollination are difficult to assess. It is estimated that 10–15 percent of all human food crops are honeybee pollinated, as are a further 10–15 percent of all animal feedstuffs, equating to a business worth many billions of dollars.

It has long been tradition that beehives are located in orchards. The nectar from a large number of flowers of just one tree type produces distinctive and tasty honey; at the same time the fruit harvest is maximized by having a guaranteed pollination brigade. So important has this fruit pollination become that the vast monocultural orchards of the USA in particular now hire huge numbers of mobile hives to visit during flowering times (see pages 70–72) for migratory beekeeping.

Honey, Wax, and Propolis

THE EVOLUTIONARY TRANSITION from carnivorous, insect-hunting wasp-like ancestors to vegetarian, nectar-feeding bees is lost in time, but has drawn many biologists to speculate on how this change occurred. One particularly persuasive theory suggested that the early bee-wasps fed on honeydew. This is the sugary liquid excrement of aphids, which is often little changed from the sugary liquid plant sap that the aphids imbibe in vast quantities and which is spilled onto the leaves as they feed. The bee-wasps also attacked the aphids, which are little more than bags of plant sap on legs (certain modern wasps still feed in this way).

There would have come a point where the larvae back in the nest were being fed so much plant sap that the protein husk of the aphids became insignificant. This was the moment in history when regurgitated plant juices started to be stored, and honey, in some primitive form, arose.

This simple outline obviously does not explain all the complexities of the transition from insect hunting to nectar feeding and storing. For example, the paper combs used by modern wasps would be poor storage vessels, compared to the waterproof and sterile beeswax of the hive. Nectar is a thin and runny liquid and would soon pour out of the horizontally aligned cells on the vertical combs. After more than 100 million years of bee evolution, however, honey is an extreme adaptation to colony life.

NECTAR TO HONEY

Nectar is a solution of sugars in water and, although concentrations can range dramatically from 5–80 percent (most are 34–45 percent), honeybees prefer the sweeter and thicker nectars: foragers returning to the hive with very dilute honey often have difficulty in passing it over to the household bees.

Maturation of honey is done inside the hive after the forager has off-loaded its nectar cargo to one of its sisters.

Even the most sugary of nectars is still a relatively runny liquid, and has to be matured by the household bees before it can be regurgitated as finished honey. The receiving bee retires to a quiet section of the comb and continually manipulates droplets of nectar in her mouthparts. She does this by regurgitating a small amount from the honey stomach and flexing her tongue before swallowing the liquid back and regurgitating some more. This continually exposes the nectar to the air, drying it by evaporation until the approximately 60 percent water content is reduced to about 20 percent. Elsewhere in the hive other bees continually fan the combs with their wings, creating a through draft. Not only does this help with temperature control, it also wafts drying air.

At the same time, enzymes in the honey stomach are changing the chemical composition of the sugary mixture. The major sugar in nectar is sucrose (the same as our familiar table sugar harvested from sugar cane and sugar beet). Sucrose is a disaccharide molecule, in that atoms are arranged in two linked saccharide rings, which are distinctive of all types of sugar compounds. The enzyme invertase splits each sucrose molecule into two smaller molecules: glucose and fructose. The solubility of these two monosaccharides is much greater than that of sucrose, further

BEES' CONSUMPTION AND PRODUCTION

It is impossible to measure precisely how much honey, wax, or brood food (see page 74) is created by a single bee, but it can be estimated. Measurements of pollen-load weight and honey-sac volume can be extrapolated to account for estimated numbers of individuals in the colony, estimated brood loads, and food stores; these can then be equated with measured wax production and known honey harvests. Latest estimates suggest that an average honeybee colony rears 150,000 bees in one year, and that this requires 45 lb (20 kg) of pollen and 130 lb (60 kg) of honey each year. It is also estimated that it takes 3 million foraging trips to accumulate enough nectar to create that amount of honey, and a further 1.3 million trips to collect the pollen. Thus, an average colony flies about 12.4 million miles in a year's food-collecting.

allowing the bee to make the highly concentrated sugar mix (about 80 percent) that is honey.

Eventually, after about 20 minutes of drying and enzyme manipulation, the bee deposits the honey into a storage cell.

Honey is an intensely energy-rich food (1,380 calories per lb or 3,040 kcal/kg), which can easily be diluted with water and fed to larvae and to the over-wintering cluster of huddling bees. Its ease of storage and readily digested chemical form make it a supremely important foodstuff for both bees and humans. It is also protected against microbial attack by its high sugar (and therefore low water) content. Any yeast or bacterial spores that do land on the honey rapidly dehydrate and die as water passes out through their cell walls by osmosis, from the relatively high water concentration inside the cells into the honey, which contains a much lower concentration of water.

WAX PRODUCTION

Underneath the worker bee's abdomen, just inside each of the overlapping armored plates (sternites) that cover the fourth to the seventh segments, are eight paired wax glands. The precise mechanism of wax production is not completely understood, but recent electron microscope studies provide some answers.

The surface of the glands, termed the wax mirrors, are smooth, shining oval plates, which lack the tough protective cement layer that gives the rest of the insect's cuticle its strength. A series of aligned secretory cells are exposed across the mirror surface; each secretes beeswax, which forms first a

A worker passing a wax scale forward.

LEFT: *Honey is deposited in the comb storage cells by household worker bees after they have matured it from nectar passed to them by the foragers.*

RIGHT: *Other bees fan their wings, helping to maintain a steady flow-through of drying air.*

blob, then, as further wax is extruded below, a flexible column or strand. The columns make a flat scale that can reach ⅛ in (3 mm) long and 1/50 in (0.5 mm) deep and sometimes be seen protruding from beneath the abdominal plates.

During comb construction the bee scrapes off the wax scales (which vary in size depending on how long each bee has been growing them) from the mirrors, using the spines of the pollen brush on its back legs, and then passes the wax scales to the front legs, where they can be held to be chewed by the mandibles. Mixing the wax with saliva makes it more malleable; it is chewed to the right consistency for a perfect building material.

When a swarm first lands at a new nest site, bees heavily laden with large scales chew and deposit them in a ridge on the roof of the cavity. Eventually there is enough of a hanging blob for bees to start constructing the first cell in what will become the regular hexagonal array. Roughly plas-

tered edges are carefully shaved to produce the six beautifully precise sides so distinctive of honeycomb cells. Further deposits of wax continue to increase the mound until it can be hollowed out from behind to make the back-to-back cell arrangement with its particularly economical use of building material. The cells can be reused again and again, and the wax cappings (covers) over the full honey storage cells and encasing the pupae are quickly reused after removal.

The wax is a tough material. It is a mixture of complex long-chain hydrocarbons, esters (combined fatty acids and alcohol-type molecules), and upwards of 300 other components. It is very strong and yet light and flexible, able to support 40 times its own weight of honey. The same wax is used to make the huge slab nests of the giant honeybee (*Apis dorsata*; see page 92).

There is a relatively precise relationship between colony forage success and the building of new

The abdomen of a worker bee, showing the wax glands.

honeycomb. At one time it was thought that an instructive message was being issued to encourage bees to stop foraging and instead to lay down wax to increase the size of the combs for honey storage. Now it is thought likely that, if individual bees are unable to find suitable cells in which to deposit their honey, they digest all of the honey instead and this additional nutrition fuels increased wax production, leading them to expand the nest.

HONEY FUELS WAX PRODUCTION

It is estimated that it takes about 1 lb (1 kg) of honey to produce 1 oz (60 g) of beeswax; that's 15½ lb (7 kg) of honey to produce the 14½ oz (400 g) of wax found in an average hive.

PROPOLIS

The final bee product is propolis. This is a tough lacquer-like substance produced from plant resins collected by the bees from tree buds and seeping wounds in plants and trees.

Using their jaws, the bees take it and transport it to the nest in their pollen baskets. Back in the colony it is used to block small gaps in the hive envelope and to strengthen the structural integrity of the combs. Its primary (and original) use may have been to coat the interior of the nest void, strengthening crumbling walls and ceiling, and blocking up all but the main access to the nest. Closing up unwanted entrances into the hive may prevent predators, parasites, and honey thieves from entering. Dwarf honeybees (*Apis florea* and *A. andreniformis*) use a sticky band of propolis as a protection against ants.

ABOVE: *Propolis (the dark resinous substance) deposited by bees on the top bars of hive frames.*

OPPOSITE: *Bees clustered within natural wax honeycomb.*

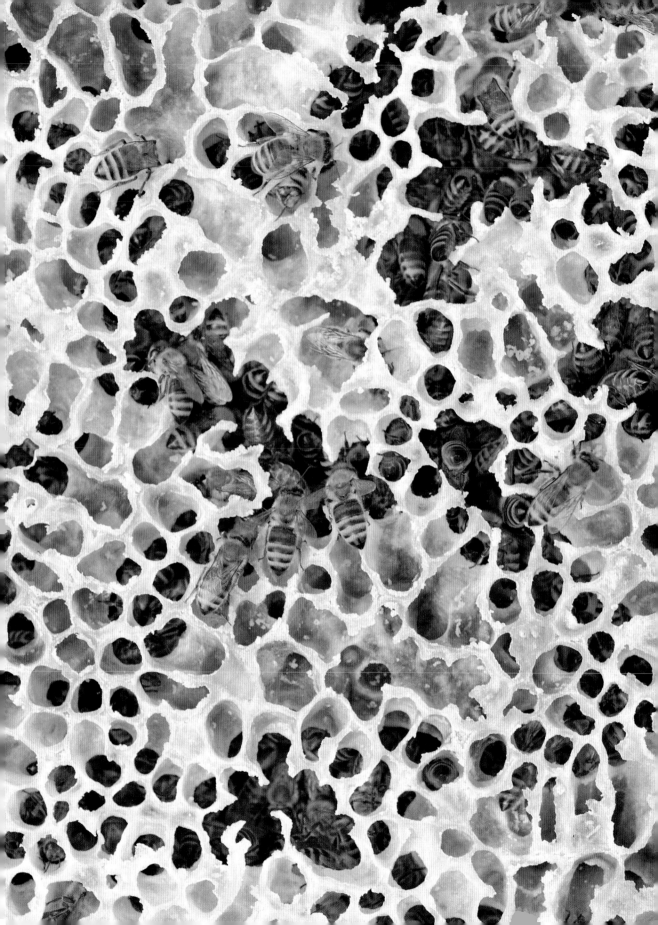

When Bees Sting

BEES HAVE INHERITED stings from the wasp-like ancestors from which they evolved. But, whereas wasps use their stings as weapons of attack—to subdue, paralyze, or kill their prey victims—bees sting only in defense. There is one exception to this: the virgin queen's fratricidal attack on her siblings after emerging from her queen cell (see page 116).

The sting is a modified part of the ovipositor, the egg-laying tube, and consequently only the females (workers and queen) possess one. The drones, like male wasps and ants, are unarmed and harmless.

The sting is a complex structure. A sharp stinging shaft of two paired, barbed lancets run backwards and forwards along a central pointed stylet. A muscular articulation controls the movement of the lancets. The venom gland, together with the venom storage sac, provides the poison, which is pushed down the shaft as the sting shaft is forced into the victim. The sac contains about $\frac{1}{50,000}$ oz (0.6 mg) of liquid venom, or about $\frac{1}{200,000}$ oz (0.15 mg) dry weight. The amount injected is usually much less than this, but depends on how long the sting is allowed to pump venom into the wound (see page 130).

When a bee stings a human (or mammal or bird), it grips the "prey" tightly with its legs and pushes the backward-pointing sting shaft into the skin of its foe, forcing the rigid stylet to angle vertically downwards. Under muscular control, the two sliding lancets move rapidly backwards and forwards, alternately, past each other, effectively sawing into the flesh. The backward-pointing barbs help the bee dig deeper into the wound. At the same time, the rhythmic pumping of the stylets pushes venom down through the venom canal formed between the stylet and lancets and injects it into the wound through a small opening near the tip of the lancets.

Honeybee venom is a complex cocktail of protein chemicals. Some of these are toxins, others are enzymes to attack and break down the proteins in the skin and flesh. Many are still poorly understood.

A major component is melittin, a small protein molecule that ruptures red blood cells and bursts other cells in the skin and blood vessels, releasing large amounts of histamine into the tissue. The histamine causes inflammation (including swelling and redness) by dilating blood vessels, which serves to increase the spread of the venom. Histamine is also a component of the venom in small amounts, but insignificant compared to the human body's own release. Even with multiple stings the histamine injected and released is not dangerous to humans, but is an important part of the venom's toxicity to other insects. An enzyme, hyaluronidase, digests and breaks down the tissue in the flesh, increasing the rate at which the venom constituents can diffuse. Apamin is a neurotoxin, interfering with the normal functioning of nerves in the area.

The sting shaft of the bee is pushed into the skin of its foe.

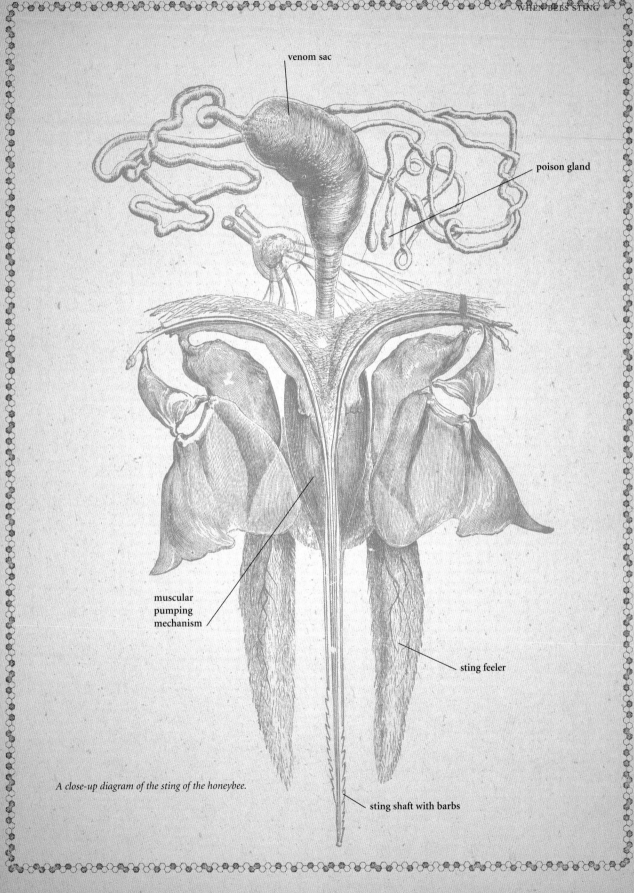

venom sac

poison gland

muscular
pumping
mechanism

sting feeler

A close-up diagram of the sting of the honeybee.

sting shaft with barbs

The bee's sting is a weapon of last resort. The barbed stylets become embedded in human skin. If the bee is brushed away, the sting, complete with venom sac and muscular pumping mechanism, remains. The insect is disemboweled and soon dies; the pump continues to push venom into the wound for 30–60 seconds, increasing the pain.

The act of stinging releases alarm pheromones (chemical scents) into the air, which alert and recruit other bees to join in the attack. Even when a bee is brushed away, the pheromones continue to be released by the disembodied sting, which leads other bees to pursue a victim unless he or she moves well away from the nest.

THE EFFECT OF A STING

The immediate local effect of a bee sting is intense pain at the site of the wound. This is followed, an hour or more later, by redness, swelling, tingling, irritation, and itching. In some cases this may last several days.

A few minutes after the initial sting, as the venom is distributed around the body in the bloodstream, a general allergic reaction may develop, with a rash, wheezing, nausea, abdominal pains, and a feeling of faintness. In the case of multiple stings, these allergic reactions are greatly amplified and, even though a single sting injects only a minute amount of toxin, 100 stings are usually enough to cause hospitalization. Even 20 stings can be fatal to some susceptible individuals.

In extreme (and luckily very rare) cases a single sting can completely disrupt the human body's allergic response; the injection of even a tiny amount of the bee's venomous cocktail starts a cascade of reactions as the body's immune system wildly overreacts to the alien chemicals.

In such cases, the victim suffers anaphylactic shock: breathlessness, wheezing, vomiting, and general confusion are accompanied by rapidly falling blood pressure and loss of consciousness. Without any immediate medical intervention—usually involving the administration of injected adrenaline (epinephrine)—death can result.

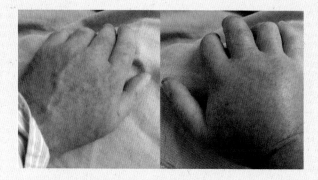

Swelling resulting from several stings to the right hand;
the left hand is shown for comparison.

Bee Habitats

IN THE NATURAL WORLD, honeybees forage wherever they can, taking advantage of local abundances of particular species at different times of the year. Almost without exception, flowering plants produce nectar and pollen—flowers having evolved to entice pollinators (see page 120)—and, in the wild, honeybees visit a vast array of native plant species. However, humans have altered the landscape over a large proportion of the globe through agriculture, and many crop plants are now cultivated in sufficient mono-culture to make them single forage sources for entire hives or even for commercial banks of hives.

Other changes to the landscape have meant that certain wildflowers have also become key forage plants.

The flower fidelity practiced by honeybees confers an advantage on both plants and bees (see page 119). It also means that beekeepers can take advantage of the bees' targeted flower visits to produce honey created from just a single nectar source. These varietal honeys have differing flavors, tones, colors, turbidities, and viscosities, and allow beekeepers to market their products much as the makers of different wines or cheeses market theirs.

A typical garden honeybee habitat.

PLANTS FOR HONEYBEES: EDIBLE HONEY: EARLY SPRING/LONG FLOWERING

The following lists of crop and wildflowers are arranged according to the flower season for temperate North America and most of Europe, with notes on flowering seasons for the southern hemisphere.

Also given are scientific names and the geo-graphical regions or zones of importance, together with approximate months of forage availability and notes on honey.

ALMOND
Prunus dulcis California, France (Feb).
Honey generally regarded as being of poor quality, mostly used in the bakery trade.

APPLES
Malus species; in particular wild crab (*M. sylvestris*) and pears (*Pyrus communis*).
Worldwide, wherever orchards are grown. Mainly Apr–Jun in northern hemisphere but highly variable, dependent on district and season; huge variety of cultivars available. Honey light amber, with good flavor, granulates quickly.

AVOCADO
Persea americana
Original range of southern and central America now extended to California, Florida, southern Spain, South Africa (Mar–Apr), Australia and New Zealand (Aug–Oct). Dark amber honey with strong flavor of caramelized molasses.

COCONUT PALM
Cocos nucifera
Hawaii, Florida, and West Indies (all year). Strong amber honey.

CLOVERS
Trifolium species: in particular alsike (*T. hybridum*) and white (*T. repens*)
Throughout the world, including North America and Europe (May–Oct) and also Australia and New Zealand (Nov–Apr), where they have been widely introduced. Barseem (*T. alexandrinum*): an important early source in Egypt (Feb–Mar).
Red clover (*T. pratense*) has a corolla rather too long for honeybees, so is visited primarily by bumblebees. Honey is pale and gently flavored with the scent of the flowers; often used in blending.

DANDELION
Taraxacum officinale (highly variable and sometimes divided into hundreds of separate species).
Important, worldwide early spring wildflower, Mar–Sept in northern hemisphere, Oct onwards in southern. Golden, coarse-grained honey.

GOOSEBERRY
Ribes uva-crispa
Europe (Mar–May) and New Zealand (Sept–Nov).
Good honey from early source.

GUM TREE
Eucalyptus species (many)
Extremely important in Australia where large proportion of honey comes from this source (Oct–Jan), and also in California (Apr–Jul). Distinctive flavor and odor. Yellow box gum (*E. melliodora*) produces pale and thin honey; honey from red box (*E. polyanthemos*) is dark amber and very dense.

HOLLY
Ilex aquifolium, I. opaca, I glabra
Western and southern Europe and southern USA (Apr–Jul). Pale color, finely flavored.

LEATHERWOOD
Eucryphia lucida
Tasmania (almost all year, but mainly summer, Nov–Apr). Honey strongly spicy, an acquired taste.

MANUKA
Leptospermum scoparium
New Zealand (most of the year but especially Sept–May). Full-bodied, herbaceous, sweet-tasting honey. Especially reputed for its antibacterial qualities.

MAPLE TREE
Acer species
Sycamore (*A. pseudoplatanus*) important in the UK, mountain maple (*A. spicatum*) in eastern Canada and northeast USA (Apr–Jun). Honey pale yellow or greenish, mild flavor, sometimes regarded as indifferent.

MESQUITE
Prosopis glandulosa
Northern Mexico and southern USA from California to Mexico (Mar–Sept). Honey varies from white to amber, with smoky scent of molasses or brown sugar.

NEEDLE BUSH (silky oak)
Hakea and *Grevillea* species
Australia (all year depending on species). Sweet, clear honey.

ORANGES (also **lemons**, **limes**, **grapefruits**, etc.).
Citrus species
Mediterranean Europe and California, Arizona, Texas, and Florida (Mar–Apr) and New Zealand (Sept–Oct).
Delicious honey, pale and dense with distinct fruit taste and echoes of blossom.

ROSEMARY
Rosmarinus officinalis
Europe, especially Mediterranean region (Mar–Nov).
Medium-bodied honey with a thick texture.

THYME
Thymus species
Mediterranean Europe, North America (May–Sept), and New Zealand (Nov–Mar).
Honey has intense aroma and aromatic flavor.

TUPELO (sour gum)
Nyssa sylvatica
Southeastern Florida and Georgia (Apr–Jun). Useful wetland forage, hives kept on high platforms or boats to avoid flooding. Honey is light amber in color, smooth and very sweet.

ACACIA (wattle, mimosa)
Acacia (large number of different species)
Europe, Canada, China, California and Arizona (May–Jul), Australia (Nov–Jul). Honey very clear, liquid and pale.

BRAMBLE (blackberry)
Rubus fruticosus
British Isles and Canada (Jun–Sept). Honey is medium colored, but rather coarse flavored.

ALFALFA *Medicago sativa* (also wild *Medicago* species)
Especially USA (west of the Mississippi) and British Columbia, dry belt, Canada (Jun–Aug). Honey is pale, white or light amber, with a scent of beeswax, also sometimes described as minty.

BUCKWHEAT
Fagopyrum esculentum
Europe, USA (California and northern states), and Canada (Jun– Sept). Honey is dark, sometimes purple-brown, and strongly flavored with hints of molasses and malt, not favored by all but recommended in "hearty" baked goods and barbecue sauces.

BLUEBERRY (bilberry, whortleberry) and **cranberry**
Vaccinium species
Western and northern Europe, Eastern Canada and Great Lakes region (usually only 2 weeks in May–Aug depending on species). Slightly fruity (more so in cranberry honey), faint buttery finish.

CANOLA
Brassica napus subsp. *oleifera*
Increasingly important in Europe and North America as this crop is extensively planted; also in Asia (May–Aug). Fine, clear honey that granulates quickly.

BORAGE
Borago officinalis
Europe and North America (Jun–Sept). Very pale or white honey. Minor nectar source from wild plants, good honey once every 5–10 years where grown for borage seed oil.

FALSE ACACIA
Robinia pseudoacacia
North America and Europe (Jun–Jul). Light honey with good density and flavor.

HAWTHORN
Crataegus monogyna and other species
Northern and western Europe, North America (May–Jun), and New Zealand (Nov–Dec). Dark amber honey has a rather nutty flavor.

RATA TREE
Metrosideros species
New Zealand (Nov–Mar).
White, pale clear honey.

KNAPWEED (hardheads, blackhead)
Centaurea nigra
Important in Ireland (Jun–Sept). Honey is golden, thin, and sharp-flavored.

SAINFOIN
Onobrychis viciifolia
North America and Europe, especially on limestone soils (Jun–Aug). Lemon-yellow honey.

LAVENDER
Lavendula species France and Spain (Jun–Jul).
Honey is pale golden, pleasant tasting, with fine granulation to a texture resembling butter. Farmed lavender, for perfume and oil, is of little use to honeybees because it is harvested at the height of its flowering period.

SWEET CHESTNUT
Castanea sativa
Southern Europe (May–Jun). Honey is dark amber, sharp, and bitter. The unrelated horse-chestnut (*Aesculus hippocastanum*) is a minor nectar source, but is locally important for pollen and the sticky resin covering its young leaf buds (propolis).

LIME TREE (linden, basswood)
Tilia platyphylos, *T. cordata*, *T. americana*, *T. x vulgaris*
Europe, Canada, and north central USA (2 or 3 weeks in Jun–Aug depending on latitude, season, and geology). Particularly important in cities where limes are often planted as street trees. Light greenish or amber honey with slight minty taste.

THISTLE
Cirsium arvense (and other species)
Worldwide, especially in uncultivated areas in Australia (Dec–Feb), Canada, California, Florida, and Great Lakes region (Jun–Aug). Pale, flavorful honey.

MELILOTS (sweet clover)
Melilotus alba, *M. officinalis* and *M. indica*
Worldwide, especially North America and Europe (Jun–Sept) and Australia (Jul–Mar). Pale greenish-yellow honey with slight cinnamon flavor.

TULIP TREE (yellow poplar)
Liriodendron tulipifera
Eastern USA, Ohio Valley, and Appalachians (May–Jun). Dark amber honey with strong flavor.

PLANTS FOR HONEYBEES: EDIBLE HONEY: LATE FLOWERING

COTTON
Gossypium hirsutum (and other species)
Southern USA (California to North Carolina), Central Asia, and Egypt (Jul–Sept). Honey is lightly flavored and pale amber in color.

HEATHER (Bell)
Erica
species Important in Scotland and other parts of UK, and Ireland (Jul–Sept). Honey is deep brown "port wine" color.

FIREWEED (rosebay, willow-herbs)
Chamerion angustifolium and *Epilobium* species
Worldwide, but especially Europe, Canada, and northern USA (Jul–Sept). Pale, white, subtle flavor with tea-like notes.

HEATHER (Ling)
Calluna vulgaris
Europe, especially Scandinavia, Scotland, and moors and lowland heaths in England (Jul–Sept). Dark golden brown honey sets to a jelly and has to be extracted with a press.

FUCHSIA
Fuchsia species
South and North America, Europe (especially noted in Britain and Ireland), and New Zealand (Jul–Sept). Because of the hanging nature of the flowers, nectar is not washed out by rain. Honey is a light color, but very mild and insipid.

IVY
Hedera helix
Europe, Asia, and North America (Oct–Dec). Honey grayish-white, with delicate odor and bitter flavor; creamy consistency and granulates quickly. Nevertheless an important later source of nectar.

GOLDENROD
Solidago canadensis
North America, although now widely planted as garden flower in Europe and elsewhere (Jul–Oct). Pale colored, slightly strong almost spicy flavor.

SUNFLOWER
Helianthus anuus
North America, central and southern Europe, Russia, and China (Jul–Oct). Light amber honey with subtle citrus undertones.

POISON HONEY:
A BOTANIST'S TALE

Frank Kingdon-Ward (1885–1958) was a British botanist, explorer, and plant-hunter of great repute. In his book *Plant-hunter's Paradise* (1937) he recounts how he and his colleague, zoologist and hunter Lord Cranbrook, traveled far up the sources of the River Irrawaddy in northern Burma, almost to Tibet. "One day the local [Tibetan] headman brought us mead, brewed from the local honey, which at this season must have been rhododendron honey. It had no ill effect, other than making us completely drunk. But when in early May he brought us real honey, it was a different story."

When Kingdon-Ward got back to the camp one evening he found Cranbrook in bed, mildly delirious and unsure how he got there. Apparently he had set out as usual that morning, with his gun, to explore the riverbank "when without warning he had collapsed and fallen into the backwater. The water had revived him somewhat, and though unable to walk he had climbed out of the river and shouted for help."

Later, Kingdon-Ward's manservant, Mano, also got hold of some. "No sooner had he eaten some—and doubtless he was immoderate—than he too felt ill, and retired to bed. The symptoms might be described as those of acute alcohol poisoning. Proud of my supposed immunity, for I had suffered the same vapours in previous years, I continued to eat popcorn impregnated with wild honey. But after a day or two, feeling listless, I began to fear chronic poisoning myself and gave up honey."

Kingdon-Ward was not entirely sure how poisonous rhododendron honey was; the local Tibetans seemed to eat it without ill effects and he concluded that they had developed some form of immunity. He then pondered, in typically scurrilous fashion:

I cannot help wondering if, with increasing cultivation of rhododendrons in the south and west of England, most of which flower between May and June, there may not presently be cases of honey poisoning in this country. That would be something for the press.

The rhododendron Wardii *var.* puralbum, *named after Frank Kingdon-Ward.*

POISONOUS OR UNAPPETIZING HONEYS

MOUNTAIN LAUREL (lambkill)
Kalmia latifolia
Eastern USA, Maine to Florida, and Indiana, introduced to Europe as an ornamental (May–Jun). All parts of the plant are toxic and the honey is poisonous to humans.

PRIVET
Ligustrum vulgare (and other species)
Europe, Asia (May–Jun), and Australasia (Nov–Dec). Honey is very strong flavored, making it objectionable and unpalatable unless it is blended with lighter honeys.

RAGWORT
Senecio species
Europe, North America (Jun–Nov), Australia, and New Zealand (Dec–May). A rank, bitter-tasting honey.

RHODODENDRON
Rhododendron ponticum and other species
Southern Europe and Asia, particularly south and west of the Himalayas (year-round flowers depending on species and location). Honey is poisonous.

SPURGE
Euphorbia species
Certain species in South Africa (where the genus is particularly diverse) produce noors (or noorsdorn) honey, which is bitter to the taste, can cause a burning sensation in the mouth and throat, and occasionally poisoning (flowers all year depending on species).

A Bee-Friendly Garden

No matter how small the domestic garden plot, there is always the potential to grow plants and create an environment that will encourage honeybees to visit. Garden flowers are often bred for their showy blooms; although this sometimes concentrates attention on abnormally expanded or multiplied petals of brighter or stranger colors, the vast majority still produce nectar and pollen, and are attractive to bees.

Despite the commercial attraction of marketing varietal honeys, each with its own distinctive aroma, texture, and consistency, most honeys are blends. Varietal honeys are usually only possible if the bees are presented with an agricultural monoculture—orchards, crop fields, or uniform woodlands—but blends are either created naturally by the bees collecting different nectars at

The globe thistle (Echinops) *is highly attractive to bees.*

different times, or are mixed commercially at the bottling plant.

Gardens are, by design, floristically diverse, so that different types, sizes, and colors of flowers can be enjoyed together, and at different times throughout the year. This diversity is also influenced by the variety of aspect, shelter from hedges, fences, and tree lines, soil type, and landscaping. Even the smallest garden is important, being part of an array of other gardens, adjoining, contrasting, and complementing each other. It is this diversity, the patchwork mosaic of small areas and corners, that makes them so attractive to bees. Wildflower or mixed floral honey, produced by bees foraging on various wild and garden flowers, is a valuable brand for small-scale honey producers.

FLOWERS TO ATTRACT HONEYBEES

Bees are after pollen and nectar, so blooms artificially bred for their double- or multiple-petaled forms (pompom chrysanthemums, for example) are less attractive to the pollinating bees, as they may be unable to penetrate through the petal barrier to get at any nectar.

Simple flowers, with petals arranged in an open form, are the best attractors of bees as they have to do little work to get at the nectar: these include asters, bramble (*Rubus*), cherry (*Prunus*), and traveler's joy (*Clematis*). Likewise, plants that have multiple small flowers making up inflorescences are highly attractive as the bees can sample many nectaries on just a single landing: examples include globe thistle (*Echinops*), goldenrod (*Solidago*), and lavender (*Lavendula*).

The best flowers for a particular garden will depend very much on where the garden is, its

It is the diversity of flowers in gardens that makes them so attractive to bees and other wildlife.

geographic location, annual rainfall and sun hours, latitude, and daily and seasonal temperature swings. Luckily, many hardy exotics can be planted in gardens throughout the world—from cold temperate to subtropical regions.

One of the major tenets of gardening—always use a variety of plant sizes, from small creeping species that overhang the path, through sprawling and upright herbaceous species, to semi-shrub plants and finally bushes and trees—will also benefit honeybees in any environment. The plants provide a variety of complementary forage sources, not just because they are different species, which might flower at different times of the year, but also because plants of varying sizes warm up in the sun at different rates each day and have nectar flows

earlier or later in the day. They are sheltered from the cold and wind by each other, which also assists the bees, allowing them to forage without buffeting even on gusty days.

There is constant debate among garden designers (and naturalists) about the competing needs of wild and garden flowers, and the respective benefits they offer to bees and other wildlife. In terms of providing nectar and pollen there is no simple answer: native weeds and exotic blooms offer both.

In wildlife terms, native plants are more important when it comes to food plants for caterpillars and other larval forms, which often tend to avoid the foreign exotics favored in grander gardens. One benefit native plants often have over garden ornamentals is that they have already adapted to the local weather and climate. As a consequence, they require much less gardening and tending and are better able to survive (or perhaps even thrive) in the extremes of heat, cold, wet, and drought, which can challenge more fragile exotics.

Leaving a wild nature corner in the garden has benefits for all wildlife, not just honeybees. While the bees (including bumblebees and many of the smaller solitary species) will visit the native wildflowers on offer, butterflies and moths will lay eggs on their caterpillars' native food plants, and birds are attracted by seeds and fruits. In addition, a non-tended part of the garden offers the sanctuary of shelter, where invertebrates (and larger animals and birds) can rest during the day or night and hibernate through winter.

Other attractions for bees

Garden ponds have greatly improved the lot of amphibians, dragonflies, and other aquatics that suffered with the decline of stock ponds in the wider landscape. Honeybees, too, benefit from the access to drinking water. They will collect water to dilute stored honey when it is being fed to the brood, and also to smear it on the comb to cool it through evaporation during very hot weather. Many flowering waterside plants will also attract foraging honeybees.

Traditionally hives have been kept in orchards, but crop-tree flowering periods are usually short, so a diversity of wildflowers between the trees and along the hedgerows allows the bees to forage successfully at different times of the year. Even arable fields, long derided as being biologically sterile and useless for wildlife, can become bee forage zones by allowing headlands (unplowed margin strips of wildflower habitat) to grow along field boundaries, hedges, and roadsides.

Bees are very susceptible to pesticides, particularly to systemic poisons, in the chemicals that enter the plant tissues to kill any insect pests eating the leaves, flowers, seeds, or fruit. Honeybees ingest these toxins with the nectar, then share their poisonous harvest with the other bees in the colony. Garden sprays and industrial crop dusting have both been blamed for honeybee deaths.

Anyone contemplating a few hives in a small back garden should be wary of the bees' potential effect on children, pets, and neighbors. Secluding the hives so that they are out of sight and out of mind is a real benefit. A tall fence also makes the bees move upwards when they set off on their foraging flights, so they are well above head height before they pass across neighboring properties. Cats and dogs may, at first, show incautious curiosity about the bees' comings and goings, while chickens, despite their propensity to peck at anything on the ground, are often regarded as suitable honeybee neighbors in the farmyard.

SOME COMMON GARDEN PLANTS SUITABLE FOR TEMPERATE NORTH AMERICA AND MOST OF EUROPE

*illustrates plant pictured

APIACEAE, carrot family
*Angelica (*Angelica archangelica*)
Lovage (*Levisticum officinale*)

ASTERACEAE, daisy family
*Cosmos (*Cosmea bipinnatus*)
Chicory (*Cichorium intybus*)
Fleabane (*Erigeron speciosus*)
Globe thistle (*Echinops ritro*)
Ox-eye (*Leucanthemum vulgaris*)
Shasta daisy
  (*Chrysanthemum maximum*)
Sunflower (*Helianthus annuus*)
Yarrow (*Achillea filipendulina*)

BORAGINACEAE, borage family
Alkanet (*Anchusa azurea*)
Borage (*Borago officinalis*)
*Viper's bugloss (*Echium vulgare*)

BRASSICACEAE, cabbage family
Candytuft (*Iberis matronalis*)
Charlock (*Sinapis arvensis*)
Golden alyssum (*Alyssum saxatile*)
Honesty (*Lunaria biennis*)
Mignonette (*Reseda odorata*)
Rockcress (*Aubrieta deltoides*)
Sweet rocket (dame's violet)
  (*Hesperis matronalis*)
*Wallflower (*Cheiranthus cheiri*)

CAMPANULACEAE, bell-flower family
*Bellflowers (*Campanula* species)

CANNABACEAE, hemp family
*Hop (*Humulus lupulus*)

DIPSACACEAE, teasel family
Scabious (*Scabiosa* species)
*Teasel (*Dipsacus fullonum*)

***ELAEAGNACEAE, Elaeagnus family**
Oleaster (*Elaeagnus argentea*)

FABACEAE, pea family
*Broom (*Cytisus* species)

GROSSULARIACEAE, goose-berry family
*Currant (*Ribes* species)

ONAGRACEAE, willowherb family
*Evening primrose (*Oenothera biennis*)

LAMIACEAE, mint family
Marjorum (*Origanum vulgare*)
*Mint (*Mentha* species)

POLEMONIACEAE, phlox family
*Phlox (*Phlox paniculata*)

LILIACEAE, lily family
*Grape hyacinth (*Muscari botryoides*)

PRIMULACEAE, primrose family
*Polyanthus (*Primula* species and cultivars)

LIMNANTHACEAE, meadow-foam family
*Poached egg plant (*Limnanthes douglasii*)

*ROSACEAE, rose family
Cotoneaster (*Cotoneaster vulgaris* and others)

LYTHRACEAE, loosestrife family
*Loosestrife (*Lythrum* species)

RUTACEAE, citrus family
*Bergamot (*Monarda didyma*)

MALVACEAE, mallow family
Hollyhock (*Alcea rosea*)
*Mallow (*Malva* and *Lavatera* species)

VALERIANACEAE, valerian family
*Valerian (*Centranthus ruber*)

Threats to Bees

❦

Fifty thousand armed warriors ready to die for their queen might sound like a formidable army not to be trifled with, but honeybees are under constant threat of attack. The stores of honey, pollen, and wax are enticing prizes to be plundered, and their own bodies, and those of their brood, are a ready source of valuable protein for hungry predators. And, living as they do, in large numbers in close proximity on the combs, they are prey to parasites and diseases.

ATTACKS FROM ANIMALS AND OTHER INSECTS

The store of honey is an attractive meal for many a sweet-toothed plunderer. In the wild of their native range, the honeybees' most serious raider is the honey badger or ratel (*Mellivora capensis*) of Africa and India (see page 147). This tough, leather-skinned predator devours comb, honey, pollen, and brood and seems impervious to the bees' stings. It is often aided by a relationship with the honeyguide birds (*Indicator indicator* and *I. variegates*), which draw attention to a bee colony by singing and displaying close to the nest. After the honey badger has taken its fill, the birds feed on the remnants of wax comb among the debris.

Elsewhere in the world vertebrate honey and brood thieves include badgers and bears. Rats, mice, hedgehogs, lizards, and other small creatures are known to enter hives, but this is usually merely for shelter in the winter. More information about dealing with pests of this nature if you are keeping hives is given on pages 234 and 242–43.

In the days of skep use, the death's head hawk-moth (*Acherontia atropos*) was a well-known visitor. Its proboscis is very short and stout, unlike the delicate curled tongues of other moths, with

which they sip nectar from flowers, and it is known to drink honey through it. Modern hives, with smaller entrance slots, are less regularly entered.

Ants are fierce honey hunters, but are usually kept at bay by the propolis seal around the combs. And honeybees themselves are not above raiding neighboring hives if given the chance. Only the constant vigilance of guard bees at the nest entrance keeps them out. Large hornets (*Vespa* species—see page 81) and some of the smaller social wasps or yellow jackets (*Vespula* and *Dolichovespula* species) actively attack and kill bees and will also raid the hive for honey and brood, causing havoc and devastation.

The honey and wax in the hive are also attractive to various insects, which readily infest the hive. The bee louse (*Braula coeca*) is a bizarre wingless fly. The larvae burrow into the wax cappings of the comb, feeding on honey and pollen, and the adults, about $\frac{1}{16}$ in (1.5 mm) long, cling to honeybees' bodies, but are not recorded as doing any harm.

The small hive beetle (*Aethina tumida*) was, until recently, a minor nuisance in hives in southern

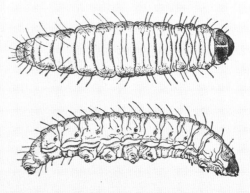

The caterpillar stage of the wax moth.

The death's head hawkmoth raids skeps and other less well-protected hives.

Africa. The African strains of honeybee had long evolved an aggressive housekeeping defense against it but, when these beetles were accidentally introduced to the USA in 1998 and Australia in 2002, it became a major problem. One of a group of beetles that usually breed on rotting fruit or fermenting tree sap, these larvae (up to 30,000 per hive) burrow freely through the wax, feeding on honey and pollen, destroying the integrity of the comb and polluting the honey with their feces. If you keep hives yourself, information about dealing with an infestation of small hive beetles can be found on page 245.

Several pollen- and nectar-eating insects will also take advantage of less-well-guarded honey stores. The large hive beetle (*Hyplostoma fuligineus*) is a sizable scarab beetle, occasionally found in African hives. *Euphora sepulchralis*, another usually flower-feeding scarab beetle, sometimes visits hives in the southern USA. At

present these are little more than occasional honey thieves.

Wax moths (*Achroia grisella* and *Galleria mellonella*) can cause major damage in hives, especially those with low populations in which large areas of the comb are not covered with bees, or in hives not regularly examined during the winter resting period and in stored frames. The caterpillars bore through the combs, feeding on honey, pollen, and brood, and leave behind thick tangled masses of messy silk tubing. Advice on dealing with this type of attack can be found on page 244.

Flying bees are always vulnerable to attack from predators in the air. Insectivorous birds of all kinds are indiscriminate in their prey-taking, but bee-eaters (as their name suggests), shrikes (butcher birds), fly-catchers, swallows, swifts, martins, and wagtails are well-known for eating bees. Woodpeckers will also gouge holes in wooden hives to get at the bees and larvae within. Suggestions for protecting bees from attacks by birds are to be found on page 242.

After birds, spiders are perhaps the most important bee killers. They use a variety of webs, snares, and pouncing tactics. There are also plenty of predatory insects that will eat bees; these include robberflies, dragonflies, and mantids. One group of small solitary wasps, *Philanthus triangulum* and its relatives, specialize in killing honeybees to feed their young, earning them the name bee wolves.

MITES AND PARASITES

Honeybees (like many other insects) are prone to parasitic mites. These microscopic animals penetrate the soft membranes between the hard armored plates of the bee's body and suck out hemolymph, the bee equivalent of blood. The tracheal mite (*Acarapis woodi*) lives inside the tracheae, the breathing tubes in the thorax and abdomen of the bee. Heavy infestations block the tubes, reducing the vigor and foraging ability of the bees; severely debilitated bees cannot fly and eventually die (see page 239 for tracheal mite disease).

Varroa mites (see page 240) have long been known as relatively benign parasites of the Eastern honeybee (*Apis cerana*). However, when they spread to the Western honeybee in Russia and Japan in the 1960s, they found a new host bee with little or no natural defenses and caused huge losses to bee numbers and honey production.

Varroa jacobsoni remains in harmless co-existence with *A. cerana* in the Malaysian peninsula and Indonesian archipelago, but its sibling species *V. destructor* has spread through Russia (1960s), Eastern Europe (1970s), South America (1971), Western Europe (1980s), the USA (1987), Canada (1989), the UK (1992), New Zealand (2000), and Hawaii (2007). At the time of going to press, the only large parts of the world free of this pest are Australia and parts of Africa. More information about treating varroa is on pages 240–41.

ABOVE: *A Japanese giant hornet* (Vespa mandarinia) *stinging a bee at the entrance of the hive.*

OPPOSITE: *Two honey badgers raiding a bees' nest.*

HONEY RATTEL

DISEASES AND VIRUSES

Bees' sociability, their habit of sharing food and bodily fluids, and their close proximity to each other in the confines of the hive help the spread of spores or germs, which is why honeybees regularly fall victim to diseases. Unlike their parasites, which at least are visible under a lens or microscope, bee diseases are identifiable only by their visible (or sometimes invisible) effects on the bees themselves.

American Foulbrood is one of the most devastating diseases. It is caused by the bacterium *Paenibacillus* (formerly *Bacillus*) *larvae*. As its name suggests, it attacks the older larvae and young pupae, which are killed and devoured by the mess of bacteria in the wax cell. It is difficult to control because the bacterium creates spores that can remain dormant until transmitted to a suitable new host larva, usually in food.

European Foulbrood, caused by *Melissococcus plutonius* (formerly *Streptococcus pluton*), is similar, but attacks mainly younger larvae. This bacterium does not form spores so is easier to control.

Chalkbrood is a fungal disease caused by *Ascosphaera apis*. The larvae die and become mummified and hard, resembling chalk.

Stonebrood is similar; the larvae appear to turn to stone. It is caused by several *Aspergillus fungi*.

Nosema is caused by an infection of the hind gut from the fungus *Nosema apis* (and *N. ceranae*, which is now found in the West). Affected bees can become unable to fly, and crawl about, apparently disoriented, on the ground. The fungus is often present in symptomless colonies, but becomes exacerbated by the stress of long periods of confinement, rapid brood increase, nutritional imbalance,

or poor weather. Treatment of this and other honeybee diseases is covered on pages 236–41.

With recent advances in microbiological techniques, many honeybee viruses have now been identified. These include black queen cell virus (which turns the queen larvae black), sacbrood virus (which reduces the larvae to little more than a sac full of liquid), several paralysis viruses (chronic, acute, and Kashmir), and the fairly simply named "deformed wing virus."

Nosema is one of the most important diseases in honeybees.

The mechanisms of viral disease development are still being unraveled. Several are associated with other problems in the hive: black queen cell virus often appears with *Nosema* infection, and *Varroa* mites are thought to be responsible for the spread of deformed wing virus and acute bee paralysis virus. Throughout the animal kingdom, viruses are well known for remaining dormant in individuals until some physiological stress trigger (such as those that encourage *Nosema*) reactivates them or diminishes the immune system that was keeping them at bay.

New diseases are still being discovered, however, and others have come and gone without a fully adequate explanation. In the first two decades of the twentieth century a mystery disease ravaged bee colonies in the UK. Isle of Wight disease, named after the island off the south coast where it was first observed, is thought to have destroyed 95 percent of British bees (see page 74).

In the 1970s large honeybee losses in 27 states across the USA caused widespread consternation

OPPOSITE: *The American tyrant flycatcher is known for eating various insects, including honeybees.*

and the term "disappearing disease" was coined; the bees were not dead or dying in the hives, they had simply disappeared. Luckily they were easily replaced and the US beekeeping industry quickly recovered.

In autumn 2006 beekeepers in the northeastern USA again reported large numbers of mysteriously empty hives. If the bees had absconded they had done so in extremely odd circumstances, leaving behind good stores of honey and pollen and with nearly mature brood capped and pupating in the combs. These disappearances were named Colony Collapse Disorder (CCD).

Research into possible causes of CCD covers a wide range of inquiry. It is known that the wax foundation used in the hive is contaminated with a large number of chemicals, including ones not used by beekeepers in the hive itself. Since the egg, larva, and pupa are being raised in close contact with contamination, research is now being focused on sublethal and synergistic effects. Nutrition is another important area of research. There have been various practical attempts to combat the disease, some of which are described on page 246.

Alarming media reports of CCD continue to appear around the world. Various suggestions have been put forward for the cause of the problem. One possibility is that it is a new viral disease, previously unknown because it used to occur only in an isolated part of the world where the local bees had developed immunity, but which has now been transported around the globe and is finding highly susceptible new hosts. The lack of genetic variability in the global commercial honeybee industry would make a new epidemic catastrophic since the bees would all now have the same high susceptibility to the disease.

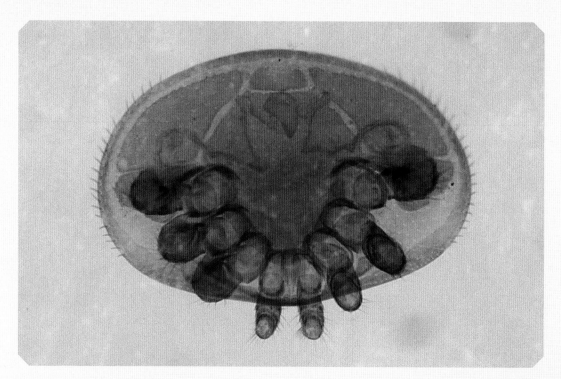

ABOVE: *A close-up of a* Varroa *mite.*

OPPOSITE: Varroa *mites feed on developing bee pupae.*

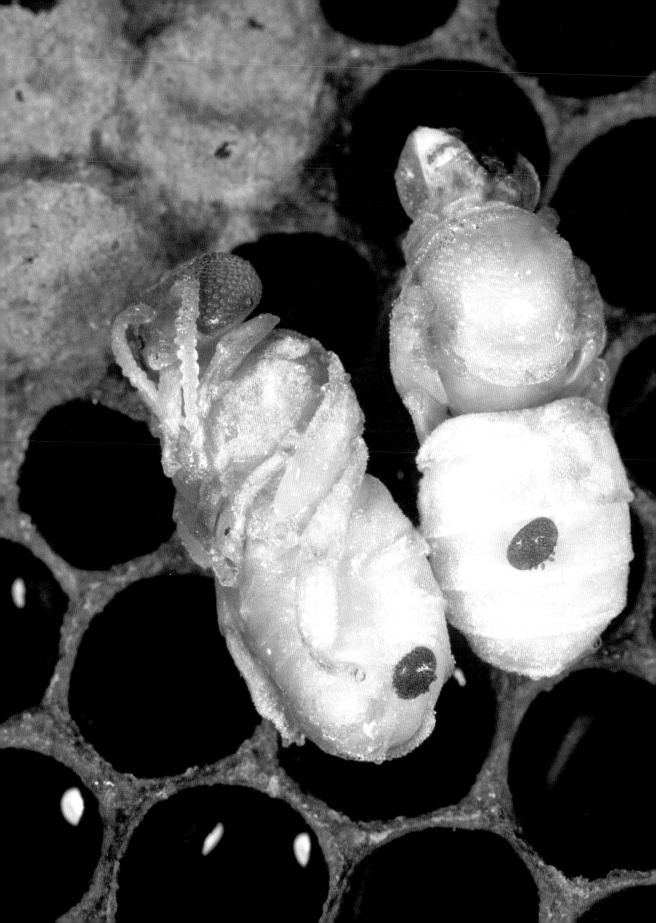

PART THREE

PRACTICAL BEEKEEPING

Choosing a Suitable Hive

Creating and Keeping Your Beehive

Where to Keep Your Bees

Clothing for Beekeeping

How Many Hives?

Keeping Records

Why Should You Try to Control Swarms?

Feeding

Hygiene and Safety

The Beekeeping Year

Diseases, Pests, and Predators

The Honey Harvest

Harvesting Beeswax

Setting Up a Commercial Apiary

The Future of Beekeeping

Choosing a Suitable Hive

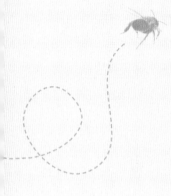

A HIVE IS the enclosed structure where bees build their beeswax, comb, store food (honey and pollen), and rear their brood (eggs, larvae, and pupae). Although bees can never be tamed, bee colonies can be encouraged to live in man-made hives.

There are two important things to remember about a man-made hive. Firstly, the internal dimensions need to be correct. Secondly, the hive must be bee-tight—the only way in must be through the entrance, with no other gaps large enough for wasps or bees from other colonies to gain access.

Most beekeepers today use hives containing movable wooden frames (see page 61 for the history of these), within which the bees are persuaded to build their combs but which can be removed for colony inspection and to extract the honey. The dimensions of the frames and their position within the hive are critical, and are based on two discoveries in the mid-nineteenth century. The Polish beekeeper Rev. Dr. Jan Dzierzon found that, for a system of practical movable frames, the center of adjacent brood combs should be 1½ in (38 mm) apart, the same as the

NESTS AND HIVES

In the wild, a honeybee colony seeks out a hollow cavity in which to build its nest. When a colony is preparing to swarm, scout bees leave the home nest, find a cavity, and measure its size for suitability by walking around inside. When the swarm arrives, the bees cluster at the top of the cavity and workers of the right age secrete wax from glands on the underside of their abdomens. They mold it into the familiar combs of hexagonal cells, starting with an attachment to the upper surface of the nest and extending downwards as the colony expands. A man-made hive is simply an attempt to mimic the conditions of a wild honeybee nest, to encourage bees to settle and make it their home.

spacing between natural combs. The "father of American beekeeping," Rev. L. L. Langstroth, further discovered that bees leave a ¼–⅜ in (6–9 mm) space between their combs and the wall of their nest. He called this the "bee space" (see below), which is the approximate gap that bees need to pass through freely. If the gap between the frame and the hive is less than ¼ in (6 mm), the bees will fill it with propolis (the resinous substance collected by bees), and they will fill any gap greater than ⅜ in (9 mm) with brace comb (bridges of wax built between adjacent surfaces in the hive). Both actions make colony inspections much more diffi-

cult because propolis and comb can stick the frames together or to the walls of the hive. Langstroth used this information to develop hanging frames with a bee space between the frame side bars and the hive walls, the basis of modern hive design today.

THE HIVE LAYOUT

There are several types of modern hive but all share the same basic structure: a cavity in which frames are placed at suitable distances apart. Starting from the bottom, the floor or bottom board of the hive is placed on a stand to reduce dampness and to raise it to a comfortable working height. The bottom board may be solid or it may incorporate a screen panel and removable tray used for monitoring levels of varroa mites in the colony (see page 240). Fillets, ⅞ in (22 mm) deep, around three sides give an entrance on the fourth side, which is formed when the brood box is placed on top.

The entrance can be reduced with an entrance block, which fits the gap and has a depression, usually in the center, measuring approximately ⅜ in (10 mm) high and 3–4 in (75–100 mm) wide. The entrance reducer should not fit too tightly as you may wish to remove and replace it.

Standing on this bottom board is the brood box, or brood chamber, within which the queen is restricted and lays her eggs, and the worker bees store pollen and honey used to feed the developing larvae. A deep box, it contains frames supported on a pair of metal or plastic runners fixed on opposite sides. The runners are narrow strips of metal or plastic, bent over along the length of

The bee space is the gap the bees need to pass through freely. In a movable-frame hive there is a bee space between the frame and the hive wall and either above or below the frame (see "top bee space" and "bottom bee space" on page 156).

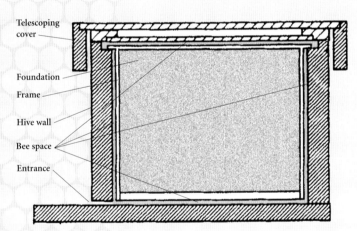

Telescoping cover

Foundation

Frame

Hive wall

Bee space

Entrance

Bee space is ¼–⅜ in (6-9 mm). A space too large (A) will be filled with brace comb, a space too small (B) with propolis. A well-designed bee space (C) will allow an easy colony inspection.

A. Space too large   B. Space too small   C. Well spaced

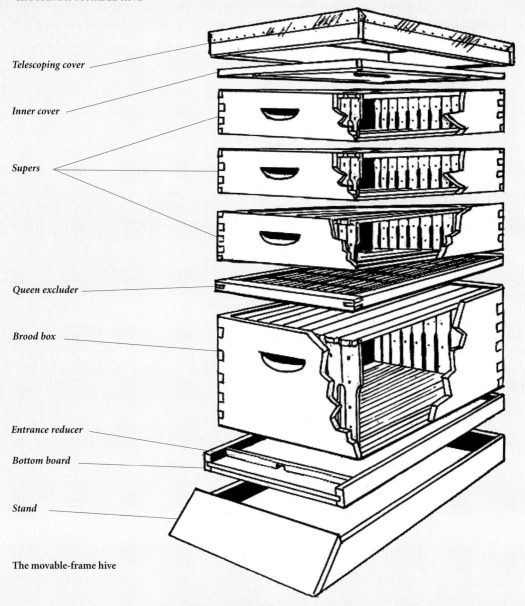

Telescoping cover

Inner cover

Supers

Queen excluder

Brood box

Entrance reducer

Bottom board

Stand

The movable-frame hive

A screen panel at the bottom of the hive is used for monitoring levels of varroa mites in the colony (see page 155).

the rabbet and nailed to the walls, with the lip curving over the thickness of the wall. A frame resting on top of the runner either has its upper surface flush with the top of the hive wall (known as bottom bee space) or its bottom bars flush with the bottom of the hive wall (top bee space).

As the queen and drones are larger than the workers, they are prevented from moving upwards from the brood

The queen is prevented from moving from the brood box (below) to the supers (above) by the queen excluder, in which the slots are only large enough to allow workers to pass through.

box into the boxes above (called supers) by a queen excluder. This is a framed grid, or a slotted metal or plastic sheet the same dimensions as the hive cross-section, in which the gaps allow only the workers to pass through.

During the active season, when the bees are collecting nectar, additional boxes are placed on top of the brood box over the queen excluder. These are the supers and are used by the bees for storing honey. Generally shallower than

the brood box, supers are of the same cross-sectional dimensions, with frames inside to fit the shallower box. By adding supers, one on top of the other, the beekeeper can ensure that the bees are always inside a cavity, which allows the colony room to expand. As a general rule, you need three or four supers for each of your colonies. The bees will fill

KEEPING FRAMES AT THE CORRECT SPACING

Frame spacers, are flat lengths of metal or plastic, with a similar function to runners but with pre-determined slots to support the frames. The slots ensure that the frames are always correctly spaced throughout the box. Frame spacers are generally used in supers as they make frame manipulations in the brood box almost impossible.

Runners are smooth metal strips used to support the frames. Castellations (shown here) have the same function but have pre-determined slots to keep the frames evenly spaced.

one with honey. The second and third give them space to expand (and store more honey) and the fourth replaces the first when you want to extract the harvest.

An inner cover, a removable flat board, fits on top of the uppermost super. There are usually one or two holes in this board, which are used for feeding bees (see page 217) and, with the insertion of a device known as a Porter bee escape (see page 249), can also be used to clear the supers of bees when the honey is harvested at the end of the summer. Over the top of the inner cover is the telescoping cover. This may have ventilation holes, usually in the top center of each side, which are protected internally with bee-proof mesh to prevent access by wasps, bees, and other insects interested in robbing the colony's honey stores. Ventilation in the telescoping cover allows air to flow

The WBC—a double-walled hive—with one side removed to show its internal structure.

through the hive, keeping it dry. The moisture level within a bee colony is high and ventilation is essential to deter mold growth. The telescoping cover is usually covered with a thin metal sheet or other suitable material to make it weathertight and durable.

WORLDWIDE HIVE STYLES

Throughout the world today one can find a number of types of hives. However, the Langstroth hive is the most popular. This hive serves very well in many climates. With the moveable frames and interchangeable parts it is a hive that is easy to inspect and to transport. Great Britain has had a number of different styles of hives but the one known as National or Modified National is favored today. It is similar to the Langstroth type hive.

With the increased interest in beekeeping in the US, particularly in urban areas, the 8-frame Langstroth has become popular with small-scale beekeepers. The 8-frame Langstroth uses the same sizes of frames as the 10-frame but the hive bodies hold just eight instead of ten frames.

The top bar hive has become somewhat popular outside of its territory in Africa. A few small-scale beekeepers are trying it. Honey production is limited, however.

Primitive hives still exist in other parts of the world. These hives can be made of logs, mud shaped into pots and pipes, corn stalks, reeds and even from scraps of wood. Generally the primitive hives do not use frames. Therefore honey is obtained from primitive hives by destruct harvest. Combs are removed and crushed and

TOP BEE SPACE AND BOTTOM BEE SPACE

In a natural colony, bees only leave a bee space between and around the combs and there are no horizontal bee spaces (horizontal divisions across the vertical sheets of comb). However, in a bee hive, horizontal bee spaces are needed by the beekeeper to make it easier to part the boxes during colony inspections. The gap is set at the bee space (see page 155) to reduce the bees' propensity to fill any smaller gaps with propolis and larger gaps with brace comb. The bee space can either be at the top of the box, over the frames (as in Langstroth, Dadant, and Smith hives), where the bottoms of the frames are in line with the bottom of the box (known as top bee space), or at the bottom (as in National, WBC, and Commercial hives), below the frames, so that the tops of the frames are level with the top of the box (bottom bee space). The two types of boxes cannot be mixed within the same hive—an important consideration if you plan to pick up second-hand equipment.

A modern decorated clay hive. Hive design © Maurice Chaudière

the honey is drained from the broken wax comb.

With the spread of beekeeping education worldwide, beekeepers are slowly changing from primitive hives to those with moveable frames.

Beekeepers around the world have invented and are still inventing various shapes and sizes of hives. However, none of them have displaced the Langstroth as the most widespread hive.

OTHER TYPES OF HIVE

Before the development of the movable-frame beehive, bees were kept in skeps and, in some parts of the world, these types of hive are still in use. In the US, however, it is illegal to keep bees in skeps because of the difficulty of inspecting for and treating disease. In the UK today, few colonies are kept in skeps but they are very useful for swarm collection (see pages 198–99).

WHICH HIVE FOR YOU?

Bees are happy in any suitable cavity and wild colonies can be found in hollow trees, cavity walls, even garbage cans. However, most beekeepers now use hives designed for a beekeeper's benefit. The type most suited to you will depend on your beekeeping plans and ambitions. If you simply want to keep two or three hives in the garden or on a rooftop you might find the 8-frame Langstroth the most suitable one. The hive bodies are lighter in weight than the usual 10-fame brood boxes.

If you wish for lighter weight but prefer the 10-frame size then it is possible to use the medium depth bodies for both the brood boxes and the honey supers. Three mediums approximate the size of

two deeps for the brood chamber. One advantage of the medium depth is that all equipment is interchangeable. Frames of honey can be put into the brood chamber for feeding the bees.

Some beekeepers may wish to experiment with a top bar hive. These are now available commercially from a few specialized suppliers. However, it may be best to learn beekeeping skills first with Langstroth hives.

All-plastic-hives were originally made for countries with limited availability of wood. Beekeepers in the US are now experimenting with the plastic hives. Some of the large equipment suppliers carry the plastic hive parts.

If you expand your apiary with the thought of moving for pollination, you may wish to consider using the migratory top instead of the telescoping cover. Hives can then be packed tightly next to each other for ease in transporting them. If you plan to use forklifts, you may also wish to put four hives on a pallet in place of individual bottom boards.

FRAMES

In the US frames are available for every depth of hive, deep, medium and shallow. Repair parts are available for wooden frames with damaged lugs. Beekeepers have a wide choice of frames and foundation.

Each wooden frame has a top bar, two side bars and a choice of two styles of bottom bars. Beekeepers fix sheets of beeswax foundation, embossed with the hexagonal honey bee cell pattern, into the frames to provide a base from which bees can draw out their comb. The foundation guides the bees to build comb within the frames, that is straight, thereby

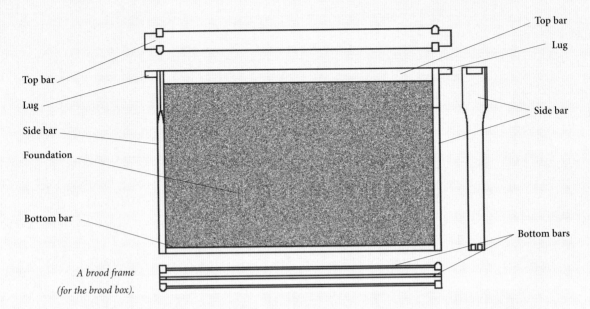

Top bar

Lug

Top bar

Lug

Side bar

Foundation

Side bar

Bottom bar

Bottom bars

A brood frame
(for the brood box).

making it possible for beekeepers to inspect the comb and harvest the honey.

Top bars

The top bar of the frame extends beyond the side bars, forming two lugs used to lift and hold the frame and from which the frame is suspended within the hive. Top bars come in several styles. One is called wedge-top. A thin slice of wood is removable to accommodate wired wax foundation. When the foundation is butted against the top bar, the wedge is nailed back in place to hold the sheet. A second type is called a grooved top bar. This is used in combination with a grooved bottom bar for plastic foundation that is snapped into place in the grooves. Recently a new style of top bar is available called the slotted top bar. It is used with slotted end bars and slotted bottom board. Wax foundation is slipped through the top bar, following the grooves in the side bars and then into the grooved bottom bar. Fitting foundation into the grooved frame is quicker than with other styles.

Side bars

In the US the Hoffman self-spacing side bar is used. These side bars have wider shoulders at the top of the side bar (widening to $1\frac{3}{8}$ in, or 35 mm) to keep the frames at the correct spacing. When the comb is drawn with 10

SPACERS

Spacers are used for keeping frames correctly spaced. There are two types made for 10-frame hives: 9-frame spacers and 8-frame spacers. Once the bees have drawn out 10 frames in brood box and honey super some beekeepers choose to reduce the brood chamber to nine frames for easier removal for inspection. Sometimes either nine frames or eight frames are used in honey supers. The bees will then draw out the comb thicker making uncapping easier. If spacers are used before the comb is initially drawn the bees will build brace comb between the frames making it impossible to remove individual frames.

frames bee space will have been preserved. The wide part of the Hoffman side bar has one flat and one tapered edge, which butt up against each other on adjacent frames, reducing the surface area available on which bees can deposit propolis.

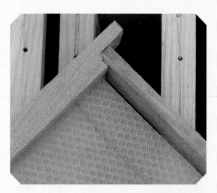

A sheet of wired foundation fitted in a frame.

Today many beekeepers, especially commercial ones, are using either plastic foundation or the combined plastic frame and foundation. These are available in either black or white and the frame/foundation combination is also available in green drone-sized comb for Varroa control.

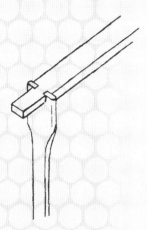

A Hoffman self-spacing frame.

Bottom bars

Bottom bars come in two different types: grooved and split bottom bar. The grooved bottom bar can be used for both wax foundation and plastic foundation. The split bottom bar can only be used with wax foundation since a nail can be driven through the two bars and the wax to hold the wax in place.

FOUNDATION

Foundation is a thin sheet of beeswax embossed on each side with the hexagonal cell pattern, offset on opposite faces, just as the bees build it in nature. It is fitted into frames to encourage bees to build straight combs, so the frames can be removed easily for inspection or to harvest the honey.

Providing foundation enables bee colonies to use fewer resources and less labor in drawing out their combs. They can develop their nest and begin brood rearing more quickly. However, it is debatable whether foundation enables them to store significantly more surplus honey than colonies given no foundation or just starter strips of wax.

Foundation can be wired to support the comb. Freshly made comb is soft and sags easily when filled with brood and/or honey and pollen, and the wires give the comb extra strength. Wiring super foundation also strengthens the comb and limits collapse and damage when the frames are placed in an extractor during the honey harvest (see page 250).

Wax foundation is available from bee equipment suppliers or you can make your own (see page 286). You can also purchase foundation made of plastic, coated with a thin layer of beeswax, which tends to give stronger combs. Before use, foundation should be stored horizontally, ideally in a warm place such as a closet, and supported top and bottom with sheets of a flat, rigid material, with a small weight on top to prevent warping.

MAKING UP FRAMES AND FOUNDATION

When making up a frame, it must be kept square at the corners, and not twisted in the vertical plane, to maintain the correct spacing in the hive. A jig may be a helpful tool for assembly.

You can make up frames at any time but it is best to fit foundation just before the frames are going to be used.

To assemble the frame, first remove the wedge under the top bar and clean out any slivers of waste wood on both wedge and frame. Push on the two side bars with the grooves facing inwards (these are not present on Langstroth or Dadant frames). Secure each side bar to the top bar with ¾ in (19 mm) black japanned (lacquered) frame nails. Push one bottom bar into place and nail up into the side bar at each end. Check that the whole frame is square and flat before nailing it together.

To put wired foundation into a frame, lay the foundation so that the bent wires face outward and the foundation is centered from side to side. Make sure the wax is butted up against the top bar. Make sure the wax is butted up against the top bar. Then replace the wedge to grip the foundation. Nail through the wedge into the top bar at an upwards angle in three or four places. Add the second bottom bar, making sure the foundation lies flat between the two, and nail it in place as before. Don't be tempted to use a single nail horizontally to secure both bottom bars at once. This is almost impossible to remove when you want to replace the foundation in the frame.

Making up a frame and inserting foundation:

1. *Removing the wedge under the top bar.*

2. *Pushing the side bar into the top bar.*

3. *Inserting the wired foundation.*

4. *Nailing the bottom bar.*

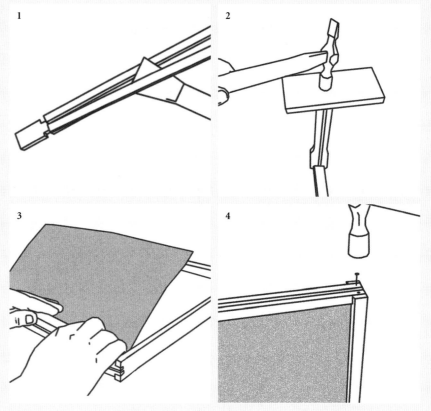

Warm way and cold way

In a Langstroth hive, if the entrance is at right angles to the frames (the usual configuration) it is called the cold way. If the entrance is on the side of the brood box, with frames parallel to the entrance, it is called the warm way. The warm way has several advantages, primarily that you can stand behind the hive during hive inspections, thus keeping out of the way of the activity at the entrance. Bees are often reluctant to build out comb on foundation just inside the entrance, so only the frame closest to the entrance will be affected if the frames are placed warm way. If the frames are placed cold way, the front corners of the foundation may not get drawn out on several frames.

Within the brood box, the brood nest (cells containing eggs, larvae, and developing pupae) is roughly spherical, surrounded by a sphere of cells containing pollen, and the rest of the comb is used for honey storage. The frames form slices through this sphere. The brood pattern on warm way frames will be symmetrical, with the brood nest area roughly in the middle. For cold way frames, the brood nest can be toward the entrance side of the frame, with the honey storage area toward the back.

In winter, bees cluster in the brood nest area and consume their honey stores to maintain colony temperature. As the stores are eaten, the natural movement of the cluster is upwards onto further honey stores. A colony does not naturally move sideways, especially if this means crossing a gap between combs. Thus, a colony on cold way frames could starve in the winter even though honey stores were available on either side of the brood nest. As a general rule, bees should always have some stores above the cluster when they have been fed for the winter, regardless of the frame orientation.

Bees starting to build out the comb on a sheet of foundation embossed with the hexagonal cell pattern.

TYPES OF HIVE

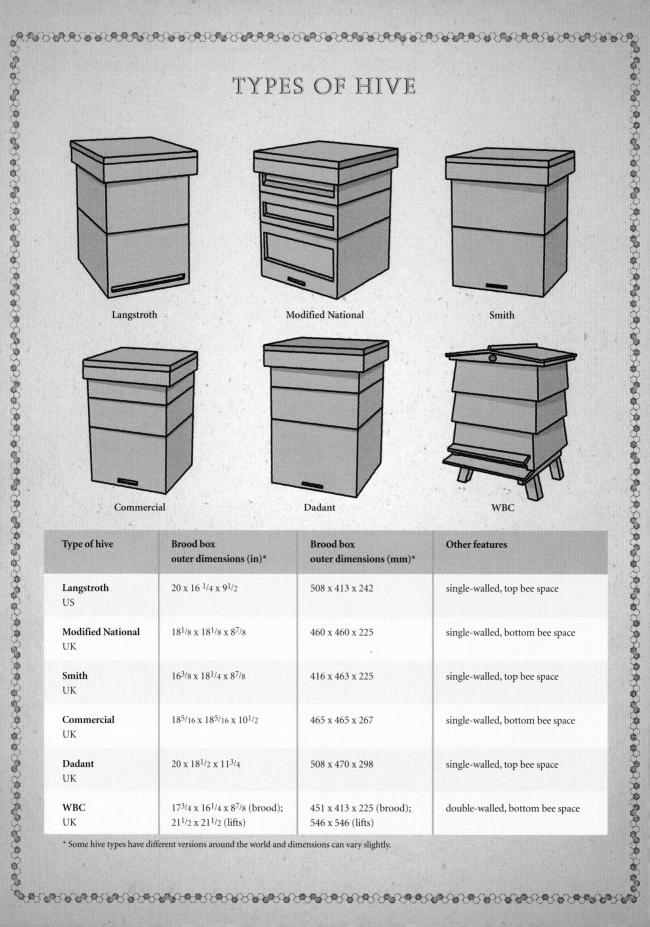

Langstroth

Modified National

Smith

Commercial

Dadant

WBC

Type of hive	Brood box outer dimensions (in)*	Brood box outer dimensions (mm)*	Other features
Langstroth US	20 x 16 $^1/_4$ x 9$^1/_2$	508 x 413 x 242	single-walled, top bee space
Modified National UK	18$^1/_8$ x 18$^1/_8$ x 8$^7/_8$	460 x 460 x 225	single-walled, bottom bee space
Smith UK	16$^3/_8$ x 18$^1/_4$ x 8$^7/_8$	416 x 463 x 225	single-walled, top bee space
Commercial UK	18$^5/_{16}$ x 18$^5/_{16}$ x 10$^1/_2$	465 x 465 x 267	single-walled, bottom bee space
Dadant UK	20 x 18$^1/_2$ x 11$^3/_4$	508 x 470 x 298	single-walled, top bee space
WBC UK	17$^3/_4$ x 16$^1/_4$ x 8$^7/_8$ (brood); 21$^1/_2$ x 21$^1/_2$ (lifts)	451 x 413 x 225 (brood); 546 x 546 (lifts)	double-walled, bottom bee space

* Some hive types have different versions around the world and dimensions can vary slightly.

Creating and Keeping Your Beehive

BEEHIVES CAN BE purchased already assembled and ready to go, even to the extent of including made-up frames fitted with foundation. While this is one of the quickest ways of getting started, it is also one of the most expensive.

Most beekeepers, therefore, purchase new hives "in the flat": a pack containing all the necessary pieces for your chosen hive design, cut accurately to size and shape. It also includes the runners, nails, and instructions.

You can, of course, produce all the wooden hive parts yourself. Plans are available on the Internet (try www.beesource.com/plans/index.htm for the WBC, Langstroth, and other hive parts) or you can get more information from organizations like the American Beekeeping Federation, www.abfnet.org.

Traditionally, hives are made from pine, which works well and lasts a long time, particularly if treated regularly with an insecticide-free preservative. Ordinary exterior paint is used or you can use linseed oil.

In one sense, the thickness of the wood used is not very important, although it should be at least ¾ in (19 mm) to make it robust, and ideally ⅞ in (22 mm). It is the inside measurements that must be accurate to maintain the bee space within the hive.

If you are building your own hive, you can buy runners for the brood box and runners for the supers. To assemble the hive, you will need 2 in (50 mm), ⅛ in (2.7 mm) gauge galvanized nails. Applying waterproof glue to the joints before nailing will strengthen them.

The corner angles of the boxes must be 90 degrees so that everything is square, and adjoining pieces of wood must be flush at the corners. All the boxes must fit with each other in any combination, with no gaps. A jig or long clamps can be used to hold the parts in place but check the angles with a square before gluing and nailing them. Pre-drilling the corner joints will help prevent the wood from splitting.

For top-bee-space hives, ensure the bottom bars are lined up with the bottom of the box.

The bottom board, inner cover, and telescoping cover must also be square and flat. If the wood here is warped or has a tendency to twist over time or when it gets damp, gaps will appear and make your hive no longer bee-tight.

Many US beekeepers now use screen bottom boards as part of their Integrated Pest Management (IPM) program to monitor and control varroa (see page 240). To make your own, purchase metal screen wire with 8 mesh per in (25 mm). This allows the varroa mites to fall through to the ground or onto a monitoring tray. The tray prevents the housekeeping bees from gaining access to the debris, but it can become a breeding ground for wax moth larvae, another menace the bees will be unable to evict (see page 245).

You will need to put ventilation holes at the center top of each side of the telescoping cover. These must be covered with mesh to prevent illegal access. The telescoping cover should be protected from the weather with a waterproof material that covers the top and folds down over the sides, where it is nailed in place.

The different parts of the hive.

Telescoping cover (A)

Inner cover (B)

Super

    Side walls (C)

    Frame spacers (D)

    Frame (made up) (E)

Brood box

    Side walls (F)

    Bottom board (G)

    Frame rests (H)

    Brood frame (made up) (I)

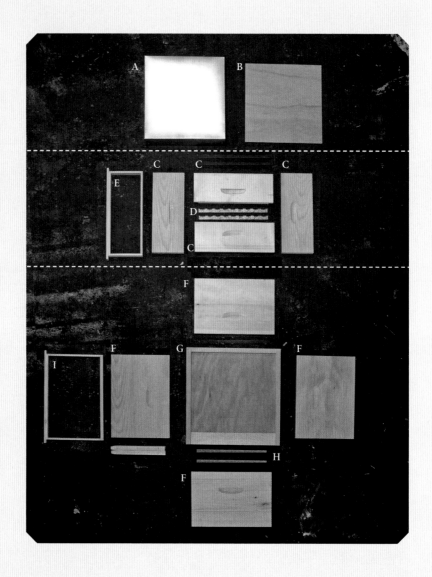

Where to Keep Your Bees

Bees can be kept almost anywhere on a carefully chosen site. In urban areas, hives are sometimes kept on flat roofs and in roof gardens; in the countryside, they are usually kept in gardens. They can also be kept away from home in out-apiaries, which are often unused corners of fields.

In the USA and some European countries, local laws sometimes regulate the keeping of bees within city limits, on certain-sized properties, or within a specified distance of the property boundary, and bees and hives may need to be registered. Ask your local beekeeping association for details.

The overall ground area required per hive must include room for you to stand when you are inspecting your

A garden apiary in London.

bees and you will need somewhere to put items such as the cover. In general, you should allow a minimum of about nine times the hive footprint area per hive, though this is not an absolute measurement as two hives can be placed side by side, reducing the overall space required.

When planning a new apiary, remember that bees have a natural tendency to swarm (see page 114) and two colonies can rapidly become four or even more. You must therefore ensure that you will have sufficient space for the maximum number of colonies you intend to keep and for any boxes temporarily accommodating bees during swarm control manipulations.

Ideally, hives should be on a south-facing slope and protected from the prevailing wind, preferably by a hedge, which slows wind speeds down dramatically. Overhanging trees are best avoided, but some amount of shade is useful in hot weather. You should also make sure your apiary is not in a frost pocket as bees won't start flying until the outside temperature warms up.

Neighbors

Bees and neighbors do not always go together amicably and you must consider this carefully when choosing your apiary site. If you are on good terms with your neighbors, talk to them before moving hives into your garden. You should explain that you may need to open colonies every seven to nine

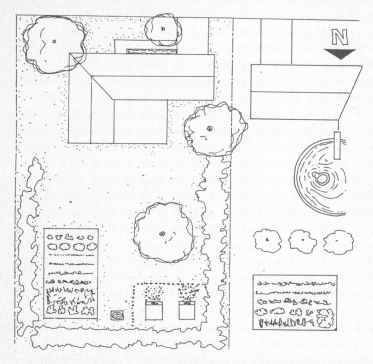

A typical garden apiary arrangement, with the hives facing south and hidden from public view. The bees' flight path is kept high by a fence around the hives and is directed away from the neighbor's garden. There is a water source provided on the property.

AVOID DAMP LOCATIONS

The inside of a beehive is a damp environment. Bees produce water vapor as part of their metabolic processes. Water vapor in a hive rises and passes out through upper ventilation, and excess moisture is usually removed from the hive by bees standing at the entrance and rapidly fanning their wings. However, if your hives are in a damp location, the bees may not be able to reduce the humidity sufficiently and mold can start to grow within the hive. You may see damp moldy patches on the woodwork and mold growing on the pollen stores. If you have to keep your bees in a damp place, make sure they are off the ground on suitable hive stands. If possible, move them to a drier location.

days, and that this may be on weekends. They have as much right to enjoy sitting in their garden as you do.

Problems can arise early in the spring. When a bee is confined to the hive, either over winter or by a cold spell, waste builds up in its rectum. One of its first actions when flying out is to defecate. Unfortunately, the yellow-brown feces can land anywhere, including on clean sheets hanging out to dry.

Another potential difficulty may arise during the swarming season. A swarm from your bees may cluster in your neighbor's garden. Worse still, it may decide to move into your neighbor's chimney or cavity wall. The best way to avoid this is to prevent your bees from swarming in the first place (see pages 198–215), but this is not always possible so prompt attendance to collect a swarm should be a priority (see pages 198–99).

If you think your neighbors will object to your bees, consider keeping the hives in a bee shed. If the entrance is in the wall facing away from your neighbors, they may never know that you are a garden beekeeper. The other option is to find a site somewhere more suitable and set up an out-apiary, away from habitation.

If bees are kept in a garden, the hives should be sited to give minimum disturbance to neighbors. Placing them a short distance from, and with their entrances facing, a head-high hedge or fence makes the bees fly up to clear the

OPPOSITE: *Bees are often kept in garden hives, and properly managed they should not bother the neighbors.*

URBAN BEEKEEPING

With society becoming ever more urban, beekeeping has also had to change. However, as people desire to get closer to nature, hives and apiaries are thriving in small town gardens and large city lots the world over. It is estimated that, in London, honeybees outnumber the city's population by 30 to 1 in the summer months. And, while you might think that cities would lead to a lack of foraging for honeybees, the parks, gardens, trees, and vacant lots actually offer a wide variety and create great honey. So, from the rooftops of London's Piccadilly to those of Brooklyn, urban beekeeping is on the rise.

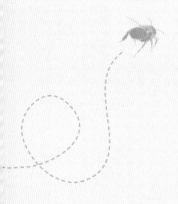

obstruction when they leave the hive. Once there, they usually continue at this altitude unless a strong wind forces them down, meaning that neighbors are rarely bothered by many bees flying across their gardens.

To get maximum pleasure from your beekeeping and to help maintain good neighborly relations, always keep gentle bees. With a gentle colony you should, as a minimum, be able to walk near your hives without a veil and not be bothered by the bees. If the bees don't come buzzing round you in this situation, they won't bother those living next door. Keep bees that don't follow you or try to defend their colony

away from the hive. For instructions on how to replace a queen heading an aggressive colony of bees, and how to breed from one that produces good-tempered bees, see page 180.

Water sources

Neighbors may become distressed to see bees drinking at their garden pond, so provide water near your hives to discourage this. You can use a shallow container, which enables the water to warm up quickly on a sunny day, but you must provide landing places for the bees from which they can drink, or many will drown. Landing places can be stones that break the water surface

An out-apiary.

long as you keep it full; if it dries out, they will search for an alternative, so don't allow this to happen.

Out-apiaries

An ideal out-apiary is easily accessible by vehicle. It is far better not to have to lift heavy boxes over fences to get them back to the vehicle. The out-apiary should have room for a reasonable number of hives to make visits economical and have extra space for spare hives and equipment. If your out-apiary is in the corner of a field with livestock, construct a fence around the hives. A beehive makes an ideal scratching post for sheep, cows, or horses, but if knocked over can have disastrous consequences.

Vandalism is relatively rare, although more commonly found in out-apiaries. Vandals can break in and turn over hives or throw stones at them. They have even been known to place lighted fireworks inside a hive. If possible, choose a secluded out-apiary away from a main thoroughfare. Hidden hives are less likely to be vandalized.

It is important to feel comfortable and safe when you are new to beekeeping. Although most beekeepers aim to breed bees that are docile and non-defensive, female honeybees can sting, and even experienced beekeepers will be stung on occasion, which can lead to an allergic reaction, and in a very few very extreme cases, a fatality (see page 130). Although you can still be stung when wearing protective clothing, it will help to reduce the chances of this occurring.

or floating pieces of wood. An alternative method is to make holes in the bottom of a container filled with soil or peat. This is then placed in a larger, watertight container, which is filled with water. The soil or peat absorbs the water and the bees can come and drink safely from the wet surface.

Bees can and should be encouraged to use the water sources you provide. Try and place a number of water sources in sunny positions as bees tend to use more than one. Keep checking to see which ones the bees are using and make sure they are constantly replenished.

A new water source can be baited with a 1:1 sugar:water syrup to encourage the bees. Trickle a spoonful of syrup over the top bars of the colonies to encourage the bees to search for its source. Once they have found the source, refill it with water, which will gradually dilute the solution. You can also add a distinctive flavor, such as peppermint essence, to help the bees find the supply. Bees will continue using the water source as

Clothing for Beekeeping

THE VEIL

On opening up a hive to inspect a colony, bees tend to be attracted to the beekeeper's face and eyes. Wearing a veil helps to protect this vulnerable area. Veils come in various types, ranging from a simple cylinder of net supported by a stiff-brimmed hat to a sophisticated hood attached to a full

A beekeeper wearing a Sherriff-type veil attached to full protective overalls.

set of overalls. All veils are designed to keep bees away from your face while allowing you to see out.

Many modern veils are of the hood type, which comprises cloth at the back and a fine, black wire mesh over the front, and two arches of stiffener providing upright support. Known as the Sherriff veil after its designer, this is now available attached to a light cotton jacket or to a full set of overalls.

An older type of veil still widely in use comprises a net cylinder attached to the brim of a hat. The hat must have a wide, stiff brim and straps, tied under the chin. The net of fine, black nylon or cotton hangs down over your face and the bottom is anchored with more straps, which tie around your chest. The bottom edge of the net should lie across the top of your shoulders—if it comes too far down, over the top of your arms, bees can get underneath when your arms are extended. Many veils of this type have one or two metal rings of the same   diameter as the hat rim, sewn onto the net to help keep it away from your face.

PROTECTIVE OVERALLS

It is advisable to use some form of protective clothing. At the very least this keeps other clothing clean; it also helps to protect you from stings as the thickness of the material, and the fact that it does not lie close to your skin, reduces the depth to which the bee's sting mechanism can penetrate.

A veil attached to the brim of a hat. In this modern version, the net cylinder is also attached to a full set of overalls.

Gloves should be lightweight and flexible.

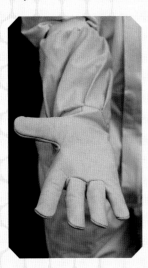

Bees tend to crawl up sleeves, under sweaters and up pants, so an all-in-one bee suit or overalls makes the ideal outfit. However, a veil and protective jacket can be worn over ordinary pants, as long as the elastic around the bottom of the jacket is drawn up tight enough to prevent access by bees at waist level.

Occasionally, a bee does get inside a veil or a suit. If this happens, you should keep calm. You can try to kill the bee by squashing it against the veil. Alternatively, move well away from the hive before lifting your veil or unzipping your bee suit to remove the intruder.

Some bees react strongly to dark colors so it is usually better to wear light-colored clothing. Avoid wearing woolly or furry garments. Wool or hair can elicit a burrowing response in the same way as the fur of a predator such as a bear does. Then, when the bee reaches a hard surface (your skin), it will sting.

PROTECTIVE GLOVES

Many beekeepers, particularly beginners, like to wear gloves when they are inspecting a colony. Some experienced beekeepers prefer not to, especially if their bees are docile, as they feel it gives them a better idea of how the bees are reacting. However, even these beekeepers will keep some gloves handy in case the bees begin to sting.

Gloves should be lightweight and flexible as you will need to be able to lift out and replace frames smoothly to avoid disturbing the bees more than necessary. Some types available from equipment suppliers have gauntlets attached to protect your forearms. An alternative is to use rubber dishwashing gloves, which don't provide as much protection but can be thrown away and replaced when they become sticky with propolis.

FOOTWEAR

Wearing calf-length rubber boots helps to protect the ankles and feet. Some bees, appropriately nicknamed "ankle-biters," have a tendency to attack at ground level. Rubber boots also give a better grip if the ground is wet or slippery.

BASIC EQUIPMENT

The two most important tools for the beekeeper are the smoker and the hive tool.

The smoker

For thousands of years, beekeepers have used smoke to calm bees during hive inspections and at other times when the bees might be disturbed. A modern smoker consists of a firebox and bellows and comes in various sizes and patterns. When buying a smoker, remember that a large one does not have to be refilled with fuel as often as a smaller type—a distinct advantage if you have many colonies to inspect. However, the larger bellows can be awkward to use if you have small hands.

The firebox is cylindrical with a grid inside to support the fuel and to allow air to flow underneath and through the cylinder. The hinged lid opens to allow you to charge and light the smoker. With the lid closed, smoke emerges through the nozzle and can be directed as required.

The bellows consist of two wooden plates, hinged at the bottom and held apart by a spring. Air exits the bellows through a small length of pipe fitted into a hole at the bottom of the plate facing the firebox, and then passes through an equivalent hole in the cylinder and into the firebox. When buying a smoker, check that the spring in the bellows is not too strong and that pumping the bellows produces a good current of air out of the nozzle.

SMOKING BEES

Bees react to smoke by going to honey storage cells and engorging, a response to their instinctive fear of fire as they prepare for flight to a new home. A bee with a crop full of honey finds it more difficult to bend her abdomen around in order to use her sting and is much less inclined to sting than a bee that is not engorged.

TOP LEFT: *A modern-day smoker with the nozzle set at an angle for ease of use.*

TOP RIGHT: *A home-made smoker.*

BOTTOM LEFT: *An old-type smoker; this design, with the straight nozzle, is rarely used today.*

BOTTOM RIGHT: *A clockwork smoker. This type emits a continuous stream of smoke but the firebox is too small to last for normal colony examinations, making it impractical.*

LIGHTING A SMOKER

THERE IS AN ART TO LIGHTING A SMOKER and keeping it lit so that it is available when required during a colony examination. A good smoker fuel burns slowly and produces a cool smoke. Soft rotten wood, wood chips, small pieces of kindling, or long pine needles are all suitable. Other useful materials are raw cotton fibers or compressed straw pellets. Avoid using any materials that have been treated with a fire retardant or chemicals such as insecticides.

You need a good ignition source to light the main fuel. Light a small piece of rotten wood or use compressed stick lighters as a primer (A). When this is alight, drop it into the firebox and add a little of the main fuel (B). Pump the bellows to push air through and the fuel should begin to burn. Keep pumping until the fuel is smoldering well, then refill the firebox and close the lid (C). If you are using a loose fuel such as shavings, you can stop these blowing out of the nozzle by twisting a handful of long grass into a circle and placing it on top of the fuel.

After use, empty the contents of the firebox out safely so they do not cause a fire. Alternatively, you can extinguish the smoker by blocking the nozzle. Some beekeepers keep a cork or wooden plug for this purpose attached to the smoker with wire or string. Others use a twist of grass. Placing a wad of material such as grass between the holes in the bellows and firebox also helps to extinguish the smoker more rapidly, thereby saving fuel.

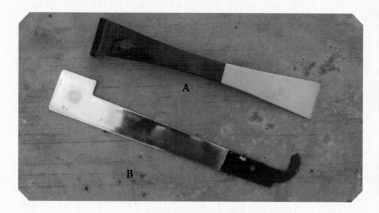

The standard (A) and J (B) hive tools.

widens out and curves over, with a beveled edge which is used as a scraper. The tail end of the J hive tool curves around to form a flattened hook and is used to lever up frames. It is more effective for this than the standard tool, any corner of which can be used to raise the frame lug. Make sure the hive tool is comfortable to hold and try out both types before deciding which one suits you best. You need to get used to keeping your hive tool in your hand at all times during colony inspections.

The hive tool
The other essential tool to have with you when examining bees is the hive tool, which is used for numerous tasks including prizing boxes apart, lifting frames, and scraping unwanted brace comb and propolis from top bars.

Hive tools come in two main types, the standard and the J. Both are made from hardened or stainless steel and both have a wider, flat, beveled end. The other end of the standard hive tool

The bee box
As a beekeeper you will accumulate various beekeeping-related objects. If you keep these in a suitable portable container you can be sure to have all the necessary equipment available when it is required. This is especially practical for visits to out-apiaries.

A bee box.

How Many Hives?

Beekeepers with only a few hives usually want to produce enough honey for themselves, their family, and friends. They may sell a few jars in the local shop or at a village fete, but honey production is not their main objective. Their real reason for keeping bees is because they enjoy looking after them.

In the USA, beekeepers with up to 50 colonies are known as small-scale beekeepers. Beekeepers with 50 to 500 colonies are known as sideliners and they make part of their income from pollination fees and honey sales. Commercial beekeepers keep from 500 to 20,000 or more colonies and make their entire living from their bees. Some but not all US states and some European countries require registration of hives and beekeepers. Check this with your local beekeeping association. If you are thinking of setting up a commercial beekeeping operation, there is more information on pages 262–63.

NEW BEEKEEPERS

It is strongly recommended that you begin beekeeping with at least two colonies (but no more than five or six until you've gained experience in managing several at once). You will need a separate hive for each colony.

Starting with only one colony makes it difficult to know whether your bees are doing well or badly. With two, you can see if they are expanding at roughly the same rate when the bees become active in the spring, and if not, you can check the weaker colony for disease

Most beekeepers operate on a small scale.

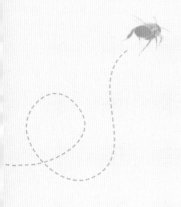

(see pages 236–47). With two colonies you can also compare their characteristics. One colony may be more ready to defend its nest and inclined to sting; one may be economical in its use of winter stores, while the other may need additional feeding. You can also compare honey crops from each colony.

A second colony will be invaluable in an emergency. If the first colony loses its queen and the bees fail, or do not have the means, to raise a replacement, a frame of brood of a suitable age can be transferred into it from the second colony. The bees from the first colony can use the young larvae from the second to rear a new queen, and it will make little difference to the second.

Ideally you should not fill all your available apiary space at the outset as colony numbers can increase rapidly if your bees swarm (see page 214). You will need space for other hives or equipment used to accommodate bees on a temporary basis—and it is very important that you keep spare equipment.

Another factor to consider when choosing your apiary and deciding the number of hives it will accommodate is the availability of suitable forage for the bees within a 2–4 mile radius. If there is sufficient forage to support four colonies, for example, putting eight on the site halves the amount available per colony. As warmer weather approaches in the spring, the brood nest expands and there are more larvae to feed. Pollen will be in great demand and, if there is too little available, less brood can be reared and colony expansion is hindered. This leads to fewer forager bees and a reduced honey crop. Find another apiary site for the extra colonies.

Buying equipment

A good source for second-hand beekeeping equipment is your local beekeeping association. Otherwise there are specialist suppliers or you can buy through advertisements in beekeeping journals. When buying equipment, make sure it is in good condition; your first experience as a beekeeper will be much more pleasurable if your equipment is in good order. Hive boxes should fit together without any gaps and you should check that the runners are positioned so that the frames sit correctly, depending on the type of hive. Make sure the brood box and supers have the correct internal measurements, allowing a bee space around the frames (see page 155). You also need to ensure that you have sufficient additional equipment on hand before the active beekeeping season starts. Avoid ordering frames, foundation, and extra boxes at the last minute during the height of the season or when you find your bees preparing to swarm—it may arrive too late.

Old hives waiting for a new owner.

Hives should be kept off the ground on a hive stand or on any sufficiently sturdy base.

PLACING HIVES ON THE SITE

Having chosen a suitable apiary site and acquired your hives, you now need to place them on the site. The hives should be kept off the ground, on sturdy stands that bring the top of the brood box level with your hands, making it comfortable to manipulate the frames during an inspection. Inspecting a large number of hives set at the wrong height can be uncomfortable and lead to back problems (see page 223).

A hive will be very heavy when the colony reaches its maximum size and has stored a large amount of honey, so avoid placing it on anything that will buckle under the weight.

Your apiary arrangement should allow you room to stand next to each hive when inspecting the frames. It is best to stand with the frames parallel to your body and to avoid standing directly in front of the entrance to the hive as that will block the bees' flight path.

When bees are old enough to start foraging, they spend some time learning the location of their hive by flying in larger and larger circles while facing toward the entrance. If your hives are in a long straight row, returning foragers find it difficult to identify their own hive and can drift into another one. A forager with a full honey crop will be accepted into the strange colony so colonies at the ends of straight rows of hives tend to collect noticeably more honey than those in the middle.

To prevent drifting, place hives with their entrances facing in different directions. The hives can be arranged completely at random, in a large circle or semicircle with the entrances facing into the center, or in groups of two on a wide square with the entrances facing outwards. The last arrangement allows you to stand behind the hives during inspections, out of the way of bees flying in and out. Two hives facing the same way can be placed on one stand without a great danger of drifting.

MAKING A HIVE STAND

You can make a hive stand simply by driving corner posts into the ground and fixing a square frame on top on which the hive is to be placed. The frame should overhang slightly at the front and back of the hive to make it more difficult for mice to climb inside, and bracing the stand between the legs makes it more stable. Alternatively, use blocks to support the ends of two substantial square rails, placed slightly closer together than the shortest side of your hive. Long rails should be supported in the middle to prevent sagging.

A

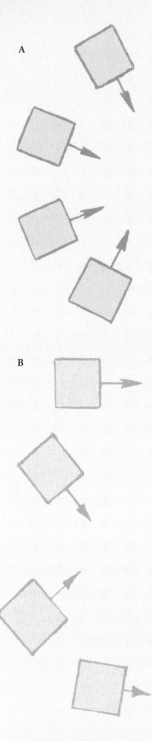

B

Configuration A is neater, but will result in more drifting than configuration B.

FINDING SUITABLE BEES

You can acquire bees at any time but it is probably best for a beginner to purchase a couple of small colonies (or "nuclei"—see page 181) in the spring. You will gain experience as your colonies expand. If you buy bees in the autumn, make sure they are well fed and have been treated properly against varroa (see page 240).

Probably the best way to find suitable bees is to obtain them from local beekeepers, perhaps through your local association. Bees can also be bought from equipment manufacturers or through classified advertisements in beekeeping journals (you will generally have to collect the colonies yourself), but always buy from a reliable source who can guarantee that the bees are healthy.

The type of bee you choose is very important, both for your own pleasure and with regard to neighborly relations. It is the duty of every beekeeper to try to keep docile bees that do not sting readily and that have a low tendency to swarm.

Colony characteristics can vary enormously. Some bees move gently on the comb during an inspection; others race around, forming clusters hanging from the bottom corner of the frame before dropping to the ground or even into your boots; others are very defensive, flying around your head emitting a high-pitched, angry buzz. Some colonies swarm readily and others make no serious swarming preparations for several years. There are colonies that will supersede their queen when she begins to fail, raising another while the original queen is still in the colony.

Take advantage of visits to local api-aries to see the differences between bee colonies. Good-tempered, local bees are those suited to local environmental conditions and are most likely to give the best performance in terms of honey yield and winter survival.

Although it is tempting to bring in good-tempered bees of a different strain from another area, it has been shown that, when the non-local queen is eventually replaced, possibly by swarming, and the new virgin queen flies out to mate with drones from local colonies, colony temper begins to deteriorate within a few queen generations. Then, to maintain docile bees, you will have to return to the original source and purchase new queens, which can be an expensive business, especially as the foreign queen may not be readily accepted and may even be killed when introduced to a colony in place of its existing queen.

If honey production is your primary objective, choosing productive bees becomes important. The amount of honey a colony collects depends on the weather and available forage within flying distance, but it also depends on the size of the foraging force, which in turn is determined by the longevity of the individual bees and/or the prolificacy of the queen. So, if, in one colony, the queen lays 1,000 eggs per day and the emerging adults eventually join the foraging force, while, in a comparative colony of the same size and with the same death rate, foragers live for one day longer, that second colony will have a greater effective foraging force and hence the potential to collect more nectar. Another factor in colony efficiency is that emerging larvae require feeding, and a very prolific queen can mean that the incoming nectar or honey stores are turned into bees, rather than your honey crop.

What comes with the bees?

If you begin by buying two nuclei of bees, each nucleus may come in a small box (a "nucleus box") containing five frames (at least two with brood and one with food stores), a queen, and a good number of bees. This small colony is often easier to handle than a full-sized one and will enable you to build your experience and confidence as it grows. However, if you purchase a full-sized colony, the bottom board, brood box, inner cover, and telescoping cover usually come with the bees. Many colonies are also sold with a queen excluder and one or two supers. If you are lucky you may get other equipment, such as a hive stand or honey extractor.

A nucleus is often delivered in a traveling box made from stiff cardboard or similar material. In this case, or if the brood box is not included in your purchase because the beekeeper wants to keep the hive, you will need to transfer your bees into your own permanent hive (see page 182).

Inspect the colony before buying

If possible you should inspect the nucleus or colony before handing over any money. When purchasing bees through an advertisement, a beekeeper from your local beekeeping association will often be pleased to go with you. You can also contact your state bee inspector for assistance. Remember always to check the colony for disease (see pages 236–39). A colony with chalkbrood may not look sick but every larva that dies before pupating is one less potential forager. Certain diseases are notifiable, which means that there is a legal obligation to notify the relevant authority of the suspected occurrence of the disease (see page 237).

Ask the vendor to guarantee that the colony has a queen. Your guarantee may also include the number of frames of brood and stores, and the fact that the colony does not have a serious (notifiable) disease. The vendor should tell you what varroa treatments have been given to the colony and when these were administered (see pages 240–41). In some countries, such as UK, beekeepers must, by law, record treatments

A traveling box for a nucleus of bees.

given to a colony, including the date the medication was purchased and administered, and the serial number on the packet. However, there is currently no equivalent legislation in the US.

TRANSFERRING THE BEES

First, prepare your new hive and position it in your apiary (see page 165). If you have bought a nucleus, ensure you have sufficient spare frames and foundation to fill up the brood box of your hive. The bees will fill any gaps with free-hanging sheets of wild comb, which can be difficult to remove if they attach it to the inner cover or build it at various angles.

Stand the nucleus box alongside the hive into which the bees are to be transferred, with the entrances of both boxes facing in the same direction. Open the entrance of the nucleus box and let the bees fly from the box for a couple of days to orient to their new position.

The bees are now ready to be transferred. Remove the cover of the permanent hive and place it upside down nearby. Take off the inner cover and put it at an angle on top of the cover or prop it against the hive stand. Remove the

frames from the brood box and prop them against the roof or somewhere out of the way but nearby. Now your brood box is ready for the bees.

Put on your protective clothing, light the smoker, and make sure it is well lit (see page 175). With gentle bees, you should have no problem transferring them but, if this is the first time you have handled bees, you may feel more confident if you smoke them first. Apply a little smoke at the entrance then take the cover off the nucleus box and blow a small amount of smoke across the top of the frames.

Stand parallel to the frames in the nucleus box so your hands fall naturally to the frame lugs. Ease the frames toward one side of the box by inserting one end of your hive tool between the side wall of the hive and the end frame and levering gently.

If the frames are stuck together with propolis, place the flat end of the hive tool vertically between each frame in turn and lever them apart. Use your spare hand to control the movement of the frame when you part it from its neighbor. If you are right-handed, separate the left-hand side of the frame

BELOW LEFT: Applying a little smoke to bees inside a nucleus box.

BELOW RIGHT: Cracking frames apart using the hive tool.

Gently lifting a frame out of the nucleus box.

first before moving across to the right-hand side to repeat the operation (use your thumb to control frame movement): do the opposite if you are left-handed. Some bees react to hand movement and this procedure minimizes movement across the box.

Now you will have a small amount of working room to lift out the first frame. Use the curved or "J" end of the hive tool to ease up the lugs at either side of the top bar so you can grasp them between thumb and first finger of each hand, and gently lift the frame

vertically out of the box. The first frame is always the most awkward as there is little room to maneuver; there will be bees in the small gaps between comb and box, and between first and second combs. As you lift the frame, the bees will roll over each other but they shouldn't suffer if you move the frame gently.

Having removed the frame with its bees, lower it gently into the new brood box. Make sure the side that was facing the wall of the nucleus box now faces the side of the brood box. When bees develop their brood nest and draw out comb on the frame, the surface may be uneven. This will be mirrored in the adjacent comb, so if there is a bulge in one side of the comb, the facing comb will be shallower to maintain the space between the comb surfaces. It is important that you maintain this nest structure by keeping the frames in sequence. The comb on one frame should not butt up against another comb. Repeat the process with the other frames, making sure they end up in the same relative positions in the brood box as they were in the nucleus box.

Shake any bees remaining in the nucleus box into the brood box. You can knock the box sharply on the ground to persuade them to let go, or use a feather or a handful of grass to brush them out. Make sure all the bees go into the brood box. The queen could be among the strays!

Fill the brood box with your extra frames and replace the inner cover and cover. You now have your first colony!

As you gain experience, the time will come when you will not necessarily open every colony on every apiary visit or go through every frame in each brood box. However, at the beginning a thorough examination is a necessary part of the learning process. Opening your first colony of bees is a very special moment. Even if you have been to an apiary meeting or helped another beekeeper, these are your bees in your hive and that makes the operation exciting and slightly daunting.

Inserting a frame into the brood box.

BUYING QUEENS BY MAIL

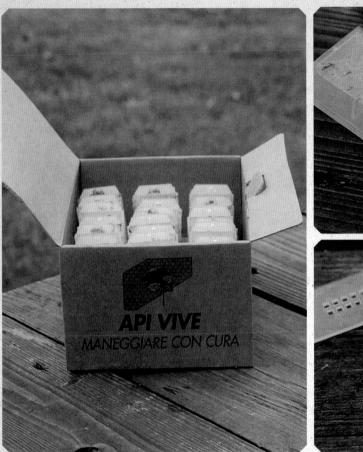

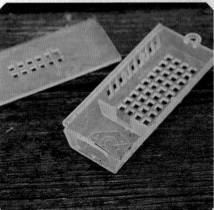

A new queen arrives through the mail in a small cage.

Wʜᴇɴ sᴛᴀʀᴛɪɴɢ ʙᴇᴇᴋᴇᴇᴘɪɴɢ you need a colony of bees rather than an individual queen (which has to be introduced to a colony). However, at a later stage there may be times when you have to buy a queen to replace an existing one that is perhaps bad tempered or if the colony's queen has been lost. Individual queens can be purchased and received by mail. They are sent in small cages, which contain a food reserve. The queen must be accompanied by about a dozen attendant workers to look after her and feed her. Queens should be bought from a reliable source, which guarantees they are healthy. They cannot be bought from overseas but you are strongly recommended to buy local queens that are

suited to your environmental, weather, and seasonal conditions.

Just before the queen cage is introduced into the colony the attendants should be removed, leaving only the queen in her cage. This improves queen acceptance. The cage has a hole which is plugged with candy. It is placed in the hive between frames in the brood nest. The bees in the hive can touch the queen in the cage with their antennae but cannot get close enough to attack her. The solid end of the cage gives the queen somewhere to retreat as colony workers will often bite at her feet. The bees eat the candy plug and by the time the queen can emerge, the bees in the colony have become used to her scent and accept her as their queen.

QUEEN MARKING AND WING CLIPPING

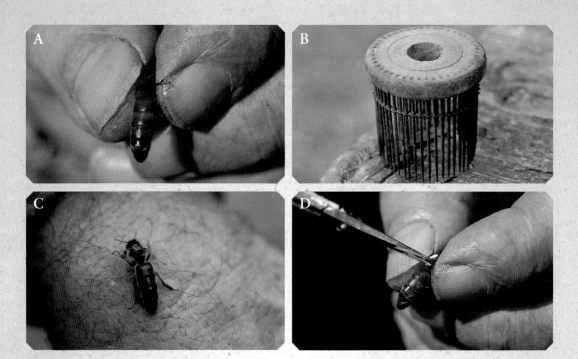

Q UEENS CAN BE MARKED to indicate the year of their birth. The convention uses five colors (white, yellow, red, green, and blue) for the years ending in 1 or 6, 2 or 7, 3 or 8, 4 or 9, and 5 or 0 respectively. As long as you remember to start with year 1, you can use a mnemonic to remember the order of the colors: What, You Rear Green Bees?

To mark a queen, first you have to restrain her. One technique, which requires considerable skill, is to pick her up from the comb between the thumb and first finger, using the second finger to give her a surface to stand on (A). Another method is to use a queen catcher, which comes in the form of a large clip, to pick her off the comb. She can then be transferred to a hand for marking. If you do not wish to hold her, she can be held against the comb with a push-in queen marking cage or captured in a queen marking tube with foam plunger to push her against the screen top for marking. Make certain that workers are not captured or injured with either device. Position the queen carefully so that she is not injured. When marking, avoid putting paint on her head, eyes and abdomen. Mark only the top of the thorax. If using numbered discs, the queen will have to be held by hand.

Once the queen is restrained, you can place a small dab of queen-marking paint on her thorax (C). Proprietary paint is available in the relevant queen-marking colors. Kits can also be obtained for gluing a numbered disc to the queen's thorax.

Having marked the queen, wait a short while to allow the paint to dry before releasing her back into the colony. It is worth watching the reaction of the workers to her return. Some bees object to the smell of the solvent in the paint and will attack the queen when they do not recognize her odor. In this case, catch her and wait a little longer before returning her to the colony again.

While the queen is restrained, the end third of one of her wings can be removed with a pair of fine scissors (D). This will unbalance her if she tries to fly out with a swarm. She will drop to the ground and the bees will return to the hive. You may have lost your queen but you still have all your bees; you can then proceed with your chosen method of swarm control.

You can mark a virgin queen if you wish but do not clip her wing before you are sure she has mated. It is usually best to leave marking and clipping until your queen is well established in the colony.

Keeping Records

Another method, especially helpful if you have many colonies, is to write down the information on a record card (see full-size record card on page 413).

BEFORE INSPECTING YOUR BEES, decide how you are going to keep a record of what you do, when you do it, and what is going on in each colony. Your records can be as simple or as complicated as you like. A stone on top of a hive could indicate the colony is rearing queen cells but it might be safer to write this information down. To start with, keeping a diary may be sufficient but for the longer term it is better to keep a consecutive record for each colony on a single card. Several simple record card designs are available, including those from Bee Craft (www.bee-craft.com) and the British Beekeepers' Association (www. british-bee.org.uk). Record cards can contain many pieces of information, ranging from the weather conditions during the inspection (and any out-of-the-ordinary reaction of the bees) to how many queen cells are being built or how much honey has been stored.

Number the hive or the queen?

When you start keeping records, it seems obvious to identify and number each hive. This is easily done by punching the number on a piece of Dymo tape, leaving on the backing and attaching the tape to the brood box with a thumbtack.

However, a better method is to assign a number to each queen. The queen number can then be punched onto tape and attached to her box.

HIVE RECORD CARD

APIARY	QUEEN						
REF. NO.	**ORIGIN**		**COLOR**		**CLIPPED**		**YEAR**
DATE	QUEEN PRESENT	QUEEN CELLS	BROOD FRAMES	STORE FRAMES	TEMPER	POLLEN	COMMENTS

A marker on top of a hive might be used to indicate a colony is rearing queen cells.

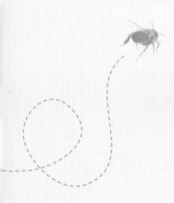

Then, when a colony prepares to swarm, for example, and you control this by removing the queen to another box (see swarming control methods on pages 198–207), then her card and the hive label move with her to the new hive. All her records are together and it is easy to look back to check on her performance. If you number your hives, you will need to note that this particular queen is now residing in a hive with a different number, which is more difficult to keep track of, especially if your colony numbers increase.

At the top of the record card note any details of the queen, such as whether she is marked (with a small dab of queen-marking paint or a numbered disc glued to her thorax) and with what color, where she came from (whether she was purchased on her own or in a colony, whether she came in a swarm or from your own queen-rearing operations), how old she is, and whether the end of one of her wings has been clipped off to prevent her flying away with a swarm.

Brood in all stages

Your records might include a column to indicate you have seen the queen, it is reassuring to do so. However, the queen can be very elusive, even if she is marked. Normally the queen is the only bee in the hive laying eggs, so if you see eggs in the cells you know she is there—or at least was three days ago.

Also include a space in the records for noting that you have seen eggs and larvae in different stages of development (see pages 99–105 for a description of the brood in its different stages). Recording the approximate number of eggs and unsealed and sealed larvae present will indicate how the colony is developing: eggs take three days to hatch into larvae; the larvae are unsealed for six days, after which the worker cells are sealed; adult workers emerge after a further twelve days. Therefore, if the queen is laying at a constant rate, there will be twice as many unsealed larvae as eggs and twice as many sealed cells as unsealed larvae. With more eggs than unsealed larvae,

It is reassuring to see the queen,
but she can be very elusive.

the colony is expanding; with fewer, the queen is either reducing her egg-laying in readiness for winter or the colony is preparing to swarm, or has already. The presence of queen cells—built out from the face of the comb and containing the developing queen bees—will confirm the latter. Estimates of the number of cells in each state and a comparison with previous inspections will tell you what is happening.

Stores

You need another column in your records to note the amount of food stores available to the bees. These stores are contained in cells around the edge of the brood nest. If they are low (see page 217 for how to tell this), you should consider feeding the colony (see pages 217–20 for what to feed them and how).

Colony behavior

Defensive behavior is worth noting in your records, as highly defensive bees do not make for enjoyable beekeeping—they will jump off the combs and become defensive as soon as you lift the cover. Others are more tolerant and need much greater disturbance before they become agitated and attack.

Honeybees defend their nest in a variety of ways. Initially they will fly around you emitting a high-pitched buzz, which you will instantly recognize as their way of telling you to go away. If you ignore this warning, their next act of defense is for a worker bee to sting you. This action pulls the sting mechanism from her abdomen and emits a pheromone or chemical scent. The sting pheromone tells the colony that it is in danger so it attracts other workers who will generally try to sting around the same area.

Another aspect of defense is where a few bees will persist in flying with you when you are walking back to the house or vehicle after an inspection. Keep records so that you can identify any colony that is behaving in this way.

Once you have identified a bad-tempered colony you can take steps to replace the queen with one that should produce better-tempered bees (see page 231).

Disease

Always keep a column in your records where you can note down any suspected disease. Some diseases, known as minor brood diseases, are probably present in most, if not all, colonies, but they do not usually give rise to great concern. Others are notifiable in some countries by law so you need to become familiar with their symptoms so you can recognize them if they ever appear (see page 236).

The most important thing when you start beekeeping is not only to be able to spot diseases but to know without a doubt what healthy brood looks like. Larvae should be pearly white and curled into a "C" shape at the bottom of the cell. The wax cappings on cells containing brood should be an even, light-brown color, dry and not perforated. In the areas of sealed brood, virtually all the cells should be sealed. Empty cells indicate either that the queen is not laying consistently or the cell contents are being removed as the worker bees detect something is wrong.

If the brood looks unhealthy, ask an experienced beekeeper for help.

OPPOSITE: *Inspecting the colony.*

Legal records

Your cards must also include space for the legal side of record keeping. The legal requirements will differ depending on where you are. In the UK, bees are classed as food-producing animals and, as such, beekeepers are legally required to record purchases of medicines and their administration to their bees. You need to note what medicine you bought and when, the number of packs and their batch numbers with the name and address of the supplier, the date you applied a particular batch of medicine, and to which colony.

COLONY INSPECTION

Having put on your overalls, rubber boots, gloves, and veil and collected your hive tool, smoker, fuel, matches, and record cards, you are now ready to undertake your first colony inspection. Before opening the hive, take time to think about what you are planning to do and why. Are you going through the colony to try to find the queen and check that the brood is normal? As you gain experience, you might be looking for signs of swarming, deciding if the colony needs a super, checking for signs of disease, or seeing if the honey is ready to be taken off the hive and extracted. What you are looking for depends on your experience, the time of year, the state of the colony, and any particular manipulations you want to perform.

Try to focus on what you aim to achieve. If you have no good reason to open a particular colony, it is often best to leave it alone. However, you can inspect all your colonies at least once every seven to nine days during the early active season—from mid-April to mid-July in temperate zones (see pages 226–34). The inspection interval

COLONY INSPECTION CHECKLIST

This is a summary of the basic things to look for in a colony inspection:

QUEEN PRESENT?

FRAMES OF EGGS?

FRAMES OF UNSEALED LARVAE?

FRAMES OF SEALED LARVAE?

BROOD LOOKS HEALTHY?

COLONY PREPARING TO SWARM
(EGGS IN QUEEN CELL CUPS)?

HONEY STORES SUFFICIENT?

TEMPER OF COLONY
(NUMBER OF STINGS)?

NEEDS A SUPER?

HONEY READY FOR EXTRACTION?

Make sure the smoker is going well before opening the hive.

Hold the smoker with the nozzle close to the hive entrance and gently blow a little smoke inside, then wait for it to take effect. Some beekeepers smoke the entrance and then continue putting on their veil and gloves.

Opening the colony
Remove the telescoping cover and place it upside down nearby. You can place other hive boxes on top, balanced on the four strips of wood making up the edge of the telescoping cover—a relatively small area where bees might be crushed. Any bees that fall out of the other boxes will collect in the upturned cover and can be shaken back inside the hive at the end of the inspection.

Next, remove the inner cover: insert the flat blade of your hive tool under one corner and pry it up a short distance. Blow a little smoke into the gap, making the bees run down between the frames. Then crack the propolis seal with your hive tool and lift the cover right off. Prop it against the side of the hive or place it on top of the upturned cover.

is linked to the life cycle of the queen (see page 231)—the time it takes to raise a new queen—since the main reason for inspection, particularly early in the active season, is to check for signs of swarming preparations.

Using the smoker
First, light your smoker (see page 175). Make sure it is going well and is filled with suitable fuel, which is not packed down so tightly that the air cannot pass through. Carry spare fuel and matches or a lighter in case the smoker goes out during the inspection.

Opening the hive:

LEFT: *Lifting the cover.*

RIGHT: *Removing the inner cover.*

How to turn the frame during an inspection:

A

B

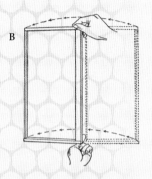

C

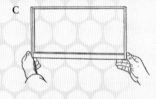

Inspecting a frame

Frames are removed from the brood box in exactly the same way as they were taken from the nucleus box when you transferred your nucleus to a brood box (see pages 182–84). Break the propolis seal between the first frame and its neighbor and ease it out gently. You now need to inspect this frame to assess what it contains. If this is a new colony starting from a nucleus, the first frame will probably contain a sheet of foundation or empty drawn comb. As the colony builds up, the bees will draw out comb on the foundation, and this will be filled with honey, pollen, or brood. Unless the queen is very prolific, the outside comb (closest to the hive wall) is likely to contain honey stores but no brood.

Many beekeepers use wired foundation in the brood frames but, even with the extra support, newly drawn comb on a warm day can be soft, so be careful.

Turning the frame

Always handle frames smoothly and gently. Sudden movement or jarring can damage the comb and may also alarm the bees, alerting them to your presence as a danger. Grasp the frame firmly by the lugs and hold it up to inspect the nearest side (A). Hold the frame you are inspecting over the hive so that, if the queen drops from the comb, she lands in the hive. To look at the other side, drop one hand down so that the top bar is vertical (B). Swivel the frame through 90 degrees until the other side is facing you before lifting the original hand to return the top bar to a horizontal position. You will now be looking at the reverse of the comb but with the top bar at the bottom rather than the top (C). Turn it back in the same manner.

BELOW LEFT, TOP AND BOTTOM: *Easing and lifting the first frame.*

PROPOLIS

Some colonies collect a great deal of propolis and others very little. It is not known why there is a variation, though it could be related to climate. Bees use propolis to fill up gaps in the hive as a kind of draft-proofing. Some colonies use it to control the size of the entrance: they can reduce this to a few, easily defended bee-sized holes in a propolis curtain that otherwise covers the entire entrance. Bees also use propolis to varnish the inside of the hive and to strengthen their combs.

Propolis is very sticky and can be difficult to remove from your fingers and clothes. Denatured alcohol will soften it and allow you to wash it off before it dries. Kitchen cream cleaners will also remove propolis but you will have to rub hard. Removing it from clothing is more difficult, but can be achieved with a suitable tar or resin stain remover. Be careful not to leave a solvent mark.

Moving to the next frame

Once you have inspected each side of the frame, prop it, resting on one of the lugs, against the hive near to the entrance so that bees can return inside easily if they wish. Alternatively, some hive stand designs enable you to place the frame at an angle between the rails to one side of the brood box. You can also buy frame supports that hang on the open box and hold the first frame. If you have a spare nucleus box, you could use that to house the first frame until you want to return it to the hive. Continue your inspection, moving the next frame into the gap you have created before lifting it from the hive. After looking at it, gently replace the frame in the gap and move it to the side away from the other frames. This effectively moves the gap along the runners by one frame. Now you can take out the next frame in the same way.

Try not to knock the frames against the sides of the hive either on the way out or when you replace them. Remember to replace the frames in the same sequence and facing the same way to retain the brood nest pattern (see page 197).

Always remain calm. If at any time you find the situation overwhelming, push the frames together, back into their original positions, replace the first frame, and close up the hive. You can resume your inspection later on.

Finding the queen

You will probably want to see the queen on your first inspection. This is easier to do if she has been marked (see page 186), but even a marked queen can be difficult to spot. If you started with a nucleus rather than a full-sized colony, she should also be easier to find because of the smaller number of bees. Remember that the queen is not that much bigger than a worker or a drone but that she has certain noticeable characteristics: some beekeepers look for her longer legs, which are often paler in color than those of worker bees; others look for her longer, more pointed abdomen (see pages 97–98).

As you lift a frame from the hive, run your eye quickly around the edge before looking in the center. Queens dislike the light and are often to be found on the side of the frame which has been in darkness during your inspection, i.e. the reverse side when you lift the frame out, so don't spend too long looking on the first side. You can come back to it later on.

If you don't see the queen, don't panic. She may be hidden by a mass of worker bees or have run onto the hive bottom. To break up a group of bees, touch them or blow on them gently so they split up and move away. Applying a small amount of smoke has the same effect. The important thing is to see eggs in worker cells, which indicate that the queen was in the hive three days ago.

Look to see if there are any occupied queen cells containing eggs or larvae. If you find this indication of swarming preparations, you need to take preventative action or you will lose a good proportion of bees as they take flight with the old queen.

Holding the frame and inspecting the reverse side.

Closing the hive

When you have finished your inspection, move the frames in the box back into their original position. Hoffman frames (see page 161) have a distinct advantage here as you can insert your hive tool between the frame and side wall and lever all of them over in one block, squeezing them together into position. If the hive is fitted with metal or plastic spacers, this is not so easy and you may have to make the maneuver in several stages.

Having regained the original gap at the side of the brood box you can now replace the first frame. If you want to remove most of the bees from this comb so they are not rolled over or crushed when you replace the frame into the narrow gap, hold it over the brood box and give it a sharp shake.

The few bees remaining will be returned to the hive on the frame. Then push the first frame up against the others to maintain the bee space.

Blow smoke gently over the top bars to persuade the bees to go back inside the box, and replace the inner cover. There are two ways to put something back on top of a brood box without crushing bees in the process. An inner cover, queen excluder, or super can either be placed in line with the side of the box—in its final orientation—or at a 45-degree angle to this. In the first case, lower it into place while checking that there are no bees between the hive parts coming together. Move any bees in the way by brushing them aside or using a little smoke. In the second case, the inner cover, excluder, or super is placed flat and then carefully swiveled into position, again making sure the bees are out of the way.

Finally, replace the telescoping cover.

The queen excluder

As the season progresses, the bees will need more space in which to store honey. When the brood box looks full of bees you will need to add a super. First, place a queen excluder on top of the brood box to prevent the queen from moving upwards and laying eggs in the combs inside the super above. Make sure that the excluder and the super both fit squarely and flush with the brood box, then replace the inner cover and telescoping cover.

At subsequent inspections, you will first have to remove the super and stand it on the upturned cover. Then lift the corner of the excluder and, if it is unframed, peel it carefully off the brood box. The bees will have stuck it down with propolis and may have built brace comb on or through it. Before replacing the excluder at the end of your inspection of the brood box, lay it on a flat surface, such as the top of another hive, and scrape off surplus material with your hive tool. Be careful not to catch the corner of the hive tool in one of the slots or you may deform the gap and make it large enough for the queen to squeeze through.

The queen excluder.

COVER CLOTHS

Some beekeepers use cover cloths to stop bees flying up from the combs during colony inspections. These are strips of heavy cotton cloth, longer than the width of the hive. A weighty rod wider than the hive is sewn into each end and this can rest across the hive walls. One cloth is placed over the opened box and parallel with the frames. It is then rolled back to reveal the frame you are about to lift out and the top bar of the adjacent frame. After the first and second frames have been inspected, and the second frame replaced and moved along the runners to create a gap, the cloth is rolled back to reveal the third frame. Another cloth is then rolled up and placed on top of the second frame. The first cloth is rolled up and the second unrolled as the inspection continues across the hive, ensuring that most of the frames are covered over during the operation.

Cover cloths tend to slow down manipulations but you can try them and decide if they help or hinder you.

CHECKING THE BROOD PATTERN

THE BROOD NEST WILL BE SITUATED toward the center of the brood box. You are more likely to find the queen here where she lays her eggs. A sphere of pollen stores surrounds the brood nest, so when you find a comb containing a large proportion of pollen you know you are close to it. The next frame should start to show what is known as the brood pattern. A queen laying a normal brood pattern will work around the comb in ever-increasing circles. The first eggs will be laid in a patch in the middle and then she will work her way outwards, using the cells that have been cleaned and polished by the house bees. As the eggs hatch and the larvae develop, you will be able to see rings of different ages, culminating in a patch of sealed brood, the final stage of development, at the center of the nest. As the first adults emerge, the workers clean the empty cells ready for the queen to lay more eggs in them, thus starting the cycle again. When the brood nest is expanding, as the active season progresses from spring into early summer, the first eggs have not completed their life cycle by the time the queen is ready to lay in those cells again, so she will move onto a new comb, thus extending the brood nest outwards. Make a note of the number of frames containing brood, however small the patch, so your records will tell you how the colony is progressing.

Why Should You Try to Control Swarms?

SWARMING IS THE natural process whereby a colony divides into two or more separate groups. The swarming season generally precedes with a major honey flow, when the nectar is flowing freely at the height of the season as the abundance of food gives the new colony the best chance of survival. Losing half of your foraging force can severely reduce your final honey crop, so if you can stop your bees from swarming, you will undoubtedly have a larger honey harvest. Preventing swarming will also reduce anxiety and potential opposition in the neighborhood. Many people are frightened by a swarm of bees. Controlling swarms will also reduce the possibility that a swarm establishes its nest in an inaccessible place such as a wall or a chimney.

Once a swarm has begun to build comb in an inaccessible place it is very

SWARM COLLECTION SERVICES

Most local beekeeping associations operate a swarm collection service. If you see a hanging swarm and are unable to collect it yourself, notify them to come and remove it before it finds a new home in an inaccessible place.

difficult to remove—either the wall or chimney has to be taken apart or the bees have to be killed in situ. Insecticides used to kill bees will poison other bees coming later to rob any remaining honey, so pest control operatives should attempt to block all access holes to the destroyed nest.

Collecting a hanging swarm in the open, while it is still clustering and deciding where to go, is vastly preferable to trying to remove one from an inaccessible place.

SWARM PREVENTION
If your colony has survived the winter and numbers are building up steadily, you need to give it room to expand; otherwise the increasing numbers of bees will lead to overcrowding, which in turn usually leads to swarming. You can help prevent, or at least delay, swarming by expanding the available cavity space. As soon as the brood box

A swarm ready to leave the hive.

looks full of bees, add a queen excluder and a super. You need to have three or four supers available for each of your colonies. The bees will need these supers for storing honey if the colony expands normally, and the extra boxes will give the bees more room to spread out, thus relieving congestion.

As the weather warms up, start regular brood nest inspections every seven to nine days to check for occupied queen cell cups. These look similar to acorn cups and all colonies build them. Only when the queen lays an egg in a cup will the bees begin to extend the sides to turn it into a queen cell and eggs in queen cell cups indicate swarming preparations. Your regular checks will ensure you become aware of these at the first opportunity.

SWARM CONTROL

Regardless of the amount of space you have given the colony, most (though not all) colonies will prepare to swarm at some time during the spring or early summer. It is important to plan in ad-

CHECKING FOR SIGNS OF SWARMING

When a colony is preparing to swarm, eggs will be laid in queen cells. The larvae will hatch in three days. These will be sealed on the ninth day after the egg is laid, and the new queen will hatch seven days later, on the sixteenth day. The old queen will leave with the swarm around the time the queen cells are sealed. Without any intervention from the beekeeper, the first virgin queen to emerge may leave the colony with an afterswarm. Further swarms will leave with the next virgin queens to emerge, until the colony "decides" that if more bees leave it will not be big enough to survive. The workers then ensure only one virgin remains and she will fly out to mate and return to head the colony.

Some strains of bee will supersede their old failing queen rather than swarm. In this case, mother and daughter will live apparently amicably in the colony for a while until the old queen dies.

All colonies build queen cell cups. When the queen lays an egg inside, the cell cup will be extended into a full queen cell.

vance so you are ready to act when you find occupied queen cell cups. You will need sufficient equipment to control swarming in each of your colonies as very often when one colony is preparing to swarm, the others will not be far behind.

If you find queen cells that are well advanced, but not sealed, there is still time to take action. You can estimate the age of the larvae and, hence, approximately how many days it will be before the cells are sealed and the swarm issues. If you find eggs in queen cells, even if one is due to hatch the

UNDERSTANDING SWARM CONTROL

The different swarm control methods can seem confusing. The main point to grasp is what is happening within the colony and when certain events take place. Once an egg is laid in a queen cell cup, the timetable for its development into a queen is fixed. It will hatch in three days, the larval stage of the queen will be five days, and the cell will be sealed on the ninth day. The larva will pupate for a further seven days and the virgin queen will emerge on the sixteenth day. These timings start on the day the egg is laid.

All methods of swarm control involve splitting the colony in one way or another. Think of the colony as consisting of three parts: the queen, the brood, and the flying bees (foragers). If you split any one of these from the other two, you will control swarming, although you will also need to take further action.

cover, and three to five frames of drawn comb or foundation. Alternatively, use a nucleus box with a full complement of similar frames. A three-frame nucleus box is adequate if you know that you are not going to keep the queen and her bees in it for very long. However, if you intend to allow her to build up her colony, a larger five-frame box gives you more time before you have to transfer this colony to a full brood box.

Another option is to place a small nucleus directly into a spare brood box but it is then best to confine the bees to the number of combs they can comfortably cover by using a dummy board. This is a narrow board with the same dimensions as a frame, which has lugs allowing it to be hung within the hive. As the small colony expands, the dummy board can be moved to give access to more combs.

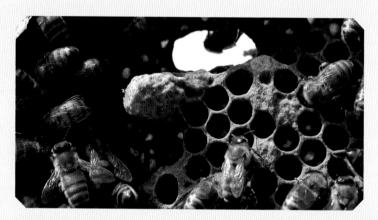

A sealed queen cell constructed by a colony that is preparing to swarm.

next day, it will still be five or six days before the queen cell is sealed. However, if you find sealed queen cells, the swarm has probably gone (see page 205).

The nucleus method: removing the queen

If you take the queen away, the first or prime swarm cannot leave the colony. One option is to pick her off the comb and remove her.

To separate the queen from the colony you will need a spare bottom board, brood box, inner cover, telescoping

WHY DISPOSE OF A QUEEN?

You may also have other reasons to dispose of a queen in addition to swarm control. For example, if the colony is highly defensive, it might be a good time to correct this behavior by replacing its queen, whose genetics strongly influence the charactersitics of the entire colony. However, if you kill the queen and the colony fails to produce a viable replacement, then the colony will not survive unless you can introduce either another mated queen acquired from elsewhere or brood of the right age on which the colony can rear a replacement (see page 203). It is prudent, therefore, to keep the old queen until you know all is well with the new one.

Assuming you find only open queen cells and no sealed ones during your inspection, you can follow these steps.

FIRST STAGE

1. Transfer a frame of food, with the bees on it, to the nucleus box and push the frame up against one side wall.

2. Find the queen and move her, on her brood frame with its bees, into the nucleus box. Before you put this frame into the nucleus box, check the comb and destroy all the queen cells that might be present (use the hive tool to break them down or pinch the cells between your fingers). Ideally the frame should contain a small patch of brood, about the size of a flat hand.

3. Put another frame of food into the nucleus, so the brood frame now has a frame of food on either side. You are most likely to find suitable food combs at the outside of the brood nest.

4. Take another two frames from the main colony in the original hive and shake or brush their bees into the nucleus. Return these frames to the original hive.

5. Fill up the nucleus box with frames but no extra bees. It is best if these contain drawn comb, but foundation will work. Then stuff some grass tightly into the entrance and replace the inner cover and cover. If you don't block the nucleus box entrance, the flying bees will leave and return to their old hive. By shutting them in, you are helping to "persuade" a number of them that they now belong to the nucleus colony.

6. Check through the original hive very carefully. Don't remove any queen cells unless they are large and nearing the point when they will be sealed, i.e. leave open queen cells, preferably containing very small larvae. To help identify these occupied queen cells the following week, stick a thumbtack into the frame top bar directly above each one.

7. Push the frames together and fill the gap(s) in the original hive with the frames of the nucleus box which you removed when you made room for the queen and the food. (The nucleus box you brought to the apiary should have a full complement of frames, enough to fill both boxes completely by the end of this operation.)

8. Place the nucleus box to one side of the original colony or on another stand in the apiary. The grass plug will gradually wilt and allow the bees out. If this

Two ways to destroy a queen cell: by pinching it between the fingers (top), or using the hive tool (bottom).

hasn't happened within 48 hours, remove the grass yourself. You now have the old queen secure in the nucleus with plenty of bees and a small patch of brood. She is insurance in case the new queen in the original colony gets lost or does not mate successfully (in this situation, you can reinstate the old queen).

SECOND STAGE

This next stage should usually take place a week later. It can be left for eight or nine days after you removed the queen, but it must be before the first of the marked queen cells you left to develop is due to hatch. This timing is because emergency replacement queens can be reared from worker larvae that are up to four days old, but seven days after the queen has gone from the main hive this opportunity has passed and there will be no more emergency queen cells (see the panel on page 203).

1. When you inspect the colony, remove all the queen cells except one. (Don't be in too much of a hurry to break down the first queen cell you find—bees are known to tear down queen cells on their own and this may be the only one they have left alone!) First, check the queen cells you marked on your last inspection, which you know contained larvae, and select one. Brush the bees off the comb(s) to see

Emergency queen cells.

A nucleus box being filled up with frames of foundation.

more closely. Avoid shaking them off as you may damage the developing queen if you jerk her up and down inside her cell. Remove any other queen cells the bees may have started to build on this comb and replace the frame carefully, making sure not to damage your chosen queen cell. Finally, check each remaining frame and remove all the queen cells and anything that looks remotely like one. You can shake or brush the bees from these frames as you check each one. Never leave more than one queen cell in the colony or you will be in danger of losing a large swarm with the first virgin queen that hatches and all your efforts will have been in vain.

2. Now the colony has only a single queen cell there will be no afterswarm as only one virgin queen can emerge. Removing the queen and bees to the nucleus is equivalent to the prime swarm leaving. You now need to leave the original colony alone for two to three weeks. This gives the virgin queen time to emerge, mature, go on her mating flight, return, and start laying eggs. After this, most colonies will make no further attempts to swarm that year. However, keep an eye on it as a small minority will try to swarm again when they have built up with their new queen.

EMERGENCY QUEENS

If a colony becomes queenless, perhaps because the queen dies unexpectedly, then the colony can raise a new queen on worker larvae that are up to four days old. These larvae are developing in worker cells but by changing their diet to one of royal jelly throughout the larval stage, they will develop into queens. In this case, the cells are elongated to house the larger insect but they are built out from a worker cell within the comb rather than from the specially constructed queen cell cups on the face of the comb (see page 114), and appear shorter than normal queen cells. They are known as emergency queen cells.

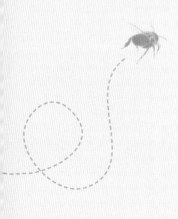

3. Now decide the fate of the old queen. She can be left to build up the nucleus into another full colony, or you can kill her and unite her bees and brood into one colony (see page 235).

The artificial swarm

Many beekeepers use another two-stage process for controlling swarming, known as the artificial swarm. This essentially fools the colony into thinking it has swarmed naturally.

If you choose this method, it is best to start swarm control as soon as you find an occupied queen cell cup. Its success hinges on the fact that there are no sealed queen cells—only eggs or very young larvae in queen cells—when you undertake the first step. If there are none containing older larvae then none will have emerged before the second stage a week later. You will need an extra bottom board, inner cover, telescoping cover, and brood box filled with frames of drawn comb or foundation.

FIRST STAGE

The first stage is taken when the queen cells contain eggs or larvae and involves separating the brood and the nurse bees from the queen and the flying bees:

1. Remove the supers from the hive (hive A) and stand them on the upturned cover.

2. Move the brood box (hive A), with its bottom board, to one side but within easy reach.

3. Put the new brood box (hive B), with its bottom board, on the hive A stand.

4. If you have your frames the cold way (see page 163) in hive B, take three out of the center. If they are the warm way, remove three from the front.

5. Find the queen in hive A. Transfer her, on her frame, into the new brood box (hive B). It is very important that you remove all queen cells on the comb you transfer with the queen.

6. If you have any spare frames of drawn comb, group them on either side of the queen's frame, and then fill up brood box B with frames of foundation.

7. Put the queen excluder on top of brood box B, add the supers and their bees, and complete the hive with the inner cover and telescoping cover.

8. Move hive A to the side, about 3 ft (1 m) away from its original position

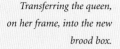

Transferring the queen, on her frame, into the new brood box.

with the entrance facing the same way as before.

9. Check the frames in brood box A. Remove any large queen cells, but leave all the remaining ones.

10. Replace the queen excluder and supers on hive B on the original site.

11. Replace the inner cover and telescoping cover on both hives.

SECOND STAGE

The second stage takes place a week later, immediately before the queen cells are due to hatch, and involves separating the new flying bees from the queen cells.

1. You should find several sealed queen cells in the original brood box (A) that was moved to one side. The virgin queens which emerge can be prevented from leaving with an afterswarm, by diverting the flying bees and reducing their numbers. To achieve this, move hive A to the other side of hive B (which contains the queen), or to any other site farther away in the apiary. Any flying bees that have learned the new position of hive A will return to find it has gone. They will enter the nearest hive, which should be hive B, and which has all the supers. The foragers will boost this colony's population and help it collect even more nectar. The colony in hive A, which has been moved twice, is now depleted of flying bees and there will be too few for it to produce any afterswarm. In all probability, the first virgin queen to emerge will kill any other developing queens while they are still in their cells.

2. As with the nucleus method (see pages 200–204), you should leave the colony with the queen cells alone for two to three weeks. Then check to see if the new queen has started laying. A good indication of this from outside the hive is when you see lots of pollen being taken inside.

3. You also need to check the queen-right colony (i.e. the one containing a queen) in hive B. If you find that the bees have started to build more queen cells, destroy them all. Then, another week later, make sure that the brood is developing normally and no further swarming preparations are being made.

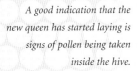

A good indication that the new queen has started laying is signs of pollen being taken inside the hive.

Can't find the queen?

Most swarm control methods require you to find the queen, but you can still take steps to control swarming without ever seeing her.

If your colony inspection reveals occupied, but unsealed, queen cells and you cannot find the queen, carry out the first three steps already outlined for the artificial swarm. After step 3, proceed as follows:

4. Find a comb in the swarming colony (hive A) with brood in all stages—eggs, larvae and sealed brood—and move it into brood box B, along with the bees on it. There must be no queen cells on this comb.

5. Check through the complete brood nest in hive A, which you have moved to one side, and make sure that all queen cells here are unsealed. It is very important that you destroy any that have been sealed.

6. Fill up brood box B with frames of drawn comb or foundation and position them around the single frame of brood. Add the queen excluder, the supers with their bees, inner cover, and telescoping cover. This hive (B) is on the original site.

A queen at the bottom of the brood box. You can still take steps to prevent swarming without seeing the queen.

7. Put brood box A, containing the rest of the brood, on another bottom board and move it to one side of the original site (where hive B is now standing) and at least 3 ft (1 m) away. Push the frames together and then fill up the gap with a spare frame of drawn comb or foundation. Make sure the entrance faces the same way as before.

8. Replace the inner cover and telescoping cover.

When you inspect the colonies a week later, you will find out which contains the queen. If there are eggs and young larvae in hive B, the queen is there and you have actually performed the artificial swarm and can proceed as before.

It is more likely, however, that she is on the brood combs in the original box that was moved (hive A). In this case, you have separated the queen and the brood from the flying bees. Any flying bees from hive A will return from foraging to the original site and join the new colony there (in hive B). The queen will not, therefore, swarm from hive A, even though it contains the queen cells. She can only swarm with bees that can fly and she is now accompanied by young bees that can't yet do so. Over the next seven to nine days, the bees should tear down the queen cells and the colony will return to a non-swarming state.

A week later, the queen will be in brood box A with eggs in worker cells and no queen cells. The bees in the queenless colony in hive B will have started building emergency queen cells. Check these frames and break down all but one of the queen cells, leaving the largest one to develop and emerge

A lost swarm. Breaking down queen cells as your only method of swarm control is not an effective solution.

to head the colony. The queenright part (in hive A) can be left to build up, and supers added if necessary. In many cases, this colony will make no further attempt to swarm. However, if there are signs of swarming preparations, remove the queen and all the queen cells.

Nine days later, remove all the queen cells again. Now queenless and unable to rear a replacement—as it has no larvae young enough to be developed into queens—this colony can be united to its daughter colony (in hive B) or to any other stock in the apiary.

Failures in swarm control

Eventually you are bound to lose a swarm. You probably missed a queen cell that was tucked away in a corner or by the bottom bars of the frames, or perhaps you failed to examine the colony at the due time. You now have to make sure your colony does not produce further afterswarms and you don't lose any more bees.

There are several clues that a colony has swarmed:

- If there are only a few eggs, or even no eggs, in worker cells and more than just a couple of sealed queen cells then you have certainly lost your queen and the swarm.
- If the bees are more bad-tempered than usual, flying up from the comb and stinging more readily, this can indicate that a colony is queenless.

If your colony has swarmed, you should proceed as follows:

❋ If all the queen cells in the colony are sealed, you should reduce them down to one so that the colony cannot lose an afterswarm.

❋ If there are both sealed and unsealed queen cells, destroy all the sealed ones. A week later, go through the colony and reduce the now-sealed cells down to one to prevent afterswarms.

If you find some of the queen cells have been chewed open at the tip and the occupant has emerged, this means that not only has the old queen left with the prime swarm, but some of the virgin queens have taken out afterswarms. In this instance you should look carefully at the remaining queen cells. Those showing a brown cocoon at the tip where the wax has been removed are "ripe" or ready to hatch. Use a penknife or the corner of your hive tool to open the tip of each queen cell very carefully. When you find one with a living queen inside, which is ready to hatch, remove the tip and let her out. You can then destroy all the other queen cells. After you have released the virgin queen into the colony, do not open it for two to three weeks. You can then check to see if she has mated and is laying eggs.

What happens if other virgin queens are already loose in the colony? This actually doesn't matter as long as there are no queen cells with a queen still inside. It is the presence of occupied queen cells in conjunction with a virgin queen and lots of bees that triggers an afterswarm. Without queen cells, this doesn't happen and the virgins will fight each other to the death, leaving just one to head the colony. If you leave two queen cells "just in case," you are very likely to lose an afterswarm if the colony is of any appreciable size.

COLLECTING A SWARM

Once news gets out that you are a beekeeper, you will probably receive calls asking you to collect swarms. (Most beekeeping associations keep a list of those willing to undertake this service and you can ask for your name to be added.)

From the beekeeper's point of view, the ideal swarm is one that is hanging from a tree branch in a large neat cluster at roughly shoulder height in a clear space with no surrounding vegetation. However, you could also find it spread out among the branches of a pine tree,

A queen cell that has been chewed open. The virgin queen has departed.

plastered up the side of a post, or hanging just out of reach on the branch of a flimsy tree.

Whatever challenges swarm collection presents, the basic principles remain the same. One of the queen's pheromones keeps the swarm together as it flies through the air and another keeps the hanging cluster stable while it decides which cavity to occupy. The key to successful swarm collection is to get the queen into your receptacle.

Equipment needed

At the beginning of the swarming season, gather together all the equipment you might need to collect a swarm so that it is ready for immediate use. You can then get to the swarm as quickly as possible. Swarms have a nasty habit of flying away about five minutes before you arrive.

You will need a receptacle in which to place the swarm. Any suitably sized container will do, from a stout cardboard box to a nucleus or brood box with an attached bottom board and appropriate cover. Many beekeepers in the UK use a traditional straw skep (see page 39). The rough inner surface gives the swarm bees a foothold and they hang in a cluster from the top.

As the swarm may be hanging in a bush or other vegetation, a pair of pruning shears is useful. You will also need a large open-weave cloth or sheet to cover the opening of your receptacle, with plenty spare to hang over the sides, and some stout string or a hive strap to secure the cloth in place. Even though swarms are not generally bad tempered, there is always an exception, so take your veil and your smoker,

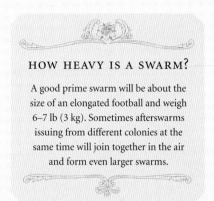

HOW HEAVY IS A SWARM?

A good prime swarm will be about the size of an elongated football and weigh 6–7 lb (3 kg). Sometimes afterswarms issuing from different colonies at the same time will join together in the air and form even larger swarms.

smoker fuel, and matches. If the swarm has settled in a tree you may also need access to a stepladder.

Some beekeepers like to use a swarm catcher for retrieving swarms that cluster on high branches. This is a bag on a metal frame, which is mounted on a pole. The bag is placed around the swarm and closed by pulling on a cord; the bees are then brought down to ground level. A swarm catcher will enable you to catch a swarm that is 13–16 ft (4–5 m) off the ground.

Catching the swarm

When you arrive at the swarm site, take a few minutes to form an action plan. A hanging swarm consisting of a well-defined cluster with some bees flying around nearby is probably the easiest to collect. Place your receptacle as far around the cluster as possible. Take a strong grip on the container and shake or strike the branch sharply to dislodge the bees. With a reasonably sized swarm, you will be surprised at the weight as it drops into the receptacle. Dislodge as many bees as possible into the container, then cover the opening with the cloth to contain the bees. Return to the ground if you have had

Collecting a swarm before it decides to move away.

ceptacle with a rough inner surface. This allows the bees to get a grip and form a cluster when the receptacle is upturned, while the flying bees are joining them.

If the bees start leaving the container to return to the clustering site, you know that you missed the queen. It is best to wait until they have clustered and start again.

Wait until all the flying bees are inside before removing the prop or closing the entrance. Then secure the cloth around the container, making sure there are no small gaps through which the bees can escape. Turn the container back over carefully so the bees can breathe through the cloth.

If you don't mind making two trips to the site, you can go home when you are sure the bulk of the swarm is in the container and then return at dusk, or when the bees have stopped flying for the day, to retrieve the container and the bees. By then, all the bees will be inside and you can simply fasten the cover and take the container away.

Unfortunately not all swarms cluster in a convenient place. If they are in the middle of vegetation, ask the owner if you can clip some of this away in order to place your container underneath the bees. If the swarm has landed and spread over a flat surface, you can brush them off into the container using a special bee brush, a large flight feather, a handful of long grass, or even your hand. If they are spread up something like a gatepost and you are able to balance the container above them, you can use smoke to persuade them to walk up inside, but don't use too much

to climb a ladder, then, holding the cloth in place, gently invert the container and put it on the ground as close as possible to the place where the swarm was hanging. This makes it easier for the stragglers to find their nest mates. If the container doesn't have an opening for the flying bees to enter, prop up the edge with a small stone or twig. You will see bees outside the container beginning to "scent" by exposing their Nasonov glands at the end of their abdomens (see page 115) and fanning their wings. Facing the entrance they waft the scent into the air, signaling to bees flying around that the colony is now in the container, so the stragglers fly down to join it. However, bees in the cluster also leave a pheromone signal where they were hanging and you will probably have to shake the branch or use the smoker a few times to persuade the bees returning there to fly into the air.

Although you can use a plastic container to catch a swarm, the bees need to be transferred immediately to a re-

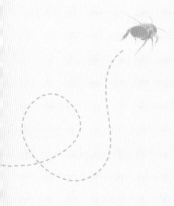

or the swarm may take off and your efforts will be wasted. Remember, the colony is looking for a suitable cavity and you have just provided a convenient one.

Swarm collection can be exciting, exhausting, and frustrating. Not only do swarms fly off before you arrive, they can fly out of your container and away, presumably to a more desirable cavity. You can leave them eagerly going into the container and return to find it quite empty. However, this does not normally so happen, once you have caught the bulk of the bees, you will usually end up taking them home.

On returning to your apiary, you need to transfer the swarm to a hive. If it is late in the evening, it is best to invert your container with the cloth underneath and place it next to where the colony will eventually be located. Undo the string or strap and prop up the edge of the container as before (see page 210) if there isn't an entrance. You can then deal with it the next day.

HIVING A SWARM

There are two main methods of hiving a swarm.

In the first, prepare the hive to receive the swarm: position it on a stand in its final position, remove the telescoping cover, inner cover, and frames, and put a queen excluder between the brood box and the bottom board. Pick up the container housing the swarm, lifting it away from the cloth and holding it over the empty brood box, and give it a hard, sharp shake to dislodge the bees. You may need another shake to get most of them out. Then place the container, with the opening facing upwards, near the new hive. Any remaining bees will soon join the colony.

Now take the frames and gently lower them into the box until they rest on the runners or on the mass of bees. Don't try to force them down into place or you will crush bees. The frames will settle into place as the bees start walking up onto them. Give them time to do this and then carefully make sure the frames are all in place. Add the inner cover and telescoping cover and you have successfully caught and hived your first swarm.

Using the queen excluder prevents the queen, and therefore the swarm, from flying from the hive entrance and absconding, and this means you can transfer the bees to a hive at any time during the day. However, you must remove the excluder two days later, especially if you have an afterswarm containing a virgin queen. If you fail to do this, she will not go on her mating flight. She will then have no sperm and can only lay unfertilized eggs, making her what is known as a drone layer. If you don't want to use an excluder, carry

BELOW AND OPPOSITE:
Hiving a swarm by getting
the bees to run up a board
into the entrance: a sharp jerk
of the container will dislodge
the bees onto the board
leading to their new hive.

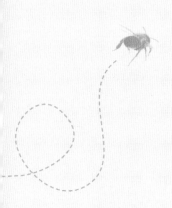

out the whole operation in the early evening. As a general rule, by morning the swarm will have decided that it wishes to remain in the hive.

The second way of hiving a swarm is to get it to run up a board into the entrance. Set up a floor and brood box, full of frames of foundation, on a stand. Find a flat board about the width of the hive and prop it up to the entrance. It helps to prop up the front of the brood box slightly, say by ½–¾ in (12–20 mm). This gives the mass of bees more room to enter.

Light your smoker and then pick up the container with the swarm and the cloth cover and place it, cloth down, on the flat board. Undo the string and spread the cloth out evenly over the board. Ideally the cloth should hang down on either side to the ground to prevent bees from crawling underneath the board. Now pick up the container and give it a sharp jerk to dislodge the bees onto the cloth. Displace the stragglers with another shake or brush them out. You can then watch the bees march into the hive. You may even be lucky enough to see the queen making her entrance. When they are all inside, remove the brood box prop if applicable, take away the flat board and cloth, and your colony is in residence.

Sometimes, because of their numbers, bees block the entrance as others are trying to get inside the hive. The remaining bees will run over them and cluster on the outside of the hive. Don't allow them to do this: brush the bees down from the outside of the hive and make sure they all go inside. Swarms hanging on the front of the hive have been known to fly away.

USING SWARMING TO INCREASE YOUR COLONIES

If your artificial swarm control is successful, you will reach a stage when you have one brood box containing the queen and another containing several developing queen cells (the swarming colony). If you want more colonies, instead of breaking the queen cells down to one at this stage, you can divide the swarming colony. When you first do this, it is probably best to stick to two divisions, but it is possible, depending on colony size, to make four or five. Simply look through all the frames of the swarming colony and decide how many queen cells there are and how many divisions you want to make. You will need a nucleus box (or spare brood box) for each one. Transfer a comb of food into each nucleus box, and follow this with two frames of brood and their bees. Each nucleus should receive at least one good queen cell on one of these frames. Remove all the other queen cells. Then push the frames together.

Distribute the remaining nurse and house bees from the original colony roughly equally between the nucleus boxes by shaking and/or brushing them off the combs. Then give all those frames containing brood to the colony in your swarm control method that houses the queen (see page 199). They will boost the colony population as the bees hatch. Place them near her brood nest and remove the equivalent number of combs without brood. Use these broodless combs to fill up the nucleus boxes, or (if they contain very little food) swap them for food combs from

Expanding your colonies by dividing the swarming colony and placing it in nucleus boxes. Here the colony is being divided into two 5-frame nucleus boxes.

other colonies and place the food combs in the nucleus boxes.

If you are using a spare brood box instead of a nucleus box, place the brood in the center of the box and flank it with the food combs if the frames are cold way, or put the brood at the front near the entrance with the food behind if the frames are warm way (see page 163). Either fill up the box with frames of foundation or use a dummy board at each side of the frames of brood and food (for cold way frames), or at the back of them (for warm way frames).

Now that the frames are in place, position the nucleus boxes at least 10 ft (3 m) away from the queenright colony. Then check the brood combs in each nucleus box and remove any very small queen cells. You don't need to remove all but one (as in the swarm control methods) as the first virgin queen to

emerge will do this for you. By removing the flying bees from the nuclei, you have prevented the virgin in each nucleus from leading off an afternoon.

Leave the divisions alone for at least two weeks. This gives the virgin queen time to emerge, mature, mate, and start laying eggs. If you keep looking to see what is happening, you could have the colony open for inspection when the queen returns from her mating flight and she could end up in a nearby (queenright) colony instead of her own. The resident bees would then kill her.

Each new small colony will begin to expand when the new queen starts laying and her new bees begin to hatch. Add further frames of foundation as required, taking the dummy board(s) out, if applicable, until the hive has its full complement of frames. When the colony becomes too big for the nucleus box, transfer it to a full-sized brood box.

Feeding

THERE ARE TIMES when your bees will need feeding, either as an emergency measure when they are short of stores and there is no prospect of a nectar flow in the area, or to supplement their stores in autumn so they have enough food for winter. At any time in the active season, a colony should have a minimum of 10 lb (4.5 kg) of stores, roughly equivalent to two full frames of food. In the autumn it needs winter stores of about 30–60 lb (13–27 kg), between six and eight full frames, some for winter survival but mainly to provide food for colony development in the spring.

When you assess your colony's stores

LEFT: Feeding bees a weak syrup solution during the early active season when food stores are low.

by checking the brood frames, you should also heft the hive so that you get used to the weight corresponding to different levels of stores. Hold it under the bottom board at the back and lift it just off the stand. When you have done this a few times you will be able to assess the colony's food situation without having to open the hive.

Bees are fed sugar, either in the form of syrup, invert sugar candy, or blocks of fondant icing. Recently, beekeeping equipment suppliers have begun selling ready-made fondant. Syrup is a solution of white, refined sugar and water. If you make your own, do not use brown, unrefined sugar as this causes dysentery among the bees. Weak syrup is fed when the bees are active and likely to use it fairly quickly. Strong, winter-strength syrup is fed when you want bees to build up their stores for the winter months. With a lower water content, the bees don't have to work as hard to reduce it to the right consistency for storage. They are also more inclined to store strong syrup than weaker solutions.

When making up syrup, the measurements do not have to be precise. Rather than weighing and measuring, pour dry sugar into the container and make it level. Stirring all the time, pour in boiling water to a line about ¼ in (5 mm) above the sugar level. Continue stirring until the sugar is dissolved. Don't worry if a few sugar crystals are left in the bottom of the container.

ROBBING

The smell of honey or sugar syrup is very attractive to bees, wasps, and other insects. If they detect your feeding operations they will make every effort to gain access and rob the syrup for their own use. It is essential that your hives do not allow robber bees and wasps inside: if there is a small gap, they can enter (remember that bee space is only ¼–⅜ in (6–9 mm). If you spot that robbers have found a way in, block the gap immediately. As a temporary measure, you can pin screen over the hole or cover it with duct tape. You can also fill holes with modeling clay or block them with plastic foam such as is found in chair cushions. Note down that you need to make a permanent repair with a wooden strip or use a wood filler when you can take the equipment away from the bees. During periods when robbing is most likely, such as in the autumn, reduce the size of the entrance to the hive as much as is practicable. If you have used a full-sized entrance during the summer to accommodate the number of bees flying in and out, insert the entrance block or fill most of the gap with plastic foam. In the autumn, most colonies can defend an entrance measuring 2 in (50 mm) wide and approximately ¼ in (6–8 mm) high.

The first time you put a feeder on a colony, do so at dusk, after the bees have stopped flying. The excitement caused when workers discover the food source inside the hive, and the risk of robbing, will both be greatly reduced. For the same reason, you also need to give the first autumn feed to all of your colonies at the same time. Feeders should be checked every couple of days and they can be refilled at any convenient time, but you must ensure that the food source does not dry up before the bees have stored all they want.

FEEDERS

Feeders come in different types and capacities. Small ones, such as a honey jar with holes in the lid, need replenishing frequently. A feeder can be emptied quickly by a strong colony with empty combs, so it is best to use a larger one, especially when feeding for the winter.

Contact feeders

Contact, or bucket, feeders work on the same principle as the honey jar. They can be purchased or you can make your own from a cylindrical plastic container with a tight-fitting lid. Rectangular containers are unsuitable as they can collapse during use and flood the colony with syrup. Punch some smallish holes in the center of the lid, about $\frac{1}{16}$ in (1–1.5 mm) in diameter. If the holes are too big, or there are too many of them, the partial vacuum formed when the feeder is inverted will be insufficient to hold in the liquid. Fill the feeder with syrup and take it to your bees, together with a spare bucket and an empty super that is taller than the feeder. Four pieces of wood of equal depth nailed together at right angles make a collar with the same dimensions as the hive. It substitutes for an empty super if one is not available.

Take off the hive cover and uncover the feed hole in the inner cover. Invert the feeder over a bucket to catch the drips that will fall before the partial vacuum takes effect, then place the feeder over the feed hole. If your homemade feeder does not have a shallow rim, place it on three or four small pieces of wood to make a bee space underneath so that the bees can reach the holes. Finally, place the empty super around the feeder and replace the cover to stop robber bees and wasps gaining access.

Pouring syrup into a full-size feeder which covers the top of the hive.

A simple jar feeder

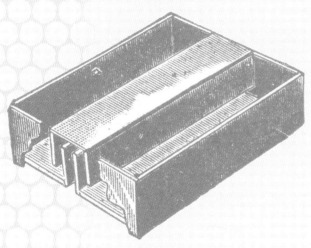

A top feeder

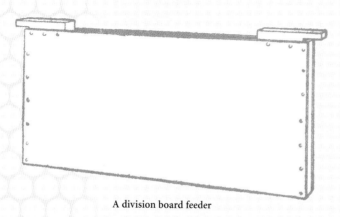

A division board feeder

Full-size feeders

Miller feeders cover the top of the hive. They work on the principle that bees have access from below and climb over a baffle to reach the syrup, which is contained in a well made by the feeder's walls. The bees' access channel has a cover which prevents them from getting to the syrup in the well and drowning there. It can be removed so that you can drip a small amount of syrup through to the frames to let the bees know the food is there. When the feeder is nearly empty, removing the cover gives the bees access to the last of the syrup to clean up. However, don't do this until the syrup level is very low and the bees have somewhere to stand without drowning. One design has the access across the center, while the other feeder has the access at one side. To use either type, remove the inner cover of the hive and put the feeder on top of the hive. It has a bee space built in below. If the hive has a slight tilt, orient the feeder so the syrup flows down toward the access and the bees can reach it at all times. Replace the inner cover and cover to make the hive bee-tight.

There are now plastic feeders that fit inside supers, which are also well worth considering.

Frame feeders

Bees can be fed using a frame or division board feeder, which goes inside the brood box. The feeder is the same size as a normal brood frame, made of plastic, and with a gap in the top to give the bees access. There is a movable float inside, slightly smaller than the feeder cross-section, to stop the bees drowning in the syrup. A frame feeder is useful for feeding just one colony in the apiary as the chances of its detection by robber bees is low. It can be placed close to the cluster, performing the jobs of both feeder and dummy frame.

CANDY

If you suddenly acquire bees that have little or no food, you can feed them during the winter by buying candy or fondant, often in large blocks, available from equipment suppliers or from a home bakery. The easiest way to cut off a large slice is to use a wet knife and keep wetting it as you cut. Remove the hive's inner cover and place the candy on top of the frames, directly over the cluster of bees. Cover it with something like a plastic food tray followed by insulation such as news-paper. Surround it with an empty super and replace the inner cover and cover.

To make your own candy, put about 1 pint (0.5 litre) of water in a large saucepan. Start to heat it and gradually add 1 lb (2.5 kg) of white, refined sugar, a little at a time. Keep stirring until the mixture starts to boil vigorously. Continue boiling for three minutes. Remove the saucepan from the heat and stand it in cold water. Keep stirring while the mixture cools, incorporating the cool syrup at the side of the saucepan into the hot mixture in the middle. As soon as it starts looking cloudy, pour it into molds or into a large metal tray where it will set. Cut the big block into smaller pieces while it is still warm and store each piece in a separate plastic bag to keep it moist until needed. When you want to use a piece, cut a hole in the bag and put the bared face of the candy over the bees.

If bees still need feeding once they have started flying in the spring, you can use syrup in a contact feeder. Any unused candy can also be used—diluted into syrup.

Candy can be bought in large blocks and is placed on top of the hives' frames.

Hygiene and Safety

APIARY HYGIENE MEASURES are designed to prevent the spread of bee diseases such as Nosema, European foulbrood (EFB), and American foulbrood (AFB), where the causative agents are easily transferred between hives. These diseases are described on pages 236–39.

The main method of transfer is the beekeeper, either via a hive tool, gloves, or protective clothing, or by moving frames between colonies. It follows that you should keep all your equipment as clean as possible. Some beekeepers

Some beekeepers place their hive tool inside the smoker as a way of disinfecting it.

wash their hive tool before use in different apiaries and even on different hives. Effective disinfectants are household bleach (sodium hypochlorite) mixed in cold water at a ratio of 100:1 parts of water to bleach, or a broad-spectrum antibacterial disinfectant available from agricultural merchants. It is best to make up a fresh solution of disinfectant on the day you plan to use it. Protective clothing can also be washed and rubber gloves disposed of when they become too dirty. Frames should not be moved to another colony unless you are confident that both colonies are disease-free (pages 236–39 give pointers for detecting disease).

Regular replacement of brood combs, either at one time or in rotation with two or three being replaced annually, reduces the risk of disease as the causative agents are removed with the

FUMIGATING COMBS

Frames taken out of use can be fumigated to kill disease organisms on the combs, and the fumigated comb can then be reused by another colony. Fumigation involves using 80 percent, or technical, acetic acid, available from an industrial chemical supplier or from bee equipment suppliers.

A brood box (see dimensions on page 164) requires about ¼ pint (150 ml) acetic acid soaked into a large cloth or felt pad. Work outside or in a well-ventilated area, and wear overalls, rubber gloves, and a mask when handling the acid as it is corrosive. Remove metal ends and metal runners as they will be corroded during fumigation. Enclose the box within a strong plastic bag, or sandwich it between solid boards top and bottom and seal the joints with tape. The fumes are heavier than air so you can apply the acid to an absorbent pad placed on top of the frames. Fumigation takes seven to ten days, after which you should air the frames until the smell has gone before using them again with bees. You can stack several brood boxes up with a pad soaked in acid on top of each as long as the joints are air-tight. Acetic acid fumes will kill the causative agents of Nosema and EFB. They will also kill the early life stages of wax moth (see page 244).

Keep the apiary tidy and, to reduce the risk of disease transfer, take away any scrapings of propolis or brace comb.

A bee sting should be removed from the skin as quickly as possible.

comb. Frames taken out of use should either be fumigated (see panel on page 221) or the comb removed and the woodwork scraped clean before being boiled in a sodium carbonate solution. Bits of comb and propolis should be scraped off empty brood boxes, ideally each time they are taken from the bees but at least annually over winter, before scorching them inside with a blowtorch until the wood begins to turn dark brown.

You should keep the apiary tidy and take away any brace comb scraped from the top bars or the queen excluder. Leaving this around, especially if it contains honey, encourages bees to investigate and possibly pick up disease.

Likewise, you should remove any varroa treatments after use and dispose of them according to the manufacturer's instructions.

Regulations regarding hygiene in honey processing are covered on page 252.

SAFETY
Take sensible precautions when you are beekeeping. Accidents can and do happen but you can do a lot to avoid them.

Stings
To reduce the risk of being stung when opening a colony, wear protective clothing, in particular a veil; a sting to the face and eyes can be dangerous.

Learn to move calmly and gently. Don't jerk the frames or knock them against the box. Rough handling can turn good-tempered bees into bad-tempered ones. Also note that bees are affected by their environment and you will observe differences in their behavior on different inspections. Temper can deteriorate when there is thunder, as bees are sensitive to electromagnetic radiation. This can also be seen in some colonies if they are sited under high-tension power cables. Bees can become more defensive if a good honey flow finishes abruptly and this is often seen when the canola finishes flowering. Being queenless can make some bees more defensive.

If you find that you are truly allergic to bee stings (for example, they cause breathing problems) but still want to take the risk and continue as a beekeeper, consult your doctor about carrying adrenaline, with you. A normal reaction to a sting can be dramatic

Moving hives in a specially designed barrow also helps to avoid back problems. The hive parts are secured with a strap.

ary, make sure your hive stands are level, secure, and sturdy. Single plastic crates, such as those used for drink bottles, are usually slippery and slightly smaller than the hive and therefore potentially unstable. Keep the apiary tidy so that there are no unexpected obstacles to trip over when carrying heavy boxes. Wear suitable footwear as the ground may be uneven.

Beekeeper's back

One problem that besets many beekeepers is "beekeeper's back." A brood box full of bees, frames, brood, and honey is heavy, and so is a super full of honey. Most hive designs do not have practical handholds for lifting. Fitting handles to the sides can make lifting easier, but if you plan to move your hives in a trailer or vehicle, external handles will make it more difficult to pack the boxes together. As you add supers, the added height makes lifting the supers on and off more difficult. All these factors can lead to back problems, so it is important to lift boxes properly. Bend your knees to lift something from the ground, rather than simply bending over and then trying to straighten up.

When moving equipment such as supers to and from the apiary, use some form of wheeled transport. Hives do not fit easily into most garden wheelbarrows, so if you use one, carry only one box at a time unless you can strap two or three together securely. Don't attempt to carry a load that makes the barrow unstable. Specially designed hive barrows are safer and generally allow you to transport more boxes at once, but again these must be tightly

and off-putting, but the swelling is usually confined to the sting area (see pages 128–30). In general, the more often you are stung, the less your reaction will be, although sometimes the reaction can become more severe. Taking antihistamine tablets can help, so if you know you are sensitive to bee stings, take one an hour before you go to the apiary so it has time to get into your system.

It is good practice to take a cell phone with you to the apiary in case of emergencies. If you go alone, always let someone know where you are and when you expect to return. In the api-

on your hive should not be above chest height, so try to extract the honey from supers when they are full in order to maintain the ideal hive height.

Transporting bees safely

If you move your bees using a vehicle or trailer, for example to forage on a particular crop, you need to ensure the safety of your bees as well as yourself. The bees must be contained within the hive by blocking the entrance, but the colony must also be well ventilated. This is achieved by replacing the inner cover with a framed wire screen traveling screen or by using a ventilated screen bottom such as the screen bottom used for varroa monitoring (see page 240). The traveling screen should be put in place during the day before the move. The colony can be protected from the weather by putting the cover back on or placing it cross-wise over the top of the hive.

The parts of the hive must be strapped together very securely. The easiest way is to use a hive strap, which can be tensioned. One strap fitted tightly around the hive should be sufficient; but to be extra sure that the boxes do not move relative to one another, use a second one at right angles to the first.

It is best to move bees at a cool time of day, i.e. in the early morning or evening. Block the entrance with plastic foam or something similar, either before the bees start flying in the morning or after activity has ceased in the evening. Remove the cover and strap the hive if this was not done when you fitted the traveling screen. If you are using a screen bottom for ventilation,

Working at the correct height reduces strain on your back.

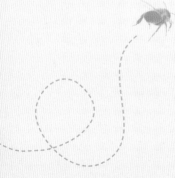

strapped together.

Set up your hive stands so that they bring the top of the brood box to a good working height for you—one where you don't have to bend or stoop to lift the frames in and out. If you are of average height, your hive stands should be in the region of $1\frac{1}{2}$ to 2 ft (45–60 cm) tall. A working platform next to the hive is ideal, so that all activities associated with a hive inspection are carried out in a fully upright, standing position, thereby reducing the strain on your back. Ideally, the supers tape over the feed hole in the inner cover to prevent bees from escaping. Check that the hive is bee-tight or you will get escapees flying around the vehicle.

Hives should be packed as tightly as practicable so they do not move around during the journey. This minimizes the possibility that the boxes shift and allow bees to get out. You should also remember to pack a cover for each hive and portable hive stands, if applicable. If you have a distance to transport the hives from the vehicle to the destination site, it is useful to take along a hive barrow or hive carrier.

If you do have to move bees during a warm day, the colony can become hot within the hive. In extreme cases, the comb will become soft and collapse, falling from the frames and killing the bees by suffocation or by drowning in the liquefied honey. If you are using a top traveling screen, you can spray a lit-

tle water over the bees through the screen to cool the colony. This is not possible with a bottom ventilation screen. Traveling with the vehicle windows open will help to keep colonies cooler. Bees on an open trailer should not overheat.

Using a smoker safely

Remember that the fuel in your smoker is alight, even if it is just smoldering. Make sure it is completely extinguished before emptying out the firebox, especially among dry vegetation. When it is out, it will still be hot, so carry a metal box or other fireproof container in which it can be laid on its side. This will stop it from falling over and spilling hot ashes onto a flammable surface.

Be particularly careful if you are putting the smoker away in the back of a vehicle where it will get jarred by the movement.

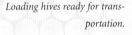

Loading hives ready for transportation.

The Beekeeping Year

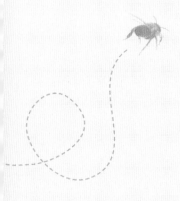

THE MAIN BEEKEEPING SEASON is usually between April and September in northern temperate zones but is dependent on ambient temperatures and the availability of local forage. Weather conditions must be suitable for bees to fly and there must be nectar and pollen available for them to collect. The rest of the year gives the beekeeper a chance to catch up with tasks such as equipment maintenance and to take the opportunity to learn and to plan for the following year.

Any beekeeping calendar can only be approximate because of climatic and environmental differences between south and north, west and east, but this will give you a guide that you can adjust to your own area.

Tasks and maintenance vary with the seasons.

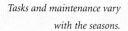

GENERAL MAINTENANCE

These are tasks for all year round.

Maintain record cards

Get into the habit of keeping records and updating them soon after completing every colony inspection (see page 187).

Keep hives tidy

Wherever your apiary is situated, it pays to keep it tidy. Cutting the grass or placing paving slabs or similar beneath each hive will enable you to see any unusual bee deaths, perhaps due to spray poisoning (see page 246). Keeping your equipment together and readily available will save you time when you find queen cells in your colony or receive an unexpected swarm call.

Clean equipment and check for disease

This is important to stop the spread of certain diseases (see page 236). Follow management techniques recommended in Integrated Pest Management (IPM) programs. Closely monitor for and, where possible, treat any disease found in the colony (see pages 236–39).

Check hives are stable and weatherproof

For particularly windy locations put weights on top of the hives to keep the covers from blowing off. The roof should be waterproof, and ideally, hives should be painted prior to wintering. Hives should be placed on stands and positioned so there is a slight slope toward the entrance and away from the prevailing wind to stop rain from being driven in. Check that livestock cannot get into the apiary and knock over the hives.

Attend local beekeepers' meetings

Your local beekeeping association should hold regular apiary or lecture meetings. Try to attend as many as possible and use the opportunity to pick the brains of experienced beekeepers.

LATE WINTER TASKS (average temperature 32–45°F / 0–7°C)

The beekeeper's main concern through the late winter and early spring months is to ensure that the colonies that survive the winter go into spring strong, disease-free, and ready to take advantage of any early nectar flow.

Visit the apiary to check the bees

During winter it is worth visiting the apiary every few weeks to check all is well. As the weather begins to improve, you will see bees flying on warmer days. Pollen loads going into the hive indicate the queen's egg-laying rate is rising, stimulated by the increasing day length, and the brood nest is expand-

WINTER BEES

During the winter months, when the temperature outside the hive falls to 57°F (14°C), the bees move closer together, and below 48°F (9°C) they form a cluster in the hive, across several combs and beneath the stored food. They consume their food stores to survive. The effect of food consumption and digestion, plus muscular movements, is to generate heat which is retained within the winter cluster by the bees' hairy bodies. As the outside temperature falls, the bees press more tightly together, reducing both heat loss and the size of the cluster. The bees in the center are very warm 70°F (21°C) but they become hungry and have to move to the edge of the cluster where the food is located. The cold bees on the edge move to the center to warm up. The bees on the outside of the cluster are just warm enough to live. As the winter progresses, the cluster moves slowly upwards, keeping contact with its stores. Warm spells during the winter help this activity as the cluster loosens and can then re-form closer to the food. The waste products of digestion accumulate inside the worker bees and they use any warm sunny spells to leave the hive and defecate.

A winter job: scraping clean the queen excluders.

to check on food supplies (see page 217) and subsequently check the weight loss on a regular (monthly) basis. Depending on your type of bee, colonies can consume 4 to 5 lb, or one full deep frame) of stores per month, but this can vary with weather conditions. If you have fed colonies properly in the autumn, they should be fine. As well as the pollen that they stored in the autumn, the bees will need water to dilute their honey/sugar stores before use, so make sure there is a water source in a sunny spot near the hives where the water will warm quickly: drinking cold water can chill a bee enough so that it is unable to fly back to the hive.

In bright, snowy conditions, shade the entrance with a board leaning against the hive
When snow covers the ground, bees are sometimes tempted by the bright reflections to fly out of the hive. Chilled bees would be unable to return and would perish. The board should discourage them from doing this.

Repair equipment and plan ahead
Use the quieter winter months to make essential repairs and prepare for the coming year.

EARLY SPRING TASKS (average temperature 37–63°F / 3–17°C)
In early spring, for every dry day when the outside temperature is around 60°F (15°C) there should be a good number of bees flying in and out of the hive, collecting pollen and nectar as more plants come into flower. Pollen is

ing. If you see a melted circle in the frost or snow on the roof, resulting from the heat generated by the cluster, this also indicates that the colony is beginning to expand.

Make sure there is sufficient ventilation in the hive
Ventilation in the hive is important during the winter to stop mold growth both on the hive walls and on the pollen stores. Air circulation can be improved by raising the inner cover slightly by putting a thin strip of wood such as a matchstick under each corner. Make sure the gap is smaller than the bee space $\frac{1}{4}$ to $\frac{3}{8}$ in (6–9 mm) to prevent robbing. If a hive has a screen bottom, this will ensure through ventilation. In snowy and icy weather, remove any snow or ice blocking the entrance to maintain a continuing flow of air through the hive.

Feed the bees if stores are low
At the start of the winter heft the hives

particularly important in the spring as it is needed to provide protein for the larvae in the rapidly developing brood nest. With more than one colony, you can compare them and note any that seem weak. The main objective for the beekeeper during the spring months is to ensure the colony is building up well, ready for the major honey flow that occurs in early or mid-summer.

Check the colony isn't starving

A colony will only starve if it had insufficient stores at the beginning of winter. The winter cluster of bees should always be touching sealed honey. To check this, go to the hive on a cold day/evening (when the temperature is below 39°F (4°C). You will need a flashlight. Gently remove the roof and inner cover. Shine the light between the frames and look down. If there is no sealed honey touching the bees you may need to feed the colony. Rearing new brood increases consumption of honey stores and you need to check

these are adequate by hefting the hive. Spring is the time when colonies are most at risk of running out of food and those that have apparently come through the winter in good shape can decline and die as stores are used up to feed the expanding brood nest. The stores may be used at a faster rate than levels of incoming nectar and pollen, leading to starvation. Don't allow the level of stores to drop below 8 to 10 lb (4-6 kg), or two full brood frames. If you think the colony is running short (and the weather conditions allow the bees to fly from the colony every few days), feed some dilute syrup (a 1:1 ratio of sugar to water) in a contact feeder directly over the cluster (see page 234). If the weather conditions prevent them from flying regularly, feed them candy or fondant instead, again placed directly over the cluster.

Make a quick, preliminary inspection

Avoid opening the hive until the temperatures rise above 57°F (14°C). Do not lift any frames from the hive; removing frames when the winter cluster is still intact will break it up and the disturbance may cause the bees to kill the queen. If the colony is doing well, it should look as strong now as it did in the autumn. If it is significantly smaller, or if you see a lot of brown/yellow streaks of excreta on the top bars and inner cover, the colony may have dysentery and/or Nosema (see pages 238–39). At this time of the year, you cannot easily treat the colony for Nosema and it may die. Dysentery will usually clear up when bees are able to fly and defecate, thus getting rid of the excess build-up of water in the rectum.

The early spring check is an important part of the beekeeping year. The bees should begin collecting nectar and pollen from early flowering plants. Food may need to be supplemented.

Regular brood nest inspections when the weather warms up will tell you how the colony survived the winter.

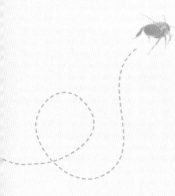

Close the hive entrances of any dead colonies to avoid robbing and to help prevent the spread of any disease that may have caused the colony's demise. Check the cause of death and fumigate the combs you have removed from the colony before giving them back to bees (see page 221). Bees in colonies that have starved to death will have their heads buried inside the cells where they were searching for food stores.

Remove ventilation props and close the feed hole

As outside temperatures increase, it is time to remove ventilation props under the inner cover and close the feed hole to help the bees maintain the colony temperature as they become more active within the hive.

Make an inspection on a warm day

As spring progresses, choose a fine day when you are comfortable outside in your shirt sleeves (temperatures of around 65°F or 18°C) to make a quick inspection. Colonies should be building up fast. As the weather warms up (above 65°F or 18°C), you can start regular (weekly) brood nest inspections.

Watch for early signs of swarming preparations

As soon as the brood box looks full of bees (don't wait for the combs to be filled with brood and/or food stores), add a queen excluder and a super. Make sure you have enough equipment for your chosen method of swarm control for each of your colonies (see pages 199–207).

Replace old frames and combs

To reduce the spread of diseases before the brood nest starts expanding at a great rate, before they contain eggs, replace old frames and combs with new foundation or fumigated comb. If you replace them with foundation, you may need to feed your colony to help the bees make wax for comb-building.

LATE SPRING/SUMMER (average temperature 63–72°F / 17–22°C)

During the late spring and summer months, if the bees have access to ample nectar supplies and the weather is warm and dry so that they can fly on a daily basis, there will be a large foraging force which should give you a good honey harvest. This is a busy time for the beekeeper, who must control swarming, monitor colony development, and extract the ripe honey from the full super combs.

Add extra supers as the number of bees increases

When the colony gets crowded, giving

it more room will help prevent swarming. So, add empty supers when the existing ones are full of bees, not honey.

Make regular inspections to control swarming

Follow your chosen swarming control method (see pages 200–207). By early summer, the swarming season should be coming to a close. If you have successfully controlled swarming or resolved the situation if a swarm did get away, it is unlikely your bees will try to swarm again. However, some queens have been known to swarm several times in a season, so although you no longer need to make strictly regular brood nest inspections, it is worth continuing to check colonies fairly frequently and many beekeepers continue this on a weekly basis.

Use swarming to increase your stocks

Swarming time gives you a chance to increase your stocks. You can use a swarming colony to produce new queens and new colonies (see page 214). To rear new queens from your favorite queen, transfer a comb of eggs and very young larvae from her colony into a queenless nucleus that does not contain any other brood of a suitable age on which the bees can rear a queen.

Requeen colonies if necessary

Queens may live for four years or more, but many beekeepers replace older queens with younger, more active ones that are in their first season (see the system of marking queens to tell their age, on page 186). Others replace queens only if they are performing poorly or are heading a bad-tempered colony or one with other undesirable characteristics.

Extract honey from supers

Depending on the nearby forage, you may find combs or supers of honey ready for extraction from early summer onwards (honey extraction is described on pages 250–56). If so, clear the supers of bees and take them away. Each full super you remove should be replaced immediately with an empty one. As summer reaches its peak, colonies begin to prepare for the winter. The queen reduces her egg-laying rate and the brood nest contracts. The workers start to pack the empty cells with honey, ready for use as food stores during the winter. When you see this happening (and your hive records will give you a clue), don't add any more supers unless you know there is a crop nearby which is going to yield a good amount of nectar. It is now time to remove the remaining supers and extract the honey.

The beekeeper has used a hive tool to remove some of the cell cappings. The honey inside is ready for extraction.

A robber bee is attacked by guards while trying to enter the hive. Reducing the size of the entrance helps the colony to defend its nest.

Help the bees defend the hive against robbers

During the honey harvest and when inspecting colonies toward the end of the summer, take care not to leave honey exposed in the combs or drips of honey on the hives and not to have the colony open any longer than necessary during an inspection because this is also the time when robber bees and wasps can be a particular problem.

As the honey season comes to an end, robbing becomes more prevalent. Help your bees to defend their colony by reducing the entrance to a width they can protect (see page 155). It is better that returning foragers line up to get in than robbers have free access to steal the honey stores.

Feed if food stores are low

In some areas, if there is a gap in available forage around mid- or late summer, and depending on whether your type of bee is prolific or not, your colonies may need feeding.

Treat against varroa

By late summer, you should start treating colonies against the varroa mite (see pages 240–41).

Begin preparing colonies for winter

In late summer you should take stock and decide how many colonies you want to take into winter. This may depend on how many you can afford to feed properly. Strong colonies are undoubtedly the key to successful wintering. If you need to reduce colony numbers, you can unite them over newspaper (see page 235). Choose which colonies you want to keep; this is the time to dispose of a queen that heads a bad-tempered colony and unite her bees to another, good-tempered one. The resulting strong colony is more likely to survive the winter and can be split into several colonies the following spring if you then want to increase. Culling is one of the best ways of improving the overall quality of your colonies, which can make your beekeeping even more enjoyable.

Check for colonies retaining their drones

In the late summer, the drones raised in spring to mate with virgin queens are now becoming surplus to requirements and are removed from the colony. When the colony shuts down for the winter, they are a drain on colony resources. Worker bees remove them by ceasing to feed them and by preventing or hindering flying drones from re-entering the hive. The drones are unable to forage for

OPPOSITE: *In summer, the number of bees increases quickly and you will need to add supers above the brood box.*

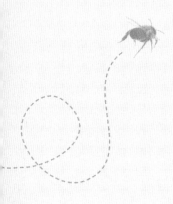

themselves and soon die. Inspect any colonies that are retaining their drones as they may be replacing their queen. Leave any queen cells to develop as these colonies do not usually swarm.

AUTUMN TASKS
(average temperature 41–66°F / 5–19°C)

The autumn is a time for ensuring the bees that go into winter are well-fed and healthy and their colonies are strong. These have the best chance of survival until the weather warms up again in early spring. As the outside temperatures fall and brood rearing has finished, the bees begin to form a winter cluster. There are certain tasks the beekeeper has to do before this happens.

Feed a colony with insufficient stores for the winter

Assess each colony's food reserves by noting the amount of stores in each brood frame. An average colony will require 30-60 lb (13-27 kg) of honey and pollen stores to survive the winter, equivalent to about eight full brood frames (each will hold around 8 lb or 4 kg of honey). During feeding, the aim is to surround the final brood nest with food, making sure in particular that there is a good amount of stores above the winter cluster. Even if you anticipate a late honey flow from something like ivy (*Hedera helix*) or Himalayan balsam (*Impatiens glandulifera*), this is never certain and you need to ensure your bees have enough to last until spring. Feed while the weather is still relatively warm (around 64°F or 18°C) and the bees are flying on fine days. They have to work to reduce the water content of the sugar syrup to a level where it will keep without spoiling, so use a thick winter syrup that requires less effort to achieve this. Feed your bees until the colony has sufficient winter stores (the brood nest is surrounded by food) or until they stop taking the syrup from the feeder.

Protect against pests and predators

Wasps are a problem toward the end of summer and into autumn. Their nests can even survive into winter in milder areas. Opportunistic robber bees from other colonies will also be trying to steal honey so make sure your hives are bee- and wasp-proof (i.e. there are no gaps). Use an entrance block or plastic foam to reduce the size of the hive entrance to one small enough for the colony to defend easily (see page 155). Avoid undertaking any extensive colony inspections. If you must, open the hive for only a very short a time: an open hive and exposed combs invite robbing. When the first frosts have killed the wasp nests, remove the entrance block or foam and pin a mouseguard in place if your hive has a deep entrance (see page 243). Hives with shallow entrances do not need a mouseguard as their shallow depth prevents mice gaining access. Having a full-width open entrance during winter is particularly beneficial if your apiary is in a damp location as it helps to maintain through ventilation. Store the queen excluder over the inner cover to prevent mice gaining access through the roof. Woodpeckers can cause problems in winter, so protect your hives by surrounding them with a small-mesh wire netting sleeve or strips of plastic (see page 242).

UNITING COLONIES

LEFT: *Strong colonies are the key to successful wintering. If necessary, unite two colonies to make one stronger colony.*

ABOVE: *Uniting colonies over newspaper.*

IN THE LATE SUMMER, when you begin preparing the bees for winter, you may decide to unite two colonies to make a stronger colony more likely to survive the cold months ahead. Probably the simplest and most effective way to unite colonies is with newspaper. First clean the wax and propolis from the top bars of the colony you wish to keep, and take any large lumps of wax and propolis from the bottom bars of the one you don't, so that the brood boxes of each colony will fit together snugly one on top of the other. Place a sheet of newspaper, large enough to cover the whole brood box surface, over the top bars of the queenright colony (the colony with the queen you want to keep). To prevent the newspaper blowing off in the wind, fold down and pin the excess paper to the outside of the brood box. Although not strictly necessary, you can make one or two small slits in the paper with your hive tool to give the bees an edge on which to start removing the paper. The process that follows involves the bees from both colonies chewing through the paper, the different hive odors blending and the bees mingling to produce one united colony.

Find the queen you want to depose, and kill her. Then place her brood box on top of the paper. If you can't find the queen, you can still unite the colonies but it will be pot luck which queen survives.

There is often a debate about which colony goes on

top of which. If you are combining a weaker colony with a stronger one, put the weaker one on top. If they are the same size, put the queenright one on top (just one queen will survive). If you have to move one of the colonies from elsewhere in the apiary to unite it with the other, then the colony that is moved should go on top. Around 24 hours later, you will see newspaper dust being thrown out of the entrance of the combined hive. This means that the bees are chewing through the paper and beginning to unite. Leave everything alone for a week.

When the colonies have mingled, you need to rearrange the frames. Separate the boxes and move all the frames containing brood into one box. In the late summer/autumn, you should be able to do this as the brood nest will have reduced in size. The total number of brood frames from the two colonies will be fewer than the full complement of frames for one brood box. Try to arrange the frames with patches of brood of the same size next to each other. Imitate the natural brood nest pattern, with large patches in the middle and smaller ones toward the outside of the nest. Arrange the food combs in the other box in a similar manner so that those containing most honey are directly above the brood nest when you put this box on top of the other. Often there will be sufficient food stores from both colonies to feed the combined one for the winter.

Diseases, Pests, and Predators

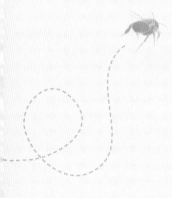

BEE DISEASES come in two types: those affecting the brood (brood diseases) and those affecting the adults (adult bee diseases). The most serious brood disease is American foulbrood (AFB). It is contagious and fatal to colonies. In states that have an Apiary Inspection Service, the local inspector should be notified to diagnose the disease and make recommendations. In those states without inspection, an experienced beekeeper can be called. It is important for all beekeepers to be familiar with the signs of this disease and take precautions to prevent its spread.

American foulbrood is the only notifiable disease in the USA and, if confirmed, colonies are usually killed and burned. The Department of Agriculture for each state can supply a copy of its appropriate bee legislation. For Continental Europe, AFB, SHB, and *Tropilaelaps* are all notifiable throughout the EU and anyone suspecting any of these diseases should get in touch with the country's apiary service. EFB is not notifiable in all EU countries and the local situation should be checked with the state apiary service. Details of notifiable bee diseases within the EU are given on the OIE

A larva that has died from American foulbrood. The larval remains have decomposed into a slimy rope.

A brood comb showing larvae that have died of chalkbrood–contaminated larvae

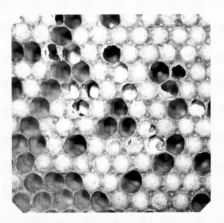

(World Organization for Animal Health) website at www.oie.int under "Animal Diseases Data."

Bees are also prey to predators such as birds, mice, rats and wasps.

BROOD DISEASES

Two of the more serious brood diseases are AFB and EFB. The other, minor, brood diseases include the fungus *Ascosphaera apis*, which causes chalkbrood, and the sacbrood virus.

American foulbrood (AFB)

AFB is caused by the bacterium *Paenibacillus larvae* subspecies larvae. This kills the larva or prepupa (the inactive stage just before the pupa) after the cell has been sealed. The cappings on the cell are sunken, darkened, may be perforated and can look greasy. The larvae are white-to-brown in colour instead of pearly white. If the cell contents are agitated with a matchstick, which is then withdrawn, they may pull out as a slimy rope. AFB may be accompanied by an unpleasant smell, described as similar to decaying insects. The larval remains dry to a dark brown scale, which sticks to the cell wall and is very difficult to remove. In the US, the bees are killed and the frames and combs are burned. The boxes and the metal parts are scorched. Other hive components may be burned or scorched, depending on the law in a given state. In the US, only two medicines are approved: Terramycin® (oxytetracycline) and Tylan® (tylosin). Neither of these cure the disease; they only suppress the spores from becoming virulent. Tylan is not approved as a prophylactic. AFB is now resistant to Terramycin.

European foulbrood (EFB)

Caused by the non-spore-forming bacterium *Melissococcus plutonius*, EFB usually kills the larva before the cell is sealed. The disease organism competes with the larva for food and effectively starves it to death. Larvae are an off-white color and lie in the cells at unnatural angles. They can look "melted" and, at a certain stage, the thin lines of the tracheae or breathing tubes are clearly visible.

In the US, EFB can be treated with Terramycin. An alternative treatment is the shook swarm method, where the colony is transferred onto clean comb.

Chalkbrood

The fungus *Ascosphaera apis* kills the larva before the cell is sealed or soon after. The larva becomes hard and chalky and can turn gray or black if fungal spores develop. Chalkbrood mummies or solid larval remains are loose in the cell and can be removed by the worker bees.

Sacbrood

The sacbrood virus prevents the bee larva from making its final moult when the prepupa turns into a pupa, and it dies before it can spin its cocoon. Generally only a few larvae in a colony are infected. The dried larvae, in their larval skins, can look as though they have died from AFB but are more easily removed from their cells. The dead larvae have pointed ends that stick up, resembling the upturned toe of a Chinese slipper.

ADULT BEE DISEASES

Of the diseases that affect adult honeybees, Nosema is the most common.

Nosema

In the US, until recently, Nosema was caused by *Nosema apis*. However, *Nosema ceranae* has also been discovered in the US. This is a disease originally affecting *Apis cerana*, the Eastern (Oriental or Asian) honeybee (see page 89), which has transferred to the Western honeybee and is proving more virulent. The spores of both types develop in the digestive system and severely interfere with the bee's ability to digest food, particularly pollen. This drastically shortens the bee's life. Nosema also reduces the worker's ability to produce brood food. Colonies with *Nosema ceranae* will not have dysentery (see box) but a colony with dysentery is not necessarily suffering from Nosema.

Nosema affects colonies year round. The colony either builds up slowly, stays the same size, or dwindles and even dies.

The active stage of Nosema can be treated with an antibiotic, fumagillin, dissolved in the autumn syrup feed. This is sold in the US as Fumadil-B®. However, the disease continues from year to year as bees pick up more spores from the comb, and it is best to replace combs with clean ones as soon as possible (see page 221 for information about fumigating comb). The symptoms gradually disappear as the colony becomes more active in the spring and bees can excrete away from the hive.

Nosema is diagnosed by crushing a sample of 20 dead bees in a pestle and mortar with a little water. A drop of the resulting liquid is examined under a 400x microscope for the presence of spores, which look like small grains of rice. Those of *Nosema apis* are only slightly different from those of *Nosema ceranae*, making visual distinction almost impossible.

Amoeba

Amoeba is caused by a protozoan, *Malpighamoeba mellificae*, which af-

DYSENTERY

Dysentery in honeybees is a condition that can develop as a result of ailments, through feeding on fermenting honey stores, or not being able to defecate outside the hive for long periods. In winter, dysentery can be a problem during prolonged periods of cold weather (usually two or three weeks of temperatures below 48°F or 9°C) when the bees are unable to perform cleansing flights. Where Nosema is present, spores are excreted in the bee feces. When the excreta is cleaned up by the house bees, they ingest the spores and become infected. With a high level of dysentery, the colony can collapse and die. Brown/yellow streaks of excreta on the top bars and inner cover are a sign of dysentery. There is currently no medical treatment, but maintaining healthy colonies that display hygienic traits will help to prevent dysentery, as will high levels of hygiene in apiary housekeeping.

fects the Malpighian tubules, the equivalent of the bee's kidneys. Transmission of the disease is similar to that of Nosema; transferring the affected colony to clean comb will help.

Diagnosis is as for Nosema; the spores of Amoeba are circular, with a diameter similar in length to that of a Nosema spore.

PESTS

Your bees may have to contend with a number of pests, which can affect their health and colony survival.

Tracheal mite

Tracheal mites can be more accurately defined as an infestation caused by the mite *Acarapis woodi*. It is known as the tracheal mite because it infests the first pair of tracheae (respiratory tubes) in a young bee. The mites breed inside and can eventually block the tubes. The bee's life is shortened and the colony does not build up properly in the spring. A severe infestation causes colony death.

To check for tracheal mites, the tracheae are examined under a low power (10x or 20x) microscope. The bee's body is pinned at an angle on a cork for easier examination and the head and thoracic collar removed to reveal the main branches of the tracheae. Badly infested tracheae are discolored, although the more normal-looking pale ivory-colored ones may also harbor plenty of mites.

The only legal treatments in the US are menthol and Apilife Var®, based on thymol; feeding grease patties has shown to be effective. Cooking oil is combined with white granulated sugar to make it oily and bind it together. A large spoonful is placed onto grease-proof paper and put on the top bars over the brood nest. Young bees are exposed to the patty as they eat it, and it masks their odor. The female tracheal mite does not then recognize them as young workers and continues questing for a new host—an ungreasy young bee. Since the arrival of small hive beetle, grease patties can only be used during winter months; the beetles are attracted to the patties and will feed on them.

The tracheal mite infests the respiratory tubes of the bee.

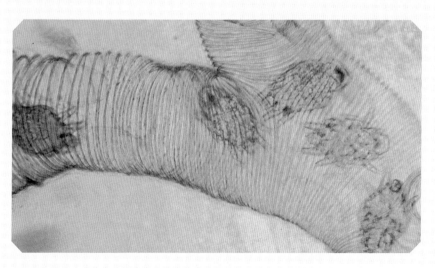

VARROA

The mite *Varroa* was originally a pest of the Eastern (Oriental) honeybee, *Apis cerana*. These bees have developed ways of keeping mite levels under control and the two live "happily" together. However, the Western honeybee, *Apis mellifera*, has not developed such controls, and without treatment mite levels increase within a colony. Because the crab-like mites pierce the skin of the prepupae, pupae, and adult bees, they act as vectors for disease organisms, particularly viruses. Several viruses are typically found in bee colonies but these do not generally become active and rarely threaten the colony. However, varroa can trigger the viruses into activity and once this happens the colony will eventually die.

Hives should be set onto bottom boards that have a panel of metal screen in the middle with holes sufficiently large for mites to fall through but not large enough for a bee to pass through to collect and remove the dead mites from the hive (see page 155). The beekeeper can monitor mite numbers, and hence infestation levels, by inserting a sticky board under the screen bottom board. Mites are collected on the tray and can then be counted at regular intervals.

Natural mortality varies throughout the year and monitoring is most effective in late spring and through the summer when mite levels are increasing. The situation can change very rapidly so you need to keep a regular check on mite levels.

Varroa treatments
Start treating colonies against the *Varroa destructor* mite in mid-summer. Management techniques involved in IPM programs include removing sacrificial drone comb (in which the mites prefer to reproduce), trapping the queen so that she can only lay eggs on a specific comb, which is then discarded when the brood is sealed (and contains the reproducing mites), and a modified artificial swarm where the new queen is confined to a frame in a cage. When the colony has no more brood, two combs of open brood are introduced from another colony. The mites then enter these cells to breed and the combs are destroyed when the cells are sealed.

In the US, you are required to use the approved treatments. These are available from a number of equipment suppliers.

The approved treatments for varroa are Apistan®, formic acid, CheckMite+® (coumaphos), Apilife

BELOW LEFT: *The varroa mite under the microscope.*
BELOW RIGHT: *A varroa mite on the back of an adult bee.*

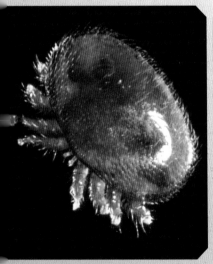

Var®, and Apiguard®. CheckMite+® is registered under an Environmental Protection Agency (EPA) Section 18 emergency use provision and the position regarding its use in any particular state should be checked with the state apiarist.

Formic acid is available in pre-packaged impregnated pads as MiteAway II®. It is placed on supporting bars on the top bars of the brood box, surrounded by a suitable box or shim, and should be applied in the temperature range 50–79°F (10–26°C). Formic acid is corrosive and protective gear must be worn during its application.

Coumaphos is an organophosphate, which is supplied in impregnated strips that are placed in the brood nest when no nectar is being brought into the hive. Mites have become resistant to coumaphos.

Apilife Var® and Apiguard® are both based on thymol. The former is supplied as an evaporating tablet, the latter as a gel. Both are effective in killing varroa mites.

Varroa treatment using pads impregnated with formic acid placed on the top bars of the brood box

A woodpeecker attack on a hive. The bees have filled the hole with propolis.

One version of the metal mouseguards available to fit over the entrance of a hive.

A mouse in an empty box in the vicinity of the apiary.

In the autumn wasps turn to robbing honey from beehives.

Birds

Swallows, martins, swifts, and titmice all eat bees but the numbers affected are low and this does not generally cause any undue stress on colonies. It is a problem if the bee that is eaten is the queen, but this happens rarely. The woodpecker can be a serious pest, especially in winter when it can smash holes through hive boxes and badly damage both the box and the frames within. Woodpeckers tend to attack occupied hives, where they will eat the bees and the brood. They have also been known to vandalize stacks of spare equipment.

Hives can be protected by wrapping them with small-mesh (1 in or 2.5 mm) chicken wire. The bees are able to fly in and out but the woodpecker cannot reach the hive surface. The wire should overlap around the hive with sufficient height to allow it to be folded over on top of the cover and cover the entire hive down to the bottom. Put the mesh in place after colonies have been fed for the winter and don't remove it until March or April.

Mice

To a mouse or other small rodent, a beehive is a very desirable residence. It is warm, protected from the wind and rain, and offers an ideal place for a nest. Mice will chew the wooden parts of equipment and consume stores, especially pollen, eating holes in the combs. When a colony is active, mice will generally be warned off by the bees' ability to sting, but as the winter cluster forms, this threat is greatly diminished and the mice will try to gain access. They can squeeze through a wide gap that is only ½ in (12 mm) high but not through a hole with a ½ in (12 mm) diameter. A bee-space-high entrance (¼–⅜ in or 6–9 mm) should prevent access. If the bottom has a deeper entrance you will need to remove the entrance block and fix a mouseguard over the entrance in the autumn. One form of mouseguard is a strip of metal the width of the hive, punched with holes of ⅜ in (9.5 mm) diameter. Bees can easily get in and out but the mice cannot. Storing your queen excluder over the top of the inner cover will prevent mice getting in from the top.

Rats

As they are larger than mice, rats can do more damage and may chew around the entrance. However, a mouseguard will prevent them from getting into the hive. If rats do gain access to your stored equipment, be aware that they can pass on diseases such as Weil's disease through the urine that they dribble as they move around. It is then probably best to destroy the combs and thoroughly scrub other equipment with a strong disinfectant.

Wasps

When adult wasps feed meat to their larvae, they are rewarded with a sweet larval secretion. In the autumn, the number of wasp larvae decreases along with the sweet rewards. Adult wasps need sugar as their energy food and turn to robbing honey from beehives if they can.

If your hive is bee-tight, it will be wasp-tight. Make sure that the cover fits closely and that all ventilation holes are covered with a fine screen. For a small colony, reduce the size of the entrance to one bee space (see page 155) if necessary, to make it easier for the colony to defend. A wasp will try to gain access through any gap but will fly away if challenged by a bee. Attacks by wasps will stimulate the colony to put more guards at the entrance, and a strong colony with a small entrance should generally be able to repel invaders. If your colony is small, move the frames containing the brood nest nearer the entrance, which will bring the guards closer to the point of attack. The guards are only really effective at the entrance and once a wasp gets inside a hive, it is generally ignored as it picks up the colony pheromone and loads up with honey. Silent robbing can quickly deplete the stores of a small colony.

If you consider your colony is too small to defend itself, think about combining it with a stronger one. This involves eliminating the queen in the colony that has been less productive and/or more defensive, before uniting the weaker colony to the stronger one, over newspaper.

Robber bees

Bees are also opportunists and try to steal honey from other colonies, particularly weak ones. Avoid leaving combs, brace comb, and sticky equipment around in the open. Make sure that there is only one way into a colony—the entrance—which can be effectively guarded by the resident colony. If you want to save a weak colony that is being robbed, move it to another location where there are no other bees.

If you open up a colony and bees

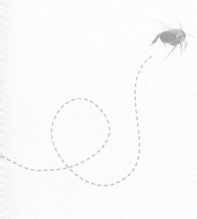

from other colonies become interested, close up the colony immediately and abandon your inspection. Inspecting a colony as late as possible in the day, when there are fewer robber bees flying around, cuts down the opportunities for robbing.

Wax moths

There are two species of wax moth whose larvae chew beeswax: the greater (*Galleria mellonella*) and the lesser (*Achroia grisella*). Both can cause severe problems in stored comb during winter. Greater wax moth larvae eat cocoons and debris, preferring to attack brood comb. They will also eat brood, honey, and pollen. Lesser wax moth larvae can tolerate a wax-only diet and damage combs in supers. Both moths leave silken webs in the damaged comb, and a severe infestation can make a considerable mess. Greater wax moth larvae often leave

tunnels below the cell cappings in active brood nests as they move from cell to cell.

Empty comb should be stored in a cold situation as this reduces wax moth activity, and therefore damage. Boxes of comb can be stacked outside with a mouse-proof mesh, such as a rigid queen excluder, on both top and bottom. Place a sound cover on top to protect the stack from rain. After preparing your colonies for winter, supers can be stored temporarily on colonies, over the hive inner cover with the feed hole open. Again, make sure the cover is sound.

Small hive beetle (SHB)

The small hive beetle (*Aethina tumida*) is a pest of African honeybees, which have developed ways of controlling its activities in the hive. The Western honeybee has no such defenses and can suffer significant damage. The small

BELOW LEFT: *A lesser wax moth and its silken web on the comb.*

BELOW RIGHT: *Greater wax moth larvae at the bottom of a hive.*

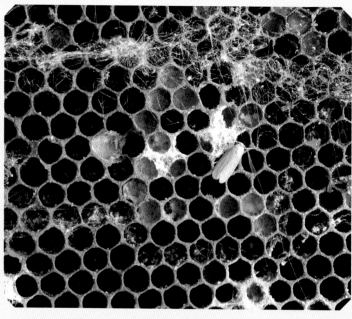

hive beetle larvae and adults prefer to eat honeybee eggs and brood but will also eat pollen and honey. Larvae burrowing through brood combs ultimately destroy the brood nest. In a severe infestation, the heat generated by the beetle larvae causes the combs to collapse and the colony to abscond. When the adult beetles and larvae defecate on the honeycomb, this causes the honey to ferment and drip out of the cells. The combs become slimy and smell like rotten oranges. In areas of small hive beetle infestation, honey must be extracted as soon as it is removed from the hive. Immediately after extraction, all equipment must be cleaned as well as the working area. Any traces of honey left will mean an infestation of SHB. If a beekeeper suspects small hive beetle infestation it is wise to have a positive identification. In states with apiary inspection, the inspector can do the identification and make recommendations. In states without inspection services an experienced beekeeper can be consulted.

The larvae have rows of spines down the back and three pairs of prolegs near the head that distinguish them from wax moth larvae. The beetles themselves are oval and measure $3/16$–$1/4$ in (5–7 mm) long and $1/8$–$3/16$ in (3–4.5 mm) wide. They are about one-third of the size of a worker bee and have clubshaped antennae. For more information, contact www.maarec.psu.edu and open up Diseases and Pests where small hive beetle will be listed. A pamphlet on small hive beetle can be printed from this site. Questions about the small hive beetle can be answered by contacting www.extension.org/pages/bee_health.

A strong colony is generally unaffected but weak colonies can suffer significant damage. CheckMite+® is authorized for beetle control within the colony and Gard Star®, a soil treatment containing permethrin, is authorized for soil treatment in an attempt to break the beetle's life cycle as it pupates in the soil. Beetle numbers can also be reduced in hives by attracting them to oil-filled traps, into which they fall and drown.

Tropilaelaps

The natural host of Tropilaelaps mites is the giant honeybee, *Apis dorsata*, but they are also found on other honeybees: *Apis laboriosa*, *Apis cerana*, and *Api florea* (see page 265). In the Far East, they have transferred to the Western honeybee, *Apis mellifera*, affecting both the brood and the adult bees. The mites are reddish brown, about $1/16$ in (1 mm) long and $1/32$ in (0.6 mm) wide, with a life cycle similar to that of the varroa mite. They feed on larval blood or hemolymph but cannot pierce the external membranes of adult bees. Damage to the colony is similar to that caused by varroa mites.

Colonies are being surveyed around the US for Tropilaelaps and other exotic pests. Information about Tropilaelaps is available in *Tropilaelaps: Parasitic Mites of Honeybees*, published by the National Bee Unit and available from http://beebase.csl.gov.uk. Other information can be found on various websites. These mites have not yet been identified outside Asia.

COLONY COLLAPSE DISORDER

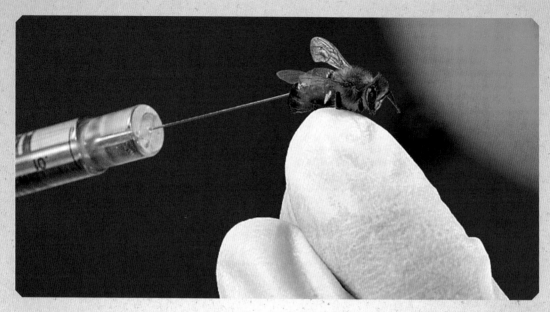

A healthy bee is injected with viruses extracted from bees in colonies showing Colony Collapse Disorder to evaluate immune responses to these viruses.

IN NOVEMBER 2006, beekeeper Dave Hackenberg discovered that a large number of his hives in Florida were virtually empty of bees. The queen might be there with a small group of young workers but the majority of bees had simply disappeared. More beekeepers began to report the same phenomenon. When a colony dwindles, other bees usually rob the honey, and wax moths will often take up residence. However, this did not seem to be happening in the affected hives. By June 2007, Colony Collapse Disorder (CCD) had been identified in 35 states and Puerto Rico. A working party of bee scientists throughout the country was established and work began on trying to identify the cause of CCD. In July, the Department of Agriculture confirmed that one-third of US honeybee colonies had been lost to CCD, and this increased to 36 percent in a survey conducted by the Apiary Inspectors of America in April 2008. Many possible causes have been proposed, with a number of them, including cell phone towers and genetically modified crops, being discounted after investigation.

The government has pledged funding, and financial support for research has also been given by various commercial companies. One research project has identified the presence of Israeli Acute Paralysis Virus in colonies affected by CCD. This was first thought to have been introduced in colonies imported from Australia to assist with almond pollination, but it was then shown to have been present in the USA for a number of years. The stress on colonies that are shipped thousands of miles to fulfill pollination contracts, particularly for the almond orchards in California, has also been put forward as a possible cause, although colonies not being used for migratory beekeeping have also been affected. Another possible cause being investigated is *Nosema ceranae* (see page 238). Some research is focusing on systemic pesticides, particularly neonicotinoids. The situation is ongoing. No cause, and therefore possible cure, has been found as this book goes to press. The only thing that is certain is that CCD does not have a single cause and that if an answer is not found quickly honeybee colonies will continue to die and their pollination activities will be severely affected, leading to food shortages.

Heavy losses of honeybees have also been reported in Europe and elsewhere. Currently, it is not thought that CCD has arrived in the UK, with colony losses there being more likely a result of *Varroa destructor* coupled with unfavourable weather conditions. Losses in Continental Europe have been blamed on pesticides and on *Nosema ceranae*.

PREPARING FOR EXTRACTION

Holding the frame over the hive and shaking it will help you determine if the honey is ripe. If a lot of liquid falls out, the frame should be returned to the hive.

BEFORE EXTRACTING, first check that the honey is ripe (with a water content of around 18 percent). Cells that are capped with wax indicate honey that is ripe and ready for extraction. If there are uncapped cells on the comb, one way to determine if the majority of the honey is ripe is to hold the frame horizontally over the hive and give it a sharp shake. If a lot of liquid falls out, the honey is not ripe and the frame should be returned to the hive. Unripe honey will ferment because of the high water content and presence of natural yeasts.

The next step is to clear the bees out of the supers. If you only have a few supers to deal with and your apiary site is away from neighbors, you can brush the bees off the combs. Take each frame from the super in turn and brush off all the bees with a large flight feather or bee brush. Place the cleared frame into an empty super and cover this to prevent bees returning to the comb(s). Remember to replace the frames in the original super with empty ones to give the bees space to store more honey.

Alternatively, you can use an inner cover. A Porter bee escape fits into the hole in the center. The bees can exit from the honey super but cannot return; therefore the super is cleared of bees. The honey super will then be ready to be removed and the honey extracted. The Porter bee escape consists of a top plate with a $^7/_8$ in (22 mm) hole in the center. A rectangular bottom part slides flush onto flanges on the underside. There are two pairs of flexible springs in the bottom part. The ends of each pair are fixed to the outside edges of the bottom part, at the center; the other ends of the springs are adjusted so that there is just less than a bee space between them.

The Porter escape is fitted into the hole in the inner cover. The supers to be cleared of bees are removed from the hive and an empty super put in their place. The inner cover is then placed on top, making sure that the hole in the Porter escape is uppermost. The full supers are replaced on top of the inner cover, followed by the inner cover and cover. The bees will naturally want to return to the lower part of the hive, either to the brood nest or to continue foraging. They will pass down the escape hole and push their way through between the springs. After 24 hours, virtually all of the bees will have left the full supers, which can then be removed for extraction. If there are any bees remaining in the supers, they can be brushed off the frames back into the hive.

The Honey Harvest

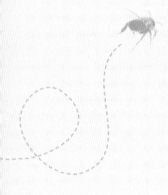

WATCHING HONEY FLOWING from the extractor gives not only the pleasure of your honey crop but also the knowledge that as the bees have foraged they have contributed significantly to the environment through their pollination activities. If the honey is for your own use, or for gifts for family and friends, you are not bound by any regulations in its production. However, if you intend to sell your honey you must abide by the appropriate laws for handling food for sale. In the US, check with local and state agencies for regulations about selling honey. Since honey is not combined with other ingredients, general food standards do not necessarily apply. See also the sections on hygiene (page 252) and other regulations for selling honey (pages 257–58).

Supers cleared of bees and ready for transporting home to harvest the honey.

BASIC EQUIPMENT

Honey can be cut straight from the comb and spread onto bread or toast, wax and all. If the honey in the cells is still liquid, you can cut the comb from the frame, break it into pieces, put it into a fine nylon mesh straining bag and let the honey drain out into a container. If you plan to do this, use unwired foundation in your super frames. However, if you want to produce more than a small quantity of honey, and store it in jars, you will need to remove it from the cells using an extractor.

For honey extraction you need a knife to uncap the sealed cells, a container in which to collect the cell cappings, an extractor, some form of straining device, and a settling/bottling tank. You also need individual honey jars or bulk storage buckets, which are generally plastic and hold roughly 30 lb (13.5 kg). Whether producing honey commercially or for yourself, it is recommended that you use food-grade equipment. This means an extractor and settling tank made from stainless steel or food-grade plastic, a nylon mesh or stainless steel sieve, and glass jars or food-grade plastic containers.

EXTRACTORS

In the US an average yield of honey per hive varies widely with regions and even within regions and states. In addition, the yield will vary with season and weather. Beekeepers use a variety of sizes for honey supers so a true comparison

A small tangential extractor suitable for a beekeeper with only a few hives.

platform or the floor puts a strain on the tank and it is better to make a wooden T-shaped support of appropriate dimensions so that the outside ends of the arms can be bolted to the three legs of the extractor. Secure a strong caster under the end of each arm of the "T" so the extractor, supported by the frame, can move freely over the floor. When you start up the extractor, allow it to move around and the strain on the tank will be reduced. You may need to tether the extractor to something solid to prevent it wandering off!

Tangential extractor

In a tangential extractor the frames are positioned in the cage so that its mesh sides each support one face of the comb at right angles to the central spindle. When the cage rotates, honey in the outer face of the comb is forced from the cells and honey in the opposite face is forced toward the midrib of the comb. The outer face of the comb is partially emptied and then the extractor is stopped and the frame turned through 180 degrees so the honey from the opposite face can be partially extracted. This is repeated until the cells are empty.

With self-spacing frames, such as Hoffman's (see page 161), the face of the comb is not supported as the comb is held away from the cage by the frame side bars. The cage must not be rotated too enthusiastically or the combs may break up.

cannot be made. You may be able to borrow or hire extracting equipment from your local beekeeping association and it is well worth taking advantage of this until you have sufficient colonies to justify buying your own.

Extractors come in two basic designs: tangential and radial. Both consist of a stainless steel or food-grade plastic tank containing a rotating cage into which the uncapped frames are placed. A lid, usually in two halves, covers the tank. Centrifugal force is used to remove honey from the uncapped cells. The honey runs down the inside of the tank and collects in a reservoir, which is emptied through a tap or honey gate.

When loading an extractor, try to balance the combs so that the machine runs evenly. An extractor can rock quite violently, particularly when it first starts rotating. As the honey is flung from the cells and the weight becomes more even, the extractor will run more smoothly. Bolting the extractor to a

Radial extractor

In a radial extractor, the circular cage holds the frames—top bars to the outside, bottom bars to the central shaft—

Frames in a radial extractor are placed in slots radiating from the central spindle.

Arrange your equipment so that you move from one task to the next in succession. Full supers should be placed next to the area where you will uncap the comb. The extractor should be close to this so the uncapped frames can be placed directly into the extractor. Then you'll need to put the empty super nearby in which to place the frames after extraction. Water sources should also be easily accessible. Uncapped and empty combs tend to drip, so try to move them over the equipment when proceeding from one stage to the next. When anything gets sticky, wipe it immediately. Use a damp cloth to wipe up honey; a wet one will spread honey into a larger sticky area.

like spokes in a wheel. As the cage rotates, honey is extracted from both sides of the comb at once. As long as the comb is securely attached to the frame, and the frame is held straight in the cage, there are few breakages. All patterns of frame can be placed safely in a radial extractor.

HYGIENE AND PREPARATIONS

Most small-scale beekeepers use the domestic kitchen for extraction, but wherever you choose to extract your honey the area must be clean and hygienic. Beekeepers should check their local and state regulations for honey houses. These regulations vary widely. The state apiarist in states with that service can supply the information or guide the beekeeper to the relevant office. In those states without a state apiarist the Cooperative Extension Service agent can provide the information needed.

UNCAPPING

Before honey can be extracted you must remove the wax cappings on the filled cells. This is usually done with a sharp or serrated-edged knife: a thin blade at least 8 in (200 mm) long is preferable, and uncapping is easier if the blade is longer than the depth of the frame. You will need a rectangular plastic container and a piece of wood about 1½–2 in (40–50 mm) wide and long enough to reach across the container. On one side of the wood cut two notches to fit over opposite edges of the container, to hold the wood steady. On the other side, cut a recess wide enough to hold the end of the frame lug.

With the wood in place across the middle of the container, the frame lug secure in the recess, and the frame held so it is pointing upwards, you can now begin to cut off the cappings. Leaning the frame toward the side being uncapped helps the cappings fall away into the container. You can cut in an upwards or downwards direction, but take care the blade does not slip.

Uncapping forks can be used instead of a knife. These lift off the cappings or simply scratch them open to allow the honey to flow out in the extractor. Any cappings collected in the extracted honey are strained out later.

For uncapping new combs, where the wax is still very soft, you can use a hot-air paint stripper. Playing the jet of air over the face of the comb melts the cappings almost instantaneously. Don't leave the heat over one area for any length of time or you will melt the comb. The uncapped comb can be put straight into the extractor.

EXTRACTING AND PROCESSING

Having uncapped your combs, put the frames into the extractor and start rotating the cage gently at first, increasing the speed gradually. Turning the cage too fast can lead to broken combs, especially with a tangential extractor. Trying to get every last drop of honey out of the combs isn't worth the effort.

Don't take the lid off while the cage is rotating. Particularly with motorized extractors there is a considerable down draft and the corner of the half lid can easily be sucked into the barrel, where it will foul the cage and probably damage the extractor. Keeping the lid on also makes the extraction process more efficient.

Cutting in an upwards direction (A)

The capping falls into the container below (B)

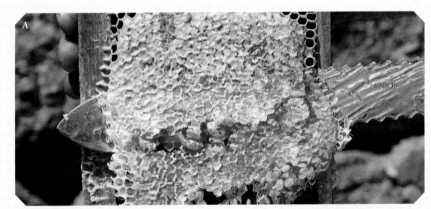

OPPOSITE: *A fully equipped commercial honey operation, with an electric uncapping machine.*

Slow and stop the cage rotation and do not attempt to remove any frames until it is stationary. Lift the frames out and store them in the empty super(s). If you want to keep the same frames for use on the same colony to prevent spread of disease, you will have to mark both the boxes and the frames.

The honey collects in the reservoir and is then run out through the tap. Depending on the depth of the reservoir, you may need to empty it before the level rises and fouls the bottom lugs of the rotating frames. The honey is generally strained at this stage so that it is ready for bottling.

STRAINING

Honey out of the extractor will contain bits of wax and other debris. These particles need to be removed by straining, before bottling the honey or putting it into bulk storage containers. There are several types of strainer you can use, including fine nylon mesh cloths or bags, single stainless steel mesh hemispherical or conical filters, and double strain-

BELOW LEFT: *As the extractor turns, the honey hits the sides of the barrel, runs down and collects in the reservoir at the bottom.*

BELOW RIGHT: *Running out the honey into a coarse sieve to remove wax and other debris.*

REUSING EMPTY SUPERS

After extraction, the empty combs can go back to the bees, which will lick them dry. Remove the Porter escape from the inner cover and put the supers, which will still be wet with honey, on top of the cover, followed by a second inner cover to stop the frames sticking to the roof. The bees will come up into the supers and remove any remaining honey. After a couple of days, replace the Porter escape and the bees will clear from the extracted supers back down into the colony. You can then reuse or store these supers, depending on the stage in the season. If there is still a honey flow in progress, put the wet supers on top of any colonies that need more space, assuming all your colonies are free of disease.

WARMING HONEY

You must be very careful when warming honey not to destroy the enzymes it contains or to evaporate the volatile aromatic compounds responsible for the unique qualities of different types of honey. Keep the temperature during processing and bottling below 95°F (35°C), and cool the honey as quickly as possible when you have finished.

An alternative to using a microwave, and especially if you have several containers to deal with, is to use a thermostatically controlled, insulated warming cabinet. If you are bottling only small amounts of honey, transfer any granulated honey into a small container and warm it gently in a water bath—standing it inside a pan containing a small amount of water, over a moderate heat. Make sure the container with the honey doesn't come into direct contact with the heat source.

ers. The last consists of a fine stainless steel mesh hemisphere with a flat circular coarser mesh that fits horizontally over the top. The coarse mesh removes large pieces of debris and the smaller pieces are strained out below in the fine mesh. Sieves can become blocked and you will need to stop the flow of honey and wait for the strainer to clear before washing it and starting again. With a double strainer, you can clear the coarse sieve several times before needing to clean the fine one.

Straining honey in a small-scale operation using a fine-mesh filter.

BULK STORAGE AND BOTTLING

If you don't want to bottle the honey immediately, you can filter it and run it into food-grade storage containers, where it will granulate or crystallize (see page 306 for information about storing honey). Any storage container or jar must be clean before use. Jars purchased from suppliers usually arrive with lids in place and can be used directly. Recycled honey jars should be washed in very hot soapy water and rinsed thoroughly. They can be dried in a hot oven and then fitted with a clean lid before storing until use. Washing them in a dishwasher is ideal. Metal lids should not be reused as the coating is easily scratched and the metal then corrodes when it comes into contact with the acid honey. Plastic lids should be washed and dried.

When ready to bottle the stored honey, you will need to warm it so that it flows and can be put into jars. This can be done in a microwave, if the container will fit inside, using short bursts of power and allowing time between

LEFT: *Bottling honey by running it off from a settling or bottling tank.*

RIGHT: *Different types of honey in a row of bottling tanks.*

each burst for the heat to dissipate. Stirring helps this process. The aim is to avoid hot spots where the honey overheats and spoils. When it has become a clear liquid, pour the honey into the settling tank.

Bottling honey

Honey in the settling tank should be left to stand. Air bubbles and small dirt particles will float to the surface and other, heavier bits will sink to the bottom. Depending on the honey's temperature, it can take between several hours and several days for it to settle.

To ensure that you bottle only clean and air-free honey, it is advisable to run some honey out from the bottom of the tank before you start bottling, and stop bottling when you see air bubbles coming out of the tap. These first and last layers of honey can be used for your own consumption. When filling the jar, run the honey down the inside to avoid entrapping air. If you are going to sell your honey, make sure that the jar is properly filled with the prescribed weight. This is usually taken to be the

case if there is no visible gap between the top of the honey and the bottom of the lid, but selected jars can be check-weighed on calibrated scales.

CLEANING UP

After extraction and/or bottling is finished, the equipment must be cleaned. Any container that has held honey is best washed with cold water; hot water will soften or melt the remaining particles and cause them to smear over the surface of the container. Extraction equipment should be covered with plastic when not in use to keep it from getting dusty. When you need it again, it should be wiped, making sure that no fibers from the cloth are left behind. Floors and surfaces should be wiped with a damp (not wet) cloth.

REGULATIONS FOR SELLING HONEY

In the US, honey is regulated by the Food and Drug Administration (FDA). As a food, it comes under the Public Health, Security and Bio-terrorism Preparedness and Response Act of 2002.

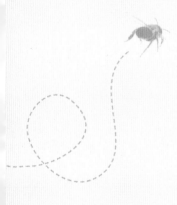

*The honey is labeled
and ready to sell.*

PRESENTING HONEY

Whether you are selling your honey or giving it away, you want it to look good. Make sure it is clean with no air bubbles on the surface. Honey naturally granulates, some types more quickly than others (e.g., canola), and incipient granulation can spoil the appearance of a bright, clear, sparkling liquid honey. Warming gently and stirring can solve the problem, at least temporarily.

Creamed honey may "frost," where lighter streaks appear on the shoulders and down the sides of the jar. Frosting can be prevented by not incorporating air into the mixture of liquid honey and seed crystal honey and by filling the jar slowly.

If you keep your bees in one place throughout the year, your honey will be from the flowers of your local area, but it can still come in a great variety of colors, viscosities, and flavors, depending on which flowers have been visited by the foragers. If some of the honey has a high water content, you can blend it with another of a high viscosity to produce a middle-range honey, which may be more suited to your taste and that of your customers.

Details can be obtained from the FDA or from the website www. cfsan.fda.gov. The United States Department of Agriculture (USDA) has issued the United States Standards for Grades of Extracted Honey, which came into effect on May 23, 1985. The full details were published in the *Federal Register* of April 23, 1985. Copies of standards and grading manuals can be obtained from the Chief, Processed Products Branch, Fruit and Vegetable Division, AMS, USDA, PO Box 96456, Rm 0709, So. Bldg., Washington, DC 20090-6456.

Various local and state laws may apply for selling your honey. It is important to be aware of any regulations in your area. In some states the state apiarist can give you the information or direct you to the correct office that handles such matters. In states without a state apiarist, beekeepers can contact their local Cooperative Extension Service agent who can provide the necessary contact.

The easiest way to ensure your labels are legal in the US is by visiting the National Honey Board website. Go to www.honey.com then open the Honey Industry box and scroll down to Labeling Information. There you will find the federal requirements. State labeling laws follow the federal ones.

Where you sell your honey will only be limited by your imagination and powers of persuasion. You can set up a stand outside your home or take one at a farmers' market. You can approach local shops or even sell it by mail order over the Internet. If you want to try shops, look for one that sells food-related items but not necessarily honey. Your honey will be much more noticeable in the butcher's shop than among all the other groceries in a convenience store.

Harvesting Beeswax

EESWAX IS ONE of the by-products of honey extraction, primarily from the cappings but also from other bits of comb strained from the extracted honey. You can also collect brace comb during colony inspections.

If you have extracted honey from only a few supers, the simplest way to clean the wax is to put the cappings on a cake-cooling rack or other grid on top of the inner cover of the hive, then give the bees access to it through the feed hole. Surround the cappings with an empty hive body to support the cover and keep out robber insects. The bees will lick any honey from the loose wax and you can then wash the cleaned cappings in soft water, and process them when you have collected more.

EXTRACTING BEESWAX

Wax can be recovered from old combs and brace comb using a solar wax extractor. This consists of an insulated box, propped up at an angle with a double-glazed glass as a cover. Inside is a large melting tray, which holds the wax. A gap at the lower edge allows molten wax to flow through into a collecting tray. The inside of the collecting tray is

LEFT: *The beeswax is melted in a double boiler, then decanted into a container, where it is allowed to settle.*

TOP RIGHT: *A mold is filled with melted wax.*

BOTTOM RIGHT: *The wax cools to form wax blocks.*

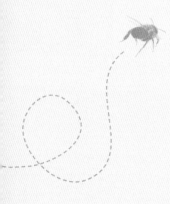

smeared with a release agent, such as dishwashing liquid, so the resulting wax block can be removed easily.

To build your own solar extractor, you can use a discarded double-glazed window as the top, and make a wooden box below to fit the window. Use weather stripping to ensure a tight seal between the window and box. At one end add some supports, roughly 6-8 in (16–20 cm) high, to set the extractor at an angle of about 40° from the horizontal. Next find a large metal tray for the melting tray. It should fit loosely into the bottom of the box and cover about three-quarters to seven-eighths of its length. Cut an opening in one of the short sides of the tray and, if possible, bend the sides to funnel the wax toward the opening. Packing insulation material under and around the melting tray will improve the extractor's efficiency. Finally, place a collecting tray at the lower end of the box, under the edge of the melting tray, so that it sits level. The edge of the melting tray must overhang the collecting tray to ensure molten wax is not spilled inside the box.

Set up the extractor to face the sun's position at midday to give maximum exposure, and make sure that it will not be shaded by surrounding vegetation. You can move it around as the sun moves but this is generally not practical.

Now place pieces of comb onto the melting tray. Crude wax pieces can be put inside a secured piece of filter cloth. This retains the cocoons and much of the dirt, while allowing the wax to melt and pass through the filter. The cloth can be discarded afterwards or used for fire lighters.

The extractor works best on clear, hot, sunny days and you will be surprised at the speed with which the collecting tray fills up. However, the wax will still melt on a duller, warm day, so keep checking the tray to see if it is full.

A solar wax extractor

Double-glazed glass top

Melting tray

Collecting tray

Insulated box

Supports

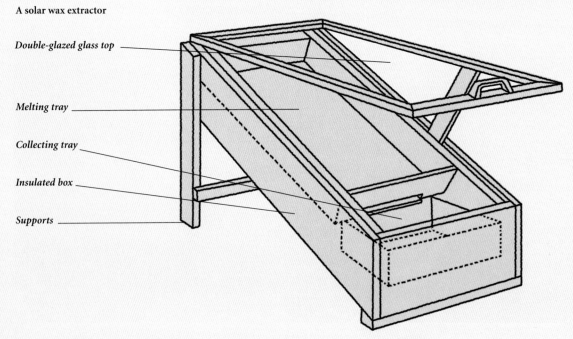

Industrial beeswax processing involves refining the wax in vats and then pouring it into molds. It is then removed in large slabs.

To stop the extractor, open the top. Wait for the wax block to cool completely and solidify before trying to remove it from the tray or you could spill molten wax and lose your harvest.

PROCESSING BEESWAX

Although your extracted wax will look clean, when you melt it there will be a dirty residue. Processing the wax cleans it, the aim being to produce a clean block of beeswax which may be pale to golden yellow in color, depending on which flowers the bees were foraging. If you only have a small amount, it is probably best to save it up until you have a reasonable quantity to process at one time.

Use a double boiler or water bath to melt the wax over a gentle heat, with the wax inside a stainless steel or glass container holding a small amount of soft water (rainwater). You can also melt wax in a microwave but, again, you must add some soft water. Don't overheat the beeswax as this darkens the wax and destroys the lovely aromas.

Small amounts of wax can be filtered through closely woven cloth secured across the top of a container. Sweatshirt material is also effective with the fluffy side up. Some of the wax will be lost on the filter but this can be reused a number of times before the filter becomes clogged.

When the clean wax has cooled slightly, pour it into small plastic containers, such as those used for margarine, and allow it to solidify. Depending on the efficiency of your filtering system, some dirt may still be present. This will collect mainly at the bottom of the final block of wax, although some may also have floated to the top. You can scrape it off and add it to the next batch to be processed.

Reclaimed beeswax can be used to make foundation (see pages 286–87) or in various other household recipes (see pages 389–95).

Setting Up a Commercial Apiary

MANY COMMERCIAL BEEKEEPERS began with a small-scale operation. If you get truly hooked and decide to turn your beekeeping into a commercial venture, there are several things you will have to consider, in addition to establishing a registered business and keeping proper accounts and records.

Firstly, are you fit and strong enough? Commercial beekeeping is very different from small-scale beekeeping. There is a lot of heavy lifting involved and you are likely to end up doing much of this yourself. The cost of lifting equipment can only be justified if you have over 300 hives.

You will need sufficient equipment for the number of hives you plan to have, and for large-scale extracting and bottling, and somewhere to store the equipment when not in use.

With hundreds of hives, you will need to find more apiary sites. The available forage at each site will have to be assessed to decide how many colonies it will support. If there is insufficient forage your honey crop will be low and you will probably need to feed your bees, adding to your work and costs. Another consideration is the distance between apiaries, in particular if you factor in the cost of gasoline. You will need a suitable vehicle and/or trailer for transporting a large number of hives at once.

A major part of your income is likely to come from pollination contracts. Here, farmers will pay for colonies to be placed on their crops. However, they

With a large number of hives, the choice of apiary sites needs careful planning.

Part of your income will be from honey sales.

will dictate where and when, and you will have to be flexible enough to remove hives at short notice. If you have mechanized your operation, you can put four hives on a pallet, each facing in a different direction. The pallet can then be lifted, with the hives in place, and deposited where required.

Your other income will be from honey and other hive products and you will need to be geared up to deal with large volumes—tons rather than pounds (kilograms). How you sell your products is another consideration. This may be through a bulk contract with a wholesaler, but you will get more added value if you pack your own honey and sell it to retailers. For both wholesale and retail markets you will have to learn how to price and market your products effectively.

For the retail market, this will probably mean establishing a brand name and a style of label and/or containers. Another decision will be what to do with your beeswax: whether you want to make your own candles, cosmetics, and other beeswax products or find a bulk outlet.

HELP FOR COMMERCIAL BEEKEEPERS

There are a number of beekeeping associations that are useful to the commercial beekeeper, including the American Beekeeping Federation (www.abfnet.org), the American Honey Producers Association (www.americanhoney producers.org), and the National Honey Board (www.honey.com). Membership of the first two is open to all beekeepers.

The Future of Beekeeping

IN MANY RESPECTS, beekeeping now is very similar to the way it was in the early twentieth century: the fundamental design of hives and other equipment has not altered, and the basic methods of beekeeping are the same. However, there are some important changes. Plastics have made a huge difference to beekeepers and you will now find plastic extractors, settling tanks, and storage buckets. You can even buy plastic frames and foundation. There is a greater use of stainless steel today rather than galvanized steel, though the basic extractor mechanism is the same.

New hive designs are introduced as the answer to a beekeeper's prayers. There are long hives to reduce the amount of lifting of heavy boxes, and hives that only use top bars rather than full frames. New methods of swarm control are proposed, though they still rely on the same principle of the stages in the development of a queen bee and the swarming timetable.

On the disease front, *Varroa destructor* is now endemic in the US. Throughout the country the mites have become resistant to two of the chemical treatments, fluvalinate and coumaphos. Although new medications are effective many beekeepers are using an Integrated Pest Management approach by determining the level of infestation first before using less harsh chemicals if necessary. Small hive beetles (*Aethina tumida*) are now present in the US.

BELOW LEFT: *New hive designs are regularly introduced. This hive design © Maurice Chaudière.*

BELOW RIGHT: *Omlet's Beehaus for urban beekeepers.*

Currently, *Tropilaelaps* mites have not been found but hives are being monitored for early detection of this pest. To combat any new pests, beekeepers must be more vigilant in their hive inspections. Different management techniques may be necessary.

As you gain experience, you will come across alternative methods of management, experimental hive models, and new and exciting discoveries. Once you have mastered the basics you can see if these work for you, whether they are better for your bees or really do double your honey crop. Bees are very tolerant of man's inventions and experiments, which is just part of what makes beekeeping such fun.

As beekeeping moves into the future and we begin to understand the importance of the honeybee, so apiary and hive technology continues to move forward.

In the UK, a company whose plastic chicken coop led to a wave of urban chicken keeping has created a plastic beehive, which it hopes will do the same for beekeeping. Omlet's horizontal plastic hive, the Beehaus (see opposite), is designed for urban beekeepers who want a way into beekeeping. The hive comes with beekeeping clothing included within the price, but bees cost extra.

Some other alternative hive models are even more experimental. Solar hives extract honey and wax from natural comb using simply the heat of the sun.

RIGHT: *This "solar hive" extracts honey and wax from natural comb using the heat of the sun. Hive design © Maurice Chaudière.*

PART FOUR

HONEY AND OTHER BEE PRODUCTS

HONEY: NECTAR OF THE GODS

MEDICINAL PROPERTIES OF HONEY

BEESWAX: NATURE'S SEALANT

CANDLES

OTHER USES FOR BEESWAX

PROPOLIS: NATURE'S MIRACLE CURE

POLLEN: NATURE'S COMPLETE FOOD

BEE VENOM: NATURAL PAIN KILLER

ROYAL JELLY: A RARE BLEND

Honey: Nectar of the Gods

ECTAR IS A THIN LIQUID consisting of water, sugars, plant pigments, and other materials. Plants produce it in order to attract pollinators such as honeybees, wasps, moths, and bats. The nectar of each plant species has a different combination of sugar and odor and, together with the flower's shape, design, and color, this helps to guide the bee from one flower to another of the same kind.

Bees make honey mainly from nectar, the sugary substance secreted by nectaries in flowers, but they also make it, though less often, from honeydew. Honeydew is a sugar-rich sticky substance secreted by aphids and other insects while they feed on plant sap. It is often deposited on leaves or plant stems. Bees usually collect it early in the morning when there is dew on the plants, which dilutes the secretion. Honeydew honey is generally very dark and strong tasting.

Honey has been valued as a sweetener since early times. In the Bible (Proverbs 24:13), Solomon exhorts: "My son, eat thou honey, because it is good; and the honeycomb, which is sweet to thy taste."

As well as being a natural sweetener, honey releases sugars into the bloodstream, providing rapid supplies of energy. Today, honey is considered a highly nutritious food. It is eaten in its natural state and used in a wide range of cooking.

COMPOSITION AND QUALITY

The composition of honey depends largely on which flowers are visited by the bees when collecting nectar. Glucose and fructose, the main simple sugars, are present in roughly equal quantities, together with a small amount of sucrose (the sugar

Bees turn nectar (and honeydew) into honey by adding enzymes and reducing the moisture content through evaporation.

we know as table sugar). The principal physical characteristics and behavior of honey are indeed due to its sugars, but other constituents—such as acids, proteins (including enzymes), pigments, and minerals—help to distinguish the individual honey types. Darker honeys, including honeydew honey, have a higher mineral content than lighter honeys.

Naturally occurring yeasts cause honey to ferment if the moisture level is too high. To avoid this, the bees reduce the amount of water in honey to around 18 percent (see page 123) before sealing the storage cell with a beeswax capping. Honeys with a higher viscosity are considered to be of a better quality than thin honeys with a higher water content.

Honey is an excellent source of energy.

HEAT AND HONEY

One quantitative measure of honey's quality is the level of hydroxymethylfurfural (HMF) that is produced as fructose breaks down. Heat accelerates this process, so a high HMF level can indicate that the honey has been overheated. As HMF production is a continuous process, however, high levels can also indicate that the honey has been around a long time. In addition, heat destroys the enzyme diastase, so a high diastase level indicates a good-quality honey that hasn't been overheated.

As heat destroys the enzymes and other properties of honey, it affects the flavor and darkens the natural honey color. Boiling honey can turn it into a hard, brittle, toffee-like substance.

Depending on the source, the amounts of sucrose, glucose, and fructose in nectar differ. During the conversion of nectar into honey, chemical changes are brought about with two enzymes being produced by the worker bee in her hypopharyngeal glands (located in the head of the bee). One, invertase, breaks down sucrose into glucose and fructose; the other, glucose oxidase, breaks down glucose into gluconic acid (the main acid in honey) and hydrogen peroxide. Gluconic and other acids are responsible for honey's low pH, which is inhospitable to bacteria, mold and fungi; the hydrogen peroxide is particularly important as it destroys bacteria. These characteristics, together with the low water content, make honey a very stable food that is able to last for years without refrigeration.

COLOR, FLAVOR, AND AROMA

Honey's color, flavor, and aroma all derive from the plant pigments and other materials secreted in the nectar, making the honey from each floral source unique. The color can range from almost as clear as water to nearly black. In between, it can exhibit tints of green and red. The flavor and aroma of honey are interlinked as both are caused by minute amounts of complex plant-derived substances.

All honey is sweet because of its high sugar content, but some honey, such as acacia, tastes sweeter than others because it has a higher proportion of fructose than glucose. However, it is not ony the fructose level that gives it its sweetness.

Dilute solutions of up to 10 percent sucrose are perceived to taste sweeter than those of the same

WHO SHOULDN'T EAT HONEY?

There is a very low risk of infant botulism resulting from honey consumption, and some bee-keepers include a warning on their honey labels that honey should not be fed to infants under 12 months of age. Each producing country has its own labeling guidelines, so it's best to check each area separately. For example, the New Zealand Food Safety Authority states that, though honey can be simply labeled, the presence of royal jelly must be specified; they are considering warnings for pollen and propolis.

Generally, diabetics are advised to avoid honey but it can be eaten in small quantities as part of a controlled carbohydrate intake. One or two tea-spoons of honey can also be used as a "quick fix" if a diabetic's blood sugar level falls too low.

concentration of the simple sugars. Strangely, at higher concentrations, the opposite is true—the fructose/glucose solution is perceived to be sweeter than an equivalent sucrose one. Hence, because of its fructose/glucose content, people generally consider that honey tastes sweeter than a solution containing the same concentration of sugar.

Most nectars have a 40–45 percent sugar content but the range can vary from 3–87 percent, with bees undoubtedly preferring those with higher sugar contents. The flavor of the resulting honey is very much a personal taste.

TYPES OF HONEY

The following colors and flavors of a number of honeys should not be taken as definitive, as honey is affected by the environmental and weather conditions when the nectar is collected. Some of the honeys listed, such as holly or fuchsia, are only available through specialist food suppliers or directly from the beekeeper (locally or online); others, such as acacia, clover, orange blossom, and meadow flower (mixed blossom), are readily available in supermarkets and grocery stores.

The color of honey is one of its most variable features and derives not from the bee but from the nectar's source.

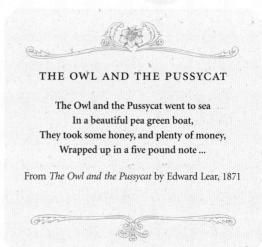

THE OWL AND THE PUSSYCAT

The Owl and the Pussycat went to sea
In a beautiful pea green boat,
They took some honey, and plenty of money,
Wrapped up in a five pound note ...

From *The Owl and the Pussycat* by Edward Lear, 1871

TYPES OF HONEY

LIGHT-COLORED HONEY

Clover honey

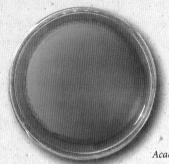

Acacia honey

Name (alternative names)	Principal countries of origin	Color, flavor, and other notes
ACACIA	California, Arizona, USA; Europe, mainly Bulgaria, Hungary, and Romania; Canada; China.	Pale golden yellow and very clear; sweet, delicate taste; vanilla and floral aroma; stays liquid for a long time.
ALFALFA	California, USA; Canada (British Columbia).	Pale, white or light amber, with a scent of beeswax; pleasing mild flavor sometimes described as minty; granulates quickly.
APPLE	Worldwide, wherever orchards are grown, especially northern and central USA; Europe.	Light amber; good flavor, may have a hint of apple in scent; granulates quickly.
BLUEBERRY (bilberry, whortleberry) and CRANBERRY	New England and Michigan, USA; eastern Canada; western and northern Europe.	Light amber; slightly fruity (more so in cranberry honey), full, well-rounded flavor; slight buttery finish.
BORAGE (starflower)	UK, Canada, New Zealand.	Pale yellow or water-white; light flavor.
CLOVER	USA, Canada, Europe, Egypt, Australia, New Zealand.	Pale amber, yellow, or white; gently flavored with the scent of the flowers; often blended (see page 278).
COTTON	Southern USA, Egypt.	Pale amber; lightly flavored.
FALSE ACACIA	North America, Europe.	Light honey with good density and flavor.
FIREWEED (rosebay willowherb)	Worldwide, but especially northern and western USA, Canada, Europe.	Pale amber to white; crystal clear; sweet and mild, subtle flavor with tea-like notes.
FUCHSIA	South and North America, Europe (especially Britain and Ireland), New Zealand.	Light color; very mild and insipid.
GOLDENROD	North America, Europe.	Light to medium gold; slightly strong almost spicy flavor; quick to granulate.

Leatherwood honey

Mesquite honey

Name (alternative names)	Principal countries of origin	Color, flavor, and other notes
CANOLA	North America, Europe, China, Southeast Asia and mid-Asia countries, including Malaysia, Thailand, and Vietnam.	White to light amber; clear honey that granulates quickly; mild taste; good source for bakery and industrial honey; often blended.
HOLLY	Western and southern Europe, southern USA.	Pale color; finely flavored.
IVY	Europe, Asia, North America.	Grayish-white or yellow, with delicate odor and bitter flavor; a creamy consistency; granulates quickly.
KNAPWEED	Ireland.	Light amber; thin consistency; mild flavor with distinctive tang.
LEATHERWOOD	Tasmania.	Light golden yellow; strongly spicy, an acquired taste; often blended.
MAPLE	Eastern Canada, northeast USA, UK.	Pale yellow sometimes with a greenish tint; mild flavor.
MELILOT (sweet clover)	Worldwide, especially North America and Europe.	Pale greenish-yellow; slight cinnamon flavor.
MESQUITE	Northern Mexico, southern USA.	Often light, but can vary from white to dark brown; viscous with a smoky scent of molasses or brown sugar.
RATA	New Zealand.	Water-white when pure; medium-bodied with a thick texture.
SAINFOIN	North America, Europe.	Lemon yellow; aromatic, good flavor.
SUNFLOWER	North America, central and southern Europe, Russia, China.	Yellow with delicate, light flavor; table honey also used in the bakery trade.

MEDIUM-COLORED HONEY

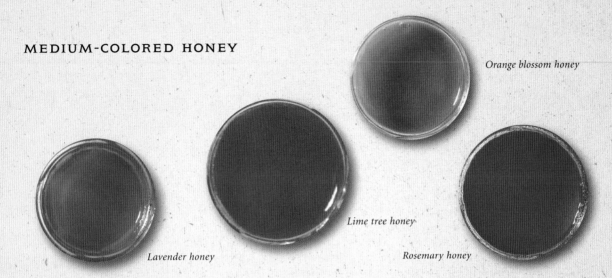

Orange blossom honey

Lime tree honey

Lavender honey

Rosemary honey

Name (alternative names)	Principal countries of origin	Color, flavor, and other notes
BLACKBERRY (bramble)	UK, Canada.	Light chestnut brown; viscous but transparent; rather coarse-flavored.
COCONUT PALM	Hawaii, Florida, USA; West Indies.	Strong amber honey.
DANDELION	Worldwide, especially UK.	Golden yellow; strong taste and flowery scent; coarse-grained.
LAVENDER	Parts of Europe, especially France and Spain.	Golden; flowery aroma, with fine granulation to a texture resembling butter.
LIME TREE (linden, basswood)	Europe, Canada, north-central USA.	Amber or light yellow with a greenish tint; strong, biting flavor with a slight vanilla taste.
MEADOW FLOWER (blossom, wild flower, multi-floral, mixed floral), such as dandelion, white, pink and red clover, wild vetch, thyme, melilot, spotted crane's bill, alfalfa, sage, etc.	Worldwide, especially Europe.	Golden yellow (varying from light to brownish); full-flavored.
ORANGE BLOSSOM (and other citrus fruits)	Florida, Texas, California, Arizona, USA; New Zealand, Asia.	Medium amber to water-white; sweet, distinctive fruit taste with echoes of blossom.
ROSEMARY	Europe, especially Mediterranean region.	Color ranging from almost white to golden with reddish tones; medium-bodied with a thick texture.
THYME	Mediterranean Europe, North America, New Zealand.	Bright amber; an intense aroma and aromatic flavor.
TUPELO	Southeastern USA.	Light amber to medium yellow with a greenish glow; mild, distinctive taste and very sweet; granulates slowly.

DARK-COLORED HONEY

Eucaplyptus honey

Manuka honey

Chestnut honey

Heather (bell) honey

Name (alternative names)	Principal countries of origin	Color, flavor, and other notes
AVOCADO	California, USA; Mexico, Australia.	Dark amber with strong flavor of caramelized molasses.
BUCKWHEAT	Minnesota, New York, Ohio, Pennsylvania, and Wisconsin, USA; eastern Canada, China, Russia.	Very dark, sometimes purple-brown; strongly flavored with hints of molasses and malt; recommended in hearty baked goods and barbecue sauces.
CHESTNUT (sweet chestnut)	Southern Europe.	Dark amber; sharp and bitter (an acquired taste); high in minerals.
EUCALYPTUS	California, USA; Australia, New Zealand, Italy.	Dark amber; distinctive, slightly caramel flavor. Yellow box gum (*E. melliodora*) produces pale and thin honey; red box (*E. polyanthemos*) honey is very dense.
HAWTHORN	Northern and western Europe, North America, New Zealand.	Dark amber; rather nutty flavor.
HEATHER (Ling)—from *Calluna vulgaris*	Europe, especially Scandinavia, Scotland, and moors and lowland heaths in England.	Bright dark amber/reddish-brown; slightly bittersweet taste and aroma of heather flowers; it sets to a jelly and has to be extracted with a press (see page 278) or sold as cut comb (see page 275).
HEATHER—from Bell, *Erica* species	Scotland and other parts of UK, Ireland.	Reddish-brown (port wine color); thick consistency; strong taste and aroma.
HONEYDEW (pine, forest honey, fir honey)	Uncultivated and forested zones of Europe, especially Greece, Turkey, and Germany; New Zealand.	Dark amber; strong, rich flavor; high mineral content. From the sugary secretions of aphids feeding on leaf sap, collected by honeybees.
MANUKA	New Zealand's coastal areas.	Dark color; full-bodied herbaceous sweet taste; rated for its antibacterial properties.
REWAREWA	New Zealand.	Dark amber; rich and malty flavor.
TULIP TREE (tulip/yellow poplar)	Eastern USA (Florida and Georgia).	Very dark amber honey with rich flavor.

HONEY TYPES AND DESCRIPTIONS

Honey can be presented in various ways. Cut-comb honey is unprocessed, straight from the comb. Extracted honey, which has been removed from the combs by centrifugal force or by pressing, is generally sold in jars or other suitable containers and described as liquid, naturally granulated or creamed. And honey from bees that have foraged predominantly on a single floral source is usually described according to the flower concerned (as with most of the honeys in the tables on pages 271–74).

Honey from a variety of floral sources may be identified by the region or country where it originated. A combination of descriptions can also be used to identify a particular honey, for example, New Zealand clover honey.

Cut comb honey is honey in its most natural state.

Cut-comb and sections

Comb honey is cut directly from the honeycomb. To produce it, super frames are fitted with thin, unwired foundation. Then, when the combs are removed from the hive, areas that are well capped are cut out and placed into suitable plastic containers with transparent lids. The trimmings can be used by the beekeeper or squashed to extract the liquid honey.

An alternative method is to provide the bees with sections to fill with honey. Equipment suppliers furnish a round section super and the replacement pieces for it. Round plastic rings are inserted into a frame that holds the thin surplus (unwired) foundation. The frames are then placed into a rack inside the honey super. On a strong nectar flow the bees will fill the round sections completely. When the honey is capped, the sections are easily removed from the frames. The edges are trimmed of excess foundation. The finished round sections are then provided with transparent, rigid plastic covers. A label—which contains the beekeeper's contact information and the weight of the section—circles the round section and holds the covers on. Although the round sections have a good market response, today cut-comb is perhaps the most popular form of comb honey sold.

Liquid honey

Liquid honey should be clear and bright with no signs of natural crystallization (see page 276). Honey extracted from the combs is strained to remove small pieces of wax and dirt. It may be warmed slightly to make it flow for easier bottling but, other than that, is as it came from the hive.

Naturally granulated honey

Over time, the vast majority of honeys crystallize or granulate naturally. This can take anywhere from several days to a few years for some types. The process involves a conversion from honey's liquid state into a crystalline solid without any deterioration of its properties.

Most honeys are supersaturated solutions of glucose, which means that they contain more dissolved glucose than can normally stay in solution. It is this sugar that begins to crystallize out from the solution as honey naturally granulates. A network of crystals forms throughout the honey, making it lighter in color because glucose crystals are white. A light honey may become almost white on crystallization, whereas a darker honey retains its brown color, though somewhat lighter than before.

The crystals that develop in different granulated honeys are varying sizes, creating either a fine or a coarse texture. A fine smooth texture is often considered more pleasant to eat.

Creamed honey has a smoother texture than naturally crystallized honey.

GRANULATED HONEY: SOLID OR LIQUID?

Although granulated honey appears solid, only about 15 percent of the honey is actually in the solid crystalline state, with the mesh of crystals holding liquid honey within it.

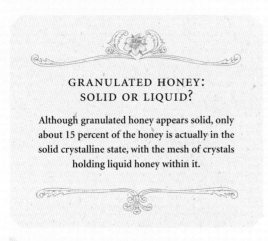

Frosting

Frosting may occur in jars of naturally granulated honey. This is seen as white patches and streaks, particularly on the shoulder of the jar and on the honey surface. Frosting is caused by the creation of small spaces within the honey when liquid glucose is drawn into crystals. The spaces fill with air during crystallization. There is nothing wrong with frosted honey although it is perhaps unsightly and may have an adverse effect on sales.

Creamed honey

Creamed honey is sometimes described as "whipped" or "spun," but these terms inaccurately imply that the honey has been processed and air added in some way. To produce creamed honey by the Dyce process, nuclei in the form of seed crystals are introduced into liquid honey and distributed evenly. This seed is chosen carefully and consists of a honey with the finest grain available. Any large crystals in the seed will spoil the finished product.

The bulk of the honey is completely liquefied by warming it to no more than 150°F (66°C). It is then cooled to room temperature as rapidly as possible. The seed, at a rate of 10–15 percent of the bulk, is warmed until it is just mobile enough to be poured into the liquid honey. Incorporating as little air as possible, the mixture is stirred thoroughly to ensure that the fine crystals of the seed are evenly distributed throughout the liquid honey. The seeded bulk is left to settle for a couple of hours to allow any large air bubbles to rise to the surface, where they are skimmed off. The creamed honey is then bottled or packed into bulk

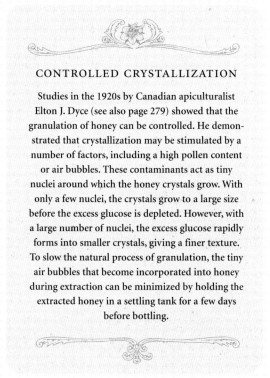

CONTROLLED CRYSTALLIZATION

Studies in the 1920s by Canadian apiculturalist Elton J. Dyce (see also page 279) showed that the granulation of honey can be controlled. He demonstrated that crystallization may be stimulated by a number of factors, including a high pollen content or air bubbles. These contaminants act as tiny nuclei around which the honey crystals grow. With only a few nuclei, the crystals grow to a large size before the excess glucose is depleted. However, with a large number of nuclei, the excess glucose rapidly forms into smaller crystals, giving a finer texture. To slow the natural process of granulation, the tiny air bubbles that become incorporated into honey during extraction can be minimized by holding the extracted honey in a settling tank for a few days before bottling.

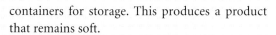

Heather moorland, an important source of monofloral honey in the UK.

containers for storage. This produces a product that remains soft.

Monofloral and multifloral honey

Monofloral honey is predominantly from one floral source—generally over 50 percent from that one source. It is generally collected by bees that are close to an extensive crop or a native species with little other forage available in the area. In the US, for example, the most identifiable monofloral honey is from clover (*Melilotus* ssp.). Colonies of bees are taken to the Dakotas to harvest the thousands of acres of sweet clover nectar. During this time nothing else is in flower so the honey produced can be labeled as clover honey.

Other common monofloral crops are from canola and borage. Canola honey is very mild and is usually sold as a blend with stronger honeys rather than on its own. Borage honey has a light flavor that is appreciated by many customers; it is sometimes sold as starflower honey.

Most honey is classed as blossom honey, multifloral, or mixed floral honey. This is honey made from a mixture of nectars by bees foraging on a range of flowers.

The honey taken from the hive will necessarily be seasonal according to the time of year it is extracted. Some beekeepers differentiate between their seasonal honeys, extracting a few combs together and selling the honey as artisinal. However, most will extract their crop in one or two sessions, allowing the honey from different times of the year to blend naturally.

Blended honey

Most honey produced in the US is naturally blended, with the bees collecting nectar from a range of floral sources. Honey can also be blended by the beekeeper to produce a uniform product to sell throughout the year. Very strong-flavored honeys can be mixed with milder ones to produce a honey that is more palatable to consumers.

COLD-PRESSED HONEY

Ling heather produces thick, almost gelatinous honey, which is too difficult to spin out of the comb using centrifugal force and is therefore generally extracted by pressing. To do this, the comb is cut from the frame and wrapped in a coarse-woven cloth. It is placed between two metal plates, which are squeezed together with a screw mechanism, causing the honey to run out from the bottom of the press.

Because of its consistency, heather honey contains air bubbles: if showing the honey at a honey show, these should be small and well distributed. Heather honey should never be overheated as this makes it muddy-looking and damages the flavor.

Honey by geographic origin

Many honeys from around the world are labeled according to the country of origin. Some also specify a geographical region within a country, describing where the bees were sited.

Toxic honey

Some honeys naturally contain compounds that are harmful or toxic to humans. One such honey comes from *Rhododendron ponticum*, which contains grayanotoxin. This can cause excessive salivation, perspiration, vomiting, dizziness, and, with higher doses, loss of coordination, bradycardia (an abnormally slow heart rate), and severe muscle weakness. However, the condition is rarely fatal and generally lasts less than a day. For poisonous honeys see pages 137–38.

HONEY STORAGE

Honey should be stored at room temperature away from direct sunlight and kept in an airtight container; otherwise it will attract water, which can

ORGANIC AND FAIR-TRADE

There is an ongoing debate whether honey can be regarded as "organic." The general consensus is that it is virtually impossible to ensure that bees collect nectar only from organically grown crops.

Much of the US and Europe's honey imports comes from Asia and Latin America. In these areas, many beekeepers are impoverished and depend on local middlemen to buy their honey. In Chile, for example, 35 beekeepers banded together to attempt to redress the situation. Now fair-trade-certified, they earn more for their honey, have built their own processing facilities, and have an improved standard of living.

lead to fermentation. The high sugar concentration gives it an almost indefinite shelf-life, although the flavor and aroma will disappear with time.

In 1932, Elton J. Dyce showed that the majority of honeys crystallize most rapidly at an average temperature of 57°F (14°C) (those with a low moisture content crystallize more quickly at a slightly higher temperature; those with a higher moisture content crystallize more quickly at a slightly lower temperature). Crystallization at above 60°F (16°C) is usually coarse and gritty (and less palatable). The rate of crystallization slows significantly at temperatures below 50°F (10°C), and before it even reaches as low as 40°F (4.5°C) it virtually halts. This is because glucose crystals cannot grow without molecular movement in the honey.

Lower temperatures increase the honey's viscosity, thereby reducing the rate of crystal growth, but they also increase glucose supersaturation (see page 276), which encourages it. At the critical temperature where these two factors balance out, crystallization proceeds most rapidly. On this basis, if you wish to keep a jar of honey in a liquid state, the best place to store it is not in the refrigerator but in the freezer!

If liquid honey crystallizes, this process can be reversed with gentle warming. Stand the jar in a pan of water, on top of a cloth or a piece of wood to prevent close contact with the heat. Gentle warming, with stirring to distribute the heat, will return the honey to its liquid state. Don't overheat the honey as this will darken it and affect its flavor.

MEAD AND OTHER HONEY DRINKS

Alcoholic honey drinks come in a number of forms:

* MEAD is essentially honey and water with added yeast, yeast nutrient, tannin and citric acid.
* PYMENT is pure grape juice sweetened with honey; adding spices to this produces HIPPOCRAS.
* Apple juice and honey are used for CYSER.

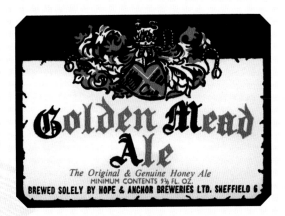

Mead comes in a range of flavors. Its alcohol content can vary from that of a mild ale to that of a fortified wine.

* Using other fruit produces MELOMEL.
* Flowers such as elderflower in mead produce METHEGLIN, and adding spices gives a spiced mead.

Other types of mead are listed on page 46.

Mead was probably the first fermented drink known to man. Many references to it can be found in Greek mythology and earlier; in Anglo-Saxon communities, it was an important part of life, but its use declined steadily after the Normans introduced wine to Britain in the eleventh century.

Mead recipes vary but one method is to heat the honey, yeast nutrient, acid, tannin, and water to no more than 150°F (65.5°C), then cool the mixture to 70°F (21°C) and add the yeast. The mixture is put in a demijohn with or airlock, and left in a warm place (at least 70°F or 21°C) for four or five days. Ideally, the temperature is then reduced to 65°F (18°C) for the remainder of the fermentation. The mead is then racked into a clean jar, corked, and stored in a cool place.

A full straightforward recipe for mead can be found on page 360.

As different honey types have different flavors, mead can be made to suit personal taste. Many consider the best meads to be those made from a light honey.

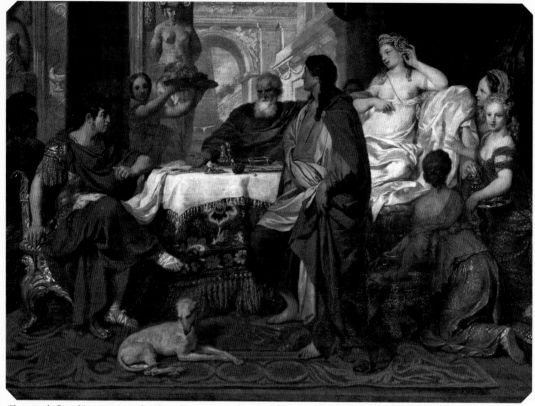

Cleopatra's fine skin was reputed to have been softened and smoothed by her baths of ass's milk and honey.

BEAUTY BENEFITS

Honey is one of man's oldest natural cosmetics, used for centuries on its own or in combination with other ingredients, including beeswax, to manufacture creams, cleansers, and tonics for the skin and hair. Cleopatra, the last Pharaoh of Ancient Egypt, was reputed to have bathed in "ass's milk and honey" and a milk and honey lotion was used by Poppea, wife of Roman Emperor Nero, to maintain her youthful appearance.

More recently, the romantic novelist Dame Barbara Cartland was a strong believer in the health benefits of honey, and the film star Catherine Zeta-Jones is a modern-day advocate of honey facials and hair treatments.

Honey is hygroscopic so it attracts and retains moisture, which may help to improve the complexion. Warm honey smoothed over the skin removes dirt from the pores and this, in combination with its natural antiseptic properties, makes it an ideal skin cleanser. See pages 369–78 for beauty recipes using honey.

EXHIBITING HONEY AND OTHER BEE PRODUCTS

LEFT: *Honey jars labeled for a competition.*

RIGHT: *The National Show of Bees and Honey held at the Crystal Palace, London, in 1934.*

Beekeepers have long competed at honey shows. They have been a part of agricultural fairs for over a century. Local, state, regional and national associations frequently hold yearly honey shows. Classes include light, amber and dark liquid honey, creamed honey, comb honey in various forms, beeswax blocks, candles and ornaments, mead of various types, cookery, and various crafts depicting bees and beekeeping.

The judges look for clean, well-presented products. Honey must be free of any dirt or other contamination and should have a clean surface. Liquid honey should have a good aroma, flavor, and viscosity, with no signs of crystallization. Creamed honey needs to be smooth, again with a good flavor.

Candles must burn without guttering or smoking. Beeswax blocks should be clean and smooth with a good aroma.

Mead, metheglin, and melomel must have finished fermentation, be free of sediment, and have a good aroma and taste.

In the cookery classes, you will find cakes, cookies, breads, pies, candies, as well as other foods (see pages 306–68 for recipes that use honey).

When entering a honey show it is important to read and follow the rules. Having a good product from your bees and presenting it immaculately will go a long way toward winning an award.

Medicinal Properties of Honey

ONEY IS AN ACID PRODUCT, with a pH of between 3.2 and 4.5, which is low enough to inhibit the growth of many animal pathogens. It is also antibacterial because of its ability to attract water: the water molecules become bound up with the sugar molecules in honey, making them unavailable for micro-organisms and thus denying microbes the water they need to survive. Some of the plant-derived components in honey are known to have anti-bacterial properties, and the enzyme glucose oxidase causes a slow release of hydrogen peroxide from honey, which also imparts antibacterial effects.

These antibacterial properties make honey effective in treating wounds of all types. Incorporated into dressings, honey rapidly helps to clean pus or dead tissue from the infected wounds, kill bacteria, suppress inflammation, and stimulate the growth of new tissue. It acts as a barrier to further contamination, and is said to reduce scarring.

Recent laboratory studies at the Belfast City Hospital, Queen's University, Belfast, and the University of Ulster, Northern Ireland, have shown that honey is effective in the treatment of Community-Associated Methicillin-Resistant Staphylococcus aureus (MRSA) skin infections, but further studies are needed to show if this has a clinical application.

Manuka honey is promoted as being particularly effective as an antibacterial. It originates from the Manuka bush (*Leptospermum scoparium*), which grows in New Zealand. The measurable activity of Manuka as an antibacterial is known as the Unique Manuka Factor (UMF), a term that has been registered and can only be used under license.

Other honeys, such as Jarrah honey from Australia, also show this trait. Human clinical trials are determining their effectiveness.

Coughs and colds

Honey is well known as a cough and cold remedy, usually drunk in combination with the likes of lemon, apple cider vinegar, or whisky. A study in 2007 at Penn State Medical College showed that honey was a more effective treatment than remedies containing dextromethorphan, the drug used in many cough medicines. A spoonful of honey also soothes a sore throat. See page 383 for a honey recipe for coughs.

Hangovers

A hangover is caused by the body's production of acetaldehyde (ethanal) in the body from the alcohol consumed. Taking honey provides the

THE TOUCH OF HEALING

Apitherapy is the term for the use of products from honeybees for therapeutic and pharmacological purposes. This can include the use of honey, pollen, propolis, royal jelly, and bee venom. Popular in Asia and Eastern Europe, the practice is also gaining a following in the West, where natural products are increasingly favored. There is a wealth of anecdotal experience to support apitherapy claims, and there is even a new journal dedicated to the therapeutic properties of bee products (www.ibra.org.uk/categories/jaas).

body with sodium, potassium, and fructose, which aid recovery. Honey is also a rapid source of energy and the fructose accelerates alcohol oxidation in the liver, thereby acting as a sobering agent. A recipe for curing hangovers can be found on page 382.

Ulcers

The probable cause of dyspepsia and peptic ulcers is the bacterium *Helicobacter pylori*, and it is possible that honey's antibacterial properties can help these disorders. Studies in New Zealand have shown that Manuka honey was effective in killing the bacterium. See page 388 for dosage.

A hot toddy with honey can help with coughs and colds.

Beeswax: Nature's Sealant

BEESWAX IS SECRETED BY HONEYBEES and used to build comb for brood-rearing and the storage of pollen and honey. The cappings on brood cells contain a large proportion of beeswax, and those on honey-storage cells are pure beeswax. Young worker bees produce beeswax from glands on the underside of the abdomen. Its production requires a temperature of 95–97°F (35-36°F), so the workers hang together in clusters. The secreted wax scales are then passed forward to the bees' mandibles, where they are molded into the hexagonal cells of the honeycomb.

The composition of beeswax is a complex mixture of about 70 percent esters, 15 percent cerotic acid, and 12 percent hydrocarbons, with traces of water, higher alcohols, minerals, and pigments. This composition appears to remain fairly constant whatever its country of origin, although there are differences in color and aroma. Because it is a mixture, it has an ill-defined melting point of 143–47°F (62–64°C). It starts to become plastic and easy to work with at 90–95°F (32–35°C), and loses its tensile strength at 105°F (40.5°C) so that, if a colony of bees overheats, possibly while confined to the hive during transportation, the combs can collapse—with fatal results for the bees.

Beeswax is a very versatile material used in a wide variety of applications, for example in wood polishes, cosmetics, and to make candles, although in some cases it has now been replaced by plastic or modern synthetic waxes. In beekeeping, it is primarily used to make wax foundation for the hives. Instructions for reclaiming beeswax from comb and for processing (cleaning) it can be found on pages 253–58.

Beeswax is used by honeybees to build comb. Here it can be seen on the top bars of the hive frames.

BEESWAX FOUNDATION

Beeswax foundation, embossed with the hexagonal cell pattern, is secured into frames, which are then put into the hive. Bees do not need foundation and frames; they are perfectly capable of building parallel combs hanging freely from the top of their nest cavity. For the beekeeper, however, foundation is a great benefit as it encourages the bees to build their comb within the frame, meaning that it can be easily inspected or removed for harvesting the honey. Some foundation is even wired to give it extra strength. With sufficient reclaimed beeswax, beekeepers can make their own foundation, although it is much easier and more convenient to purchase foundation from equipment suppliers.

Commercially produced foundation

Thick, continuous beeswax lengths are produced on a water-cooled roller rotating through a vat of liquid beeswax. The lengths of wax are then passed between warmed rollers, reduced to the required thickness for foundation, and embossed with the hexagonal cell pattern on both sides. From this thin roll, sheets are cut to the correct size—according to the type of hive and to whether the sheets are for brood or super frames. For unwired foundation, the sheet is rolled thinner than for wired foundation. For cut-comb production, it is rolled even thinner to make thin foundation.

In the USA, foundation is wired with vertical lengths of crimped wire. The sheets are placed on a slowly moving conveyor at a set distance apart. Continuous, crimped wires are fed in from overhead and pressed into the wax with a heated roller. The crimp helps to hold the wires in place and prevent them from being pulled out of the foundation. The spacing between the sheets leaves short lengths of bare wire. The wires are cut at one side of the sheet. Longer wires stick out from the other edge; these are bent at right angles for securing the foundation in the frame using the wedge.

Foundation sold in other countries may have each sheet wired individually using a continuous strand of stainless-steel $\frac{1}{32}$-in (0.44-mm) frame wire. Longer loops are left at the top of the sheet and go under the frame wedge; shorter loops at the bottom fit between the bottom bars to hold the foundation within the frame. The foundation is wired by placing it on a wiring board, which has pegs on either side of the long edge, a short distance away from the wax. The wire is secured at one corner and then run between the pegs forming "V" shapes, up and down the sheet. At the other corner, the wire should be made taut and secured. A low-voltage electric current is then applied to the wire to warm it and embed it into the wax.

Using a foundation press

Beekeepers can make their own foundation from reclaimed beeswax using a foundation press. This creates sheets that are larger than the final dimensions of the frame, but which can then be cut to size. The molds are traditionally made with metal surfaces, but are now available with silicone rubber cell formers, the pattern on one mold being offset against the other to mimic natural comb construction. The beeswax is filtered to remove most of the dirt and debris, although it does not have to be as clean as the wax used for making cosmetics, polish, or candles.

A large sink is first filled with cold water and a

WAX CONVERSION

Some equipment suppliers offer an exchange system where clean beeswax is swapped for an agreed amount of ready-to-use foundation. This is a much easier, if less interesting, way of converting reclaimed beeswax.

small amount of dishwashing liquid. The press is placed within a sizable shallow tray containing a little soft water to catch any excess wax. To melt the block of wax more easily it is broken into smaller pieces, which is easily done by placing the wax in a plastic bag or wrapping it in a cloth and smashing it with a hammer. The pieces of wax are melted over gentle heat using a stainless steel double boiler or water bath (beeswax should never be melted over direct heat), and the molten wax is then poured into the press until it is nearly full. The press is closed and squeezed shut, allowing excess wax to collect in the tray below. It is then submerged in water in the sink for 20–30 seconds. It is opened, while still under water, and the wax sheet removed. The press is taken out of the water and allowed to drain. The dishwashing liquid prevents water droplets from forming on the mold, which would make small holes in the next sheet of foundation. The process is repeated as required, and the sheets of foundation stacked on a flat surface with a flat piece of wood and a small weight on top to keep the sheets flat.

If the first few sheets of foundation do not come out properly, they can be remelted, together with the overflow in the tray.

For cutting the foundation to the correct size, a template can be easily made from plywood, with a small block of wood nailed to one side for use as a handle. Place a sheet of foundation on a flat cutting surface, position the template, and cut around it, using a sharp knife. Surplus wax should be returned to the melting pot.

Using a foundation mill

The only equipment available in the US for making foundation by hand is a foundation mill, a rather expensive piece of equipment. The foundation will have to be cut to the appropriate size in both length and width. A foundation mill can have two sets of rollers. One is to reduce the thickness of the sheet and the second will emboss the honeycomb pattern.

To produce foundation, a wooden board, width approximately the depth of the finished sheet, is dipped several times into a vat of melted, clean

A foundation press in use.

beeswax. While the wax is still slightly warm and flexible the two sheets are peeled off the board. They are then ready to be passed through the rollers. While still somewhat soft, the sheets are trimmed to size with a knife. The sheets can then be stacked with thin paper, such as tissue paper, between each sheet.

Some beekeepers are now using a wooden frame with a special top bar to allow the bees to construct comb within a frame. Unless drawn on a strong nectar flow, the bees may not fill the frame with comb, thus making rather fragile unsupported comb. Sometimes the bees will draw comb across several frames which makes it impossible to remove individual frames.

Wiring foundation

Home-made foundation is generally thicker than commercially made foundation, but the bees seem to prefer it, possibly because it has not been refined so closely. As it cannot be used for cut-comb, it is usually wired. This is done with a wiring board, as described on page 286, or by wiring the frame and then embedding the wax foundation, using a transformer to electrically heat the wire and embed it into the wax.

Care must be taken not to overheat the wire or it will melt right through the wax. However, any spoiled sheets can simply be recycled.

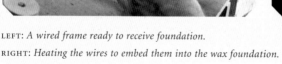

LEFT: *A wired frame ready to receive foundation.*
RIGHT: *Heating the wires to embed them into the wax foundation.*

Candles

A CHIEF USE FOR BEESWAX since before the Middle Ages has been the manufacture of candles, both for domestic and religious purposes. Pure beeswax candles are still highly valued. They burn with a warmer and brighter flame than the common paraffin wax candles sold today, and solid beeswax candles will burn for up to twice as long as those of the same size made from other waxes. They also smoke the least, as long as the correct size of wick is used for the candle diameter and the wax has been thoroughly cleaned: clean wax is important to prevent sputtering during burning. A wick for a beeswax candle must also be much thicker for the same candle diameter than a wick for a candle made from paraffin wax and stearin.

When the candle is lit, the flame melts the wax, which is then drawn up the wick by capillary action to the point where it contacts the air, vaporizes, and is burned. If the wick is too thin for the candle's diameter, not all of the melted wax can be absorbed by the wick and the surplus overflows and runs down the side of the candle. When the pool of liquid wax formed in the top of the candle builds up too much, it can extinguish the flame. If the wick is too thick, however, there is insufficient molten wax for it to draw on and the candle burns away rapidly with a very smoky flame, like a taper.

The four traditional methods for making candles are pouring, dipping, molding, and rolling. See page 390 for recipes for making both hand-dipped and rolled beeswax candles.

Candle pouring

This was the original way of making candles and is still the only way to make very long candles by hand. Using a ladle, wax (at just above its melting point) is gently poured down the wick into a reservoir. The wick is rotated during pouring to ensure an even coating. If the wax is too hot, it melts wax already on the wick. If it is too cool, it will not pour evenly. After the wax on the wick has cooled, the process is repeated until the candle is the required diameter. A knot is then tied in the bottom of the wick to keep the wax from sliding off. After every two or three pourings, the candle is rolled on a laminate surface to keep it straight. When finished, the "pont" or "icicle" of wax at the base is cut off with a warm knife. The candle is left to cool in a draft-free room.

ABOVE: *Candle pouring as depicted in a nineteenth-century woodcut.*

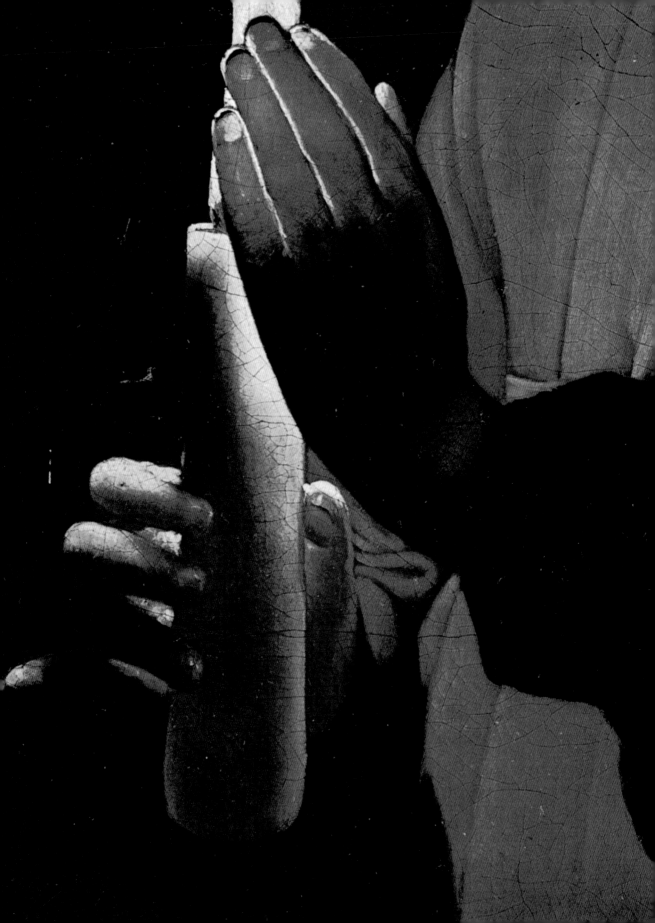

Candle dipping

Dipped candles are made in a similar way to poured ones: layers of wax are built up by dipping the wick into a pot of liquid wax that is taller than the final candle length, holding it there for a few seconds and then removing it smoothly. The pot stands in a heated water bath to keep the wax at the desired temperature: initially around 170°F (77°C). After cooling slightly, the wick is dipped again. Once the wax begins to build up on the wick, the temperature of the water bath can be lowered to around 155–160°F (68–71°C) to help build up more wax on each dip. Dipping is repeated until the candle is the required diameter and it is finished off in the same way as a poured candle. Hot wax on the final dip gives a smooth finish. Instructions for hand-dipping beeswax candles at home can be found on page 390.

Molded candles

Candles can also be cast in molds, and there are many attractive mold designs available from candle mold and wick suppliers. Glass, polycarbonate, or thin latex molds can be used, but these need a release agent and the results can be very variable.

Most people today use silicone rubber molds, which don't require a release agent and give more predictable results. Externally, these molds are cylindrical with a split at one side for easy removal of the contents. They usually come with a recommended wick size. The wick is cut slightly longer than the finished candle. It is threaded through the hole in the bottom of the mold (the candle is molded upside down) and secured at the other end between two thin sticks, which rest on top of the mold; the wick is pulled taut.

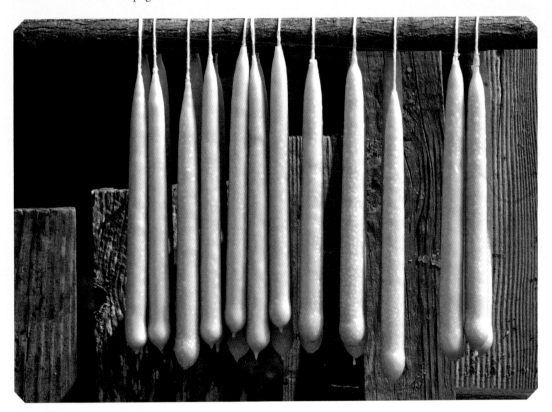

ABOVE: *Hand-made candles cooling on a rack.*

OPPOSITE: *Beeswax candles emit a warm, bright light.*

Rubber bands are used to hold the mold tightly together with the two sides of the split being matched carefully to minimise the "seam" on the finished product. The mold is filled slowly with wax at about 167°F (75°C), making sure that no air is trapped in any corner as this will spoil the candle's finished appearance.

Wax in molds takes longer to cool than the wax of a dipped candle. The wax also shrinks as it solidifies, so the mold may need to be filled up. If necessary, a hole can be made in the base of the candle for wax to be added. When the candle is cold, the rubber bands and the wick support are removed and the mold can be carefully opened.

Rolled candles

One of the easiest ways to make attractive candles of varying designs is with sheets of unwired brood or super foundation. These can be rolled in either direction, making a tall, thin candle or a shorter, fatter one. To do this, the wick is cut to size and placed just inside the edge of the foundation; the sheet is then gently rolled up by hand. Any slight bends can be straightened by rolling the candle backwards and forwards once or twice.

For a tapered top, the side of the sheet forming the top of the candle is trimmed to an angle (a straight line running from one corner, at one end of the sheet, to a little way down from the corner at the other end): the steeper the angle, the greater the taper.

Two different-colored sheets (purchased already dyed) can be cut and rolled together for attractive variations.

To make fatter candles, a second sheet is butted up to the edge of the first sheet, and rolling continues across both sheets. A thicker wick is required for the larger candle. Instructions for making a rolled beeswax candle can be found on page 390.

MELTING BEESWAX

For all candlemaking, clean beeswax must be melted in a double boiler or a stainless steel container in a hot water bath. The wax should not come into direct contact with the water or the heat. The temperature of the wax can be checked with a meat thermometer: different candlemaking methods require different temperatures for the best results. The melting point range for beeswax is 144–147°F (62–64°C).

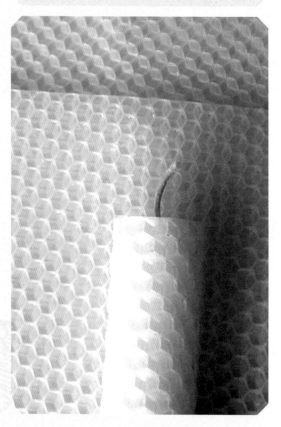

Rolled candles made from sheets of foundation will burn faster than solid beeswax candles.

Other Uses for Beeswax

PHARMACEUTICALS AND COSMETICS

Beeswax has long been used as an inert base for cosmetics and a base for medicinal salves and pharmaceutical products. On the Indian sub-continent, where Ayurvedic medicine has been practiced for centuries, beeswax is often used as a "carrier" for herbs, minerals, and other materials considered to have medicinal value; the beeswax ensures an even distribution over the skin and helps to keep the skin from drying out.

Beeswax has also been used for coating pills and in pharmaceutical production. The shelf-life of penicillin is extended when it is mixed with beeswax and peanut oil, and this mixture has a less violent and more sustained effect, enabling doctors to give larger doses at less frequent intervals.

In addition, hand, face, and foot creams and lip salves can be made with beeswax—see recipes for some of these on pages 369–77.

AS A SEALANT AND LUBRICANT

In the Middle Ages, sealing wax for letters was typically made from beeswax combined with Venice turpentine, a resinous extract from the European larch tree (*Larix decidua*). (Originally uncolored, sealing wax was later stained red with vermilion.) Beeswax was replaced in the sixteenth century by shellac, turpentine, resin, chalk, and coloring, which made a harder seal. A beeswax recipe for sealing wax can be found on page 393.

Before the advent of airtight lids, a jar's contents would be preserved by sealing it with a thin layer of beeswax. A beeswax coating over the edge of a lid served the same purpose.

A thin layer of beeswax will lubricate drawers and zippers that are sticking, and it is also applied to the underside of skis to reduce friction. Lace makers use it to lubricate their pins, and seamstresses for waxing sewing thread to make it waterproof, more supple, and easier to handle. Stabbing a nail into a cake of beeswax before use makes it easier to drive the nail into the wood, which is also then less liable to split.

A wax seal, stained red.

IN POLISHES

Today, polishes are made with synthetic paraffin waxes, but originally beeswax was the only suitable material available and it is still preferred for use on unsealed or unvarnished wooden furniture.

Beeswax polish comes in two forms: hard and paste. Both are made by combining beeswax and turpentine, but the paste version also contains water and soap or borax. In each case, the other ingredients act as a carrier for beeswax, ensuring that it is spread evenly and thinly over the surface. The only difference is that the hard polish requires more elbow grease to get a good shine! A recipe for hard beeswax furniture polish can be found on page 391.

WAX MODELS

In Egypt, beeswax models were used in rituals, as the wax models could be easily broken or burned in fire. The most famous beeswax models are probably those of Madame Tussaud (1761–1850)—see page 48. Her first wax model, made in 1777, was of the French Enlightenment writer and philosopher Voltaire (1694–1778).

IN MODELING AND CASTING

Beeswax is an ideal medium for modeling and carving, allowing very precise, intricate details to be inscribed. It can be used alone or combined with other waxes when the addition of beeswax can help to increase the softness and workability of the material.

In liquid form, beeswax will take even the minutest impressions of a mold and, once set, it is not influenced by ordinary climatic conditions. It can also be colored easily. As such, beeswax has been used in the lost-wax technique of casting for centuries (see page 46). In this process the very fine detail carved in the beeswax is transferred to the finished model. Models, including beeswax flowers, can be made from shapes cut from beeswax sheets, both plain and colored, and welded together using a modified soldering iron.

IN BATIK

Batik is thought to originate from Asia, particularly Java, although some think it began in Egypt. The word "batik" originates from the Javanese *titik* and means "dot." Wax is used to cover areas of cloth not intended to be dyed, in order to prevent dye from

Using wax to create a batik design.

reaching them. Originally, beeswax was probably used on its own, but now it is mixed with about 30 percent dammar resin.

Batik works best on natural fabrics, which take dye well. The cloth is dyed with successively darker colors, with the wax mixture being applied between dyeings to cover increasing areas of the fabric in the desired pattern. The wax must be hot enough to penetrate the cloth, producing an identical pattern on both sides.

The first application goes over areas where the original fabric color is to be retained. The cloth is then dipped in the first dye color. When the cloth is dry, more wax is added over areas that are to retain the color of the first dye (in other words, they are not to take the color of the second dye). The cloth is then dyed with the second color, and so on until the design is completed. The dyes are chosen carefully as the colors influence each other. For example, if the first dye is yellow and the second blue, unwaxed parts will turn green. On completion, the wax is removed, usually with boiling water.

IN ENCAUSTIC PAINTING

Encaustic painting uses hot beeswax into which colored pigments are infused. In ancient Greece, Rome, and Egypt, colored waxes combined with dammar resin were kept liquid on a metal sheet placed over a pan of hot coals, and then applied to the surface using a brush or palette knife-like tool. The finished work was "burned in" by subjecting it to low heat (the sun's rays in the case of Egyptian painters), melting the entire surface. This did not disturb the image, but did create a smooth finish that could be burnished. Wax portraits on sarcophagi lids were fashioned this way.

Today, colored encaustic waxes are available and an electric iron similar to a travel iron is used for painting. The wax is melted onto the sole plate of the iron and then transferred to special card with a high-quality, non-absorbent surface. Four basic

strokes produce different effects and the finished picture is polished with a soft tissue pad. Heated pens, brushes, and other tools can be employed to produce delicate effects. Heat lamps are used to "burn" the surface of the painting.

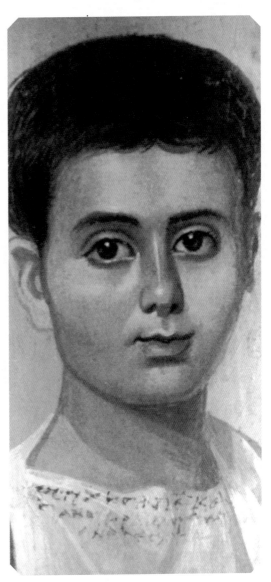

Encaustic painting was used in second-century Egypt to create funerary portraits on wooden panels, which were then attached to mummies.

AND FINALLY...

Beeswax has an extraordinary range of other uses, including as a coating for cheese and a glazing agent for chewing gum and sweets such as gummy bears.

Rastafarian dreadlocks can be styled with various substances, including beeswax, which can also be found in moustache wax. Pure beeswax blocks are particularly good for achieving the required shine on army boots and beeswax has been used for dubbin since medieval times to soften, condition, and waterproof leather.

Gardeners also use beeswax in grafting wax, printers use it in their ink, and archers use it on their bow strings. Detailed watermarks are engraved on beeswax before the pattern is transferred to the metal plates used for papermaking.

In the musical world, a beeswax mouthpiece for a didgeridoo softens during use and forms a tighter seal, and accordion makers blend beeswax with

CHEESE COATINGS

Beeswax is still used as a natural coating for some cheeses, to protect the food as it ages. Many traditional cheesemakers prefer it to modern plastic coatings as it avoids the possibility of contamination of the cheese with unpleasant flavors from the plastic.

pine rosin to make an adhesive to fix reed plates within the instrument.

In Eastern Europe colored beeswax is used to decorate eggs, particularly for Easter.

Dutch Gouda cheeses sealed within wax casings.

Propolis: Nature's Miracle Cure

PROPOLIS IS A STICKY RESINOUS SUBSTANCE that worker bees collect from conifers and other types of tree. They use it as a thin coating on the inside walls of the hive and to strengthen their combs. Larger amounts are used to block cracks and reduce the size of the hive entrance by constructing propolis "curtains."

If a mouse or another small animal has gained access to the hive and has died or been killed, the bees are unable to remove the body, so they will embalm it with propolis to inhibit microbial and mold growth. Propolis is a complex mixture of many components; the elements and consistency vary according to the source. Generally it is brown, but its color can range from red to yellow depending on the particular sources of resin in the locality of the hive.

ANTIBACTERIAL AND ANTIFUNGAL PROPERTIES

The antibacterial and antifungal properties of propolis have long been known. Egyptian priests used it many thousands of years ago, the Greek

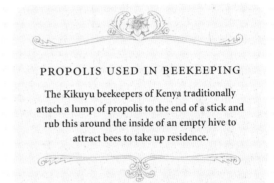

PROPOLIS USED IN BEEKEEPING

The Kikuyu beekeepers of Kenya traditionally attach a lump of propolis to the end of a stick and rub this around the inside of an empty hive to attract bees to take up residence.

Propolis (the reddish-brown deposit) on the side of a frame.

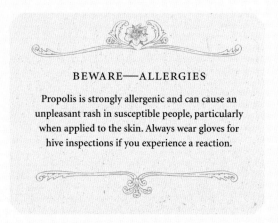

Propolis recovered from the hive.

physician Hippocrates (c. 460–370 BC) prescribed
it for the treatment of sores and ulcers, both internally and externally, and Celsus, a Roman medical
writer, described its use in poultices in the first
century AD. In the sixth to eighth centuries, propolis was described in Koranic manuscripts as a
blood purifier and treatment for skin problems
and bronchial catarrh. In the twelfth to fifteenth
centuries, it was advocated for the treatment of
toothache, and propolis was placed on the navel of
a newborn baby, presumably for its antiseptic
properties.

The South American Incas, who controlled
regions that extend through modern-day Peru,
Chile, and Ecuador during the twelfth to fifteenth
centuries, are believed to have used propolis to
treat inflammations, and hunter-gatherer groups
are also reported to have used it as an adhesive.

Today, propolis is used mainly in North America
and Europe, particularly eastern Europe, as a natural supplement or in herbal medicines. It is combined with pollen, royal jelly, or other products in
tablet form and it is also supplied as a tincture, dissolved in around 70 percent alcohol, for the treatment of cuts and skin rashes. Some of its other
varied uses include as an additive in cosmetic
products, such as skin lotions, beauty creams, lip-
sticks, toothpastes, soaps, and shampoos, and in
products such as mouthwashes and chewing gum.

Since 1992, James Fearnley, a British authority
on propolis, has been undertaking research into its
pharmacological and clinical properties, commis-
sioning a number of studies at the University of
Oxford and Manchester University Dental School.
In 2002 he established BeeVital Ltd to offer propolis and pollen health products to consumers and
the company was awarded a government research
and development award to investigate the antibiotic, anti-inflammatory, antifungal, and antiviral
properties of propolis. The chemical composition
of samples from around the world was analyzed to
identify any particular advantages of propolis from
specific regions and the project included clinical
trials for the treatment of mouth, gastric, and
stomach ulcers, dermatological problems, wound
healing, and immune deficiency diseases.

Pollen: Nature's Complete Food

POLLEN IS PRODUCED in the anthers of flowering plants (the terminal lobes of the stamens—see page 120). This fine powdery substance is made up of microscopic grains, each containing a male gamete capable of fertilizing the female ovule or seed. Pollen is transported to the female ovule by bees visiting flowers of the same species, and also by wind, other insects, and animals.

Together with honey, pollen forms a large part of the diet of developing bee larvae and is also eaten by adult bees. Its composition varies not only between flower species but from season to season, making a precise analysis difficult. The major constituent is protein (nearly 24 percent), with hydrocarbons making up 27 percent and fats around 5 percent. However, most of the hydrocarbons are fructose and glucose, which are added by the bee to help bind the pollen load together to bring it back to the hive. Pollen is also known to contain substantial quantities of potassium, calcium, and magnesium, together with high levels of zinc, iron, manganese, and copper.

POLLEN AS A FOOD SUPPLEMENT

Humans primarily use pollen as a nutritious food. Originally it was consumed in honey but the filtering process for commercially produced honey now removes most of the pollen grains, hence the demand for pollen as a food supplement, usually as tablets, granules, or candy bars. There is no scientific evidence that it actually increases stamina but, given its composition, it seems likely that its use, in small quantities, could be beneficial.

Bees in a colony usually collect pollen from a number of different plant species flowering at the same time, giving them the range of proteins and vitamins they need. The resulting multi-colored pollen store reflects the different colors of the pollens from different flowers. For human consumption, too, it is generally considered that pollen of different colors is better than pollen of a uniform color.

Pollen is essential to bees as a source of protein; for humans it is used as a food supplement.

BEWARE—STOMACH UPSETS

There have been reports of people suffering adverse reactions (generally stomach or gastrointestinal upsets) after eating pollen. It is therefore best to begin by eating only small amounts.

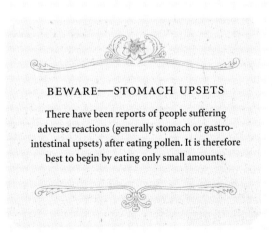

MEDICAL BENEFITS OF POLLEN

The medical use of pollen was first recorded in the early Middle Ages by Arabs and Jews in Islamic Spain, although this may not have been pollen collected by bees. Its use as an astringent was recommended by Maimonides (1135–1204), a Jewish physician to the Sultan of Egypt. Ibn el-Beithar, a Spanish apothecary, botanist, and essayist who became Saladin's doctor, described its use as an aphrodisiac in the early 1200s. He also claimed that it was good for the stomach, bowels, and heart.

In more recent times, Abraham Lincoln, as a child in Indiana, is recorded as asking for bread and honey when he visited a neighbor's house, and being especially pleased if it contained a good deal of pollen.

Pollen has been demonstrated to be beneficial in the treatment of chronic prostatitis (inflammation of the prostate gland), reducing the inflammation, discomfort, and pathology of patients, possibly because of the high levels of zinc. It has also been shown to protect against the harmful effects of radiation during X-ray.

Eating local honey regularly, especially the wax cappings from honeycomb or unprocessed honey such as cut-comb, is believed to alleviate the symptoms of hay fever, from which up to 25 percent of people suffer. There is little scientific evidence for this, although some research does seem to suggest there may be a beneficial effect.

ABOVE: *A worker bee collects pollen.*

OPPOSITE: *A 500x magnification of pollen grains from a variety of common plants.*

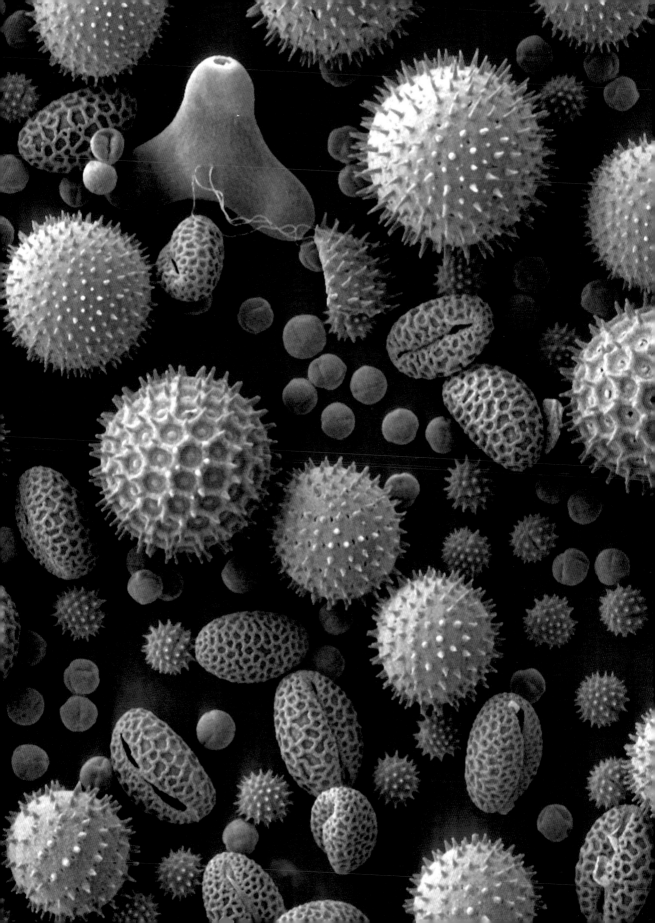

Bee Venom: Natural Pain Killer

BEE VENOM IS PRODUCED by two glands associated with the bee's sting mechanism and used during defense of the colony (see pages 128–30). It is 88 percent water, with glucose, fructose, and phospholipids (fats) as the other main components. Venom also contains pharmacologically active elements, enzymes, peptides, and amines.

BEE VENOM THERAPY

Bee venom has been used in the treatment of arthritis for many years. The medical writings of Hippocrates (c. 460–359 BC), Pliny the Elder in *Historia Naturalis* (c. AD 77), and Galen (c. AD 148) refer to the use of bee stings for this. In modern-day bee venom therapy, the venom should only be administered by trained therapists either using a direct bee sting (the insects are placed next to the skin) or an injection of venom extract. Beekeepers should refuse to give this type of therapy as they could be subject to litigation if things go wrong.

Bee stings are also thought to help ease the symptoms of multiple sclerosis, tendonitis (inflammation of the tendons), and fibromyalgia (muscle and joint tenderness). The peptide melittin in bee venom stimulates the production of cortisol, an anti-inflammatory. It also causes the pain when someone is stung. Several clinical trials indicate possible benefits of bee venom therapy, but more controlled research is needed.

BEWARE—ALLERGIC REACTION

Patients should be tested for allergic reactions before starting any form of bee venom therapy. The allergic reaction caused by a bee sting can range from mild irritation and swelling to an anaphylactic shock, which, without urgent treatment, could result in death.

A Chinese man receives treatment with bee venom for rhinitis (an inflammation of the nasal membranes).

Royal Jelly: A Rare Blend

ROYAL JELLY IS SECRETED by the hypopharyngeal and mandibular glands in nurse bees between the ages of 5 and 15 days. It is fed to all larvae for the first three days after hatching. Those destined to develop into queen bees are fed royal jelly until they pupate, whereas the diet of the workers changes at about three days to ordinary brood food (see pages 102–3).

Royal jelly was first identified in the 1770s when scientists showed that it was different from brood food. The first chemical analysis appears to have been commissioned by Rev. L. L. Langstroth in 1853, but a detailed analysis was not possible until techniques had been improved in the 1940s. It has been shown to contain 67 percent water, 12.5 percent proteins, 11 percent sugars, 5 percent fatty acids, and 1 percent ash, leaving 3.5 percent undetermined, with traces of vitamins and enzymes.

China is the world's largest producer and consumer of royal jelly; it has for a long time played a central role in traditional Chinese medicine. It is used today in the treatment of a wide range of ailments, including anxiety, arterio-sclerosis, asthma, depression, fatigue, hair loss, impotence, insomnia, stomach ulcers, high and low blood pressure, and several skin conditions.

Royal jelly is widely used as an ingredient in moisturizers and skin creams, and as a supplement in health foods. In Asia, it is also found in beverages.

In spite of claims for the therapeutic benefits of royal jelly, these have not been demonstrated with scientifically controlled medical studies. In 1940, it was shown that royal jelly has no sex-enhancing effects—the levels of testosterone it contains are extremely small, about the same as those normally circulating in male blood plasma. The effect on humans is therefore likely to be negligible. Any benefits are most likely to be those due to royal jelly's antibacterial properties—it is reported to be beneficial in wound healing and skin cleansing—but these need further investigation.

Royal jelly for human use is collected from queen cells when the queen larvae are about four days old.

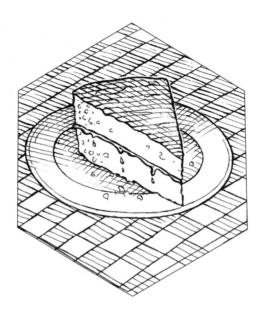

PART FIVE

RECIPES AND HOME CRAFTS

FOOD AND DRINK:
Soups and Appetizers
Main Dishes
Desserts, Cookies, and Sweets
Cakes and Breads
Dressings and Chutneys
Drinks

BEAUTY RECIPES
HEALTH RECIPES
DOMESTIC CRAFTS

Food and Drink

Honey is an endlessly versatile ingredient in both sweet and savory dishes, and is used extensively around the world as a naturally healthy and nutritious food. The most authentic dishes use honey produced in the country of origin of the recipe and are made with the best type of honey.

DIFFERENT TYPES OF HONEY

There are a bewildering variety of honeys on offer, so when cooking with honey, it is useful to understand the properties of the different types.

Specialist and monofloral, or single-flower, honeys (made by bees feeding on just one species of flower or plant) tend to be lighter in color and more delicately flavored. Honey such as acacia, orange blossom, and Scottish heather are well-known examples, but there are many others. It is best to use these raw (uncooked), or toward the end of the cooking time, as their particular flavor will almost certainly be lost if overheated.

Some monofloral honeys are darker in color, such as chestnut and eucalyptus, and more intensely flavored. They keep their flavor better than lighter honeys when heated, so are good for marinating meats and for use in dark fruit cakes.

Multifloral, or polyfloral, honey is made from a mixture of nectars, a result of bees foraging on several different species of plants. Popular types include pure wildflower honey and alpine floral honey, which also has a delicate floral fragrance.

For most recipes a standard blended honey (made by combining different types of honey to produce a particular consistency and taste) will suffice, adding sweetness and flavor without overwhelming other ingredients. It is the clear, liquid honey often used in recipes; creamed honey is best for spreading on toast or scones rather than for cooking.

Comb honey hasn't been processed or heat-treated so is one of the more nutritious ways of eating honey. It is best to eat it raw.

COOKING WITH HONEY

As honey can burn at high temperatures, which affects the flavor, it is best to cook dishes containing honey on a low or medium heat where possible. When using honey to replace refined sugar, remember that honey is sweeter than superfine or granulated sugar, so as a general rule, allow one part of honey to replace one and a quarter parts of sugar. Also remember that honey will keep cakes and cookies moist, so they may need slightly longer cooking times—but this also means they will tend to last longer.

MEASURING AND STORING HONEY

The easiest way to measure honey is to dip a measuring spoon into boiling water first, which helps the honey slide easily from the spoon rather than stick to it. When weighing larger amounts of honey, place a measuring cup on the scales and weigh it alone before adding the honey. Honey will last many years if kept in a cool, dark place, in an airtight jar.

CURRIED HONEY SWEET POTATO SOUP

The natural sweetness of sweet potatoes is complemented by roasting them with honey. Use a robust dark honey for roasting the potatoes and a light single-flower honey to drizzle over just before serving. The sweet potatoes can also be served as an accompaniment to roast meat or grilled fish, or added to winter salads.

SERVES: 4
VEGETARIAN

2$^{1}/_{4}$ LB SWEET POTATOES, PEELED AND CUT
   INTO 1-IN CHUNKS
4 TBSP SUNFLOWER OIL
1 TBSP TANDOORI CURRY POWDER
2 TBSP DARK HONEY, SUCH AS CHESTNUT,
   EUCALYPTUS, OR HEATHER HONEY
1 CLOVE GARLIC, PEELED AND CRUSHED
1-IN PIECE GINGER, PEELED AND FINELY
   CHOPPED
1 RED CHILE, SEEDED AND FINELY CHOPPED
1 RED ONION, PEELED AND FINELY CHOPPED
4 CUPS VEGETABLE STOCK
$^{1}/_{4}$ CUP COCONUT MILK
SEA SALT AND BLACK PEPPER, FRESHLY
   GROUND

TO SERVE:
4 TBSP THICK YOGURT
1 TSP CHILI POWDER
BLACK PEPPER, FRESHLY GROUND
2 TSP LIGHT HONEY, SUCH AS ACACIA OR
   ORANGE BLOSSOM

1 Preheat the oven to 375°F and place a heavy baking sheet in the middle of the oven to heat.

2 In a large bowl, toss the sweet potatoes with 2 tablespoons of the oil, the curry powder and dark honey, and stir to coat the sweet potato pieces.

3 Stir in the garlic, ginger and chile.

4 Spread the sweet potatoes and the flavored oil and spices on the hot baking sheet and roast for 15 minutes.

5 Turn the sweet potatoes once, cook for 10 to 15 minutes, until they are soft, then cool.

6 Meanwhile, heat the remaining 2 tablespoons oil in a nonstick skillet and add the onion; cook over low heat for 20 minutes, or until the onion is soft and translucent.

7 Transfer the cooked onion and the sweet potatoes, along with the sticky contents of the baking sheet, into a food processor and add half the stock. Process until well combined.

8 Add the remaining stock and the coconut milk, pulse again until smooth, and return the soup to a clean saucepan. Season to taste with salt and pepper and gently reheat the soup.

9 Serve in warmed bowls and top each bowl with 1 tablespoon yogurt, a pinch of chile powder, a grind of black pepper, and a drizzle of light honey.

ENDIVE, PANCETTA, AND HONEY-BROILED FIG SALAD

Figs and honey are a classic combination and one much favored in Greek cooking. This recipe uses two different honeys: orange blossom to give the dressing a floral fragrance, and a medium blended to drizzle over the figs before grilling. For a vegetarian version, substitute feta cheese for the pancetta.

SERVES: 4

JUICE OF ¹/₂ ORANGE

2 TSP ORANGE BLOSSOM HONEY

3 TBSP EXTRA-VIRGIN OLIVE OIL

SEA SALT AND BLACK PEPPER, FRESHLY
    GROUND

9 OZ BELGIAN ENDIVE

1 SMALL HEAD RADICCHIO, SHREDDED

1 TSP PEANUT OIL

16 SLICES PANCETTA, CUT INTO LARDONS

8 RIPE BLACK FIGS, HALVED

4 TBSP MEDIUM-BLENDED HONEY (AMBER
    COLORED)

1 In a small bowl, whisk together the orange juice, orange blossom honey, and olive oil and season with salt and pepper. Set aside.

2 Pull off the outer leaves of the endive, quarter the cores, and place them all in a large bowl along with the radicchio.

3 Heat the peanut oil in a nonstick skillet over medium heat. When hot, add the pancetta and fry for 4 to 5 minutes, until crisp and brown. Drain on paper towels and set aside.

4 Preheat the broiler to high. Place the figs, cut side up, on the broiler pan and drizzle with 2 tablespoons of the medium-blended honey. Broil the figs for 1 to 2 minutes, until they begin to caramelize.

5 Rewhisk the dressing and pour it over the salad leaves; toss gently and divide among 4 serving plates.

6 Scatter the pancetta over the salad and place 4 fig halves on each plate. Drizzle the remaining honey over the figs and serve.

HONEY-BROILED SCALLOPS

Scallops are best cooked simply and are particularly delicious drizzled with rosemary or clover honey and broiled. Honey can lose some of its beneficial qualities when cooked, so consider spooning a little extra raw honey over the scallops before serving. For an even quicker version, leave out the leek and top the cooked scallops with thinly sliced scallions.

SERVES: 4

1 TBSP UNSALTED BUTTER

1 TBSP OLIVE OIL

1 LEEK, TRIMMED AND CUT INTO THIN
   STRIPS

SEA SALT AND BLACK PEPPER,
   FRESHLY GROUND

8 SCALLOPS, CORALS REMOVED

1 TBSP PEANUT OIL

1/4 CUP HONEY, SUCH AS ROSEMARY

TO SERVE:

1 LEMON, CUT INTO QUARTERS

2 TBSP FRESH CHERVIL LEAVES

1 Melt the butter with the oil in a nonstick skillet over medium heat and gently sauté the leek for 5 minutes, or until soft. Season with salt and pepper and set aside.

2 Slice each scallop in half horizontally.

3 Prepare 4 cleaned scallop shells or gratin dishes by brushing the inside with the peanut oil. If using shells, wrap a strip of foil around the edges to protect them, and remove before serving.

4 Place the shells/dishes in the broiler pan, divide the leeks among them, then top the leeks with 4 scallop halves on each shell and drizzle with the honey.

5 Preheat the broiler to high. Broil for 2 to 3 minutes until the scallops are just beginning to caramelize.

6 Serve in the shells or dishes with a lemon wedge, and the chervil scattered over.

SPICED HONEY HUMMUS

Honey is an unusual addition to hummus, but it gives this rich chickpea dip an extra dimension and enhances the spices without making it noticeably sweet. Honey is used extensively in Middle Eastern and North African cooking and it has a natural affinity with spices such as cumin, coriander, and chiles, so try hummus in warmed pita bread with falafel and salad, or use as a dip for lamb kebabs. To make the hummus more substantial, add a few whole chickpeas to the finished dip and drizzle with some good olive oil.

SERVES: 4
VEGETARIAN

1 CUP DRIED CHICKPEAS
1/2 CUP TAHINI
1/2 TSP GROUND CUMIN
1/2 TSP GROUND CORIANDER
1/2 TSP GROUND CHILE
3 TBSP LEMON JUICE
1 TBSP EXTRA-VIRGIN OLIVE OIL
1 CLOVE GARLIC, PEELED AND CRUSHED
1 TBSP HONEY
SEA SALT AND BLACK PEPPER,
    FRESHLY GROUND

TO SERVE:
1 TSP HONEY
WARM PITA BREAD OR CRUDITÉS

1 Place the chickpeas in a colander and rinse thoroughly with cold water, then drain and transfer to a large bowl. Cover them with cold water and soak for at least 12 hours.

2 Drain the chickpeas, rinse again with cold water, and transfer to a large saucepan. Cover with plenty of cold water and place over medium heat; bring slowly up to a boil, then simmer for 2 hours, skimming off any foam that accumulates.

3 When the chickpeas are tender, drain and reserve the cooking water. Rinse the chickpeas briefly under cold water and drain well, then transfer them to a food processor and pulse until coarsely chopped.

4 Add the tahini, spices, lemon juice, oil, garlic, and honey and process until smooth—add a little of the cooking water if you want a smoother hummus.

5 Season with salt and pepper and spoon into a serving bowl. Drizzle with a little more honey and serve with warm pita bread or crudités.

HONEY ROAST HAM

Honey roast ham has become a Christmas and Thanksgiving staple, along with turkey, as it is delicious served as part of a cold buffet or in sandwiches, with pickles and relish. It is best to buy a boned, rolled, and tied ham with a good layer of fat. The ham is cooked at relatively high heat, so use a dark honey such as chestnut; keep an eye on the meat as it cooks and baste regularly.

SERVES: 6 TO 8

1 (3¹/₂- TO 4-LB) BONED AND TIED
   UNSMOKED HAM
¹/₃ CUP APPLE CIDER OR APPLE JUICE
¹/₄ CUP DARK HONEY, SUCH AS CHESTNUT
2 TBSP DEMERARA SUGAR
¹/₂ TSP DRY MUSTARD POWDER
¹/₂ TSP GROUND CINNAMON

TO SERVE:
FRESH BAY LEAVES

1 Preheat the oven to 350°F.

2 Place a large piece of foil (that will wrap around the ham) in a roasting pan big enough to fit the ham comfortably. Wipe the ham dry with paper towel and place it in the roasting pan with the rind side up.

3 Pour the cider around the ham and loosely wrap the ham in the foil, making sure the edges are sealed so the cider does not leak out. Roast in the center of the oven for 1 hour 15 minutes, and then remove. Increase the oven temperature to 400°F.

4 Open the foil and lift the ham onto a board; empty the juices from the pan.

5 With a sharp knife, carefully cut away the rind from the meat, leaving the fat intact. Score the fat into a diamond pattern and return the ham to the roasting pan; pull the foil up to protect the sides of the ham but leave the fat uncovered.

6 In a small bowl, mix together the honey, sugar, mustard powder, and cinnamon and, using a palette knife, spread the mixture evenly over the fat.

7 Return the ham to the oven for 15 to 20 minutes, baste, then roast for 10 to 15 minutes longer, until golden brown.

8 Let cool to room temperature in the pan, then remove the ham and carve it. Garnish with fresh bay leaves and serve.

CITRUS HONEY AND BALSAMIC PORK STIR-FRY

Asian cooking often uses honey as a substitute for refined sugar as it not only adds sweetness but also depth of flavor. Using honey as a marinade gives this dish a wonderful sticky quality; you could also try the marinade with chicken or shrimp. This low-fat and tasty supper dish could be served with sticky rice in place of noodles.

SERVES: 4

1 LB PORK LOIN, TRIMMED OF ALL FAT AND
    CUT INTO STRIPS
FINELY GRATED ZEST AND JUICE OF 1 LIME
FINELY GRATED ZEST AND JUICE OF
   1/2 LEMON
3 TBSP HONEY
1 TBSP BALSAMIC VINEGAR
2 TBSP PEANUT OIL
9 OZ FINE EGG NOODLES
5 OZ SNOW PEAS
4 SCALLIONS, SLICED
1 LB BOK CHOY, THICKLY SLICED
4 OZ BEAN SPROUTS

TO SERVE:
2 TBSP CHOPPED FRESH CILANTRO LEAVES

1 Place the pork in a large nonreactive bowl. In a small bowl whisk together the lime and lemon juice and zest, the honey, vinegar, and 1 tablespoon of the oil.

2 Pour the mixture over the pork, mix well to coat, cover and chill for 15 minutes.

3 Bring a large pot of water to a boil and add the egg noodles, cook for 3 minutes, then drain and keep warm.

4 Heat a wok over high heat until the oil that is ingrained in the wok's metal surface begins to smoke, then add the remaining 1 tablespoon oil, the pork and the marinade; stir-fry over high heat for 2 to 3 minutes.

5 Add the snow peas and half of the scallions and stir-fry for 2 to 3 minutes, until the pork is cooked through.

6 Add the bok choy and bean sprouts and toss together. Cook for 1 minute.

7 Divide the noodles among 4 serving bowls and spoon the pork and vegetables over. Top with the remaining scallions and the cilantro and serve.

STICKY HONEY RIBS

Wonderfully aromatic, sweet, hot, and sticky, these ribs are best made with a pungent dark honey. They are perfect for a barbecue, as they can be left to marinate in the sauce for up to 24 hours. Roast in the oven as directed in the recipe, then transfer the ribs to a hot grill for the last 20 minutes of cooking, turning regularly until dark brown and sticky. Serve with baked potatoes and coleslaw.

SERVES: 4 TO 6

3¹/₂ TO 4 LBS PORK SPARE RIBS
¹/₂ TSP SEA SALT
5 TBSP DARK HONEY
1 TSP DRY MUSTARD POWDER
1 FRESH RED CHILE, SEEDED AND FINELY
   CHOPPED
1 TBSP WHITE WINE VINEGAR
¹/₄ CUP DARK MUSCOVADO SUGAR
1 TBSP TOMATO PASTE
¹/₂ TSP GROUND GINGER

1 Preheat the oven to 375°F. Line a large roasting pan with a double layer of foil and place the ribs in the pan.

2 In a small bowl, whisk together the remaining ingredients and pour the mixture over the meat; using your hands, rub the marinade into the ribs to make sure they are thoroughly covered.

3 Cover the pan with foil and bake in the center of the oven for 30 minutes, then remove the top foil and turn the ribs over. Re-cover the pan and return to the oven. Cook for 30 minutes longer, basting occasionally.

4 Increase the oven temperature to 425°F and roast uncovered for a final 5 to 10 minutes, until the ribs are dark brown and sticky; keep an eye on them so they don't burn. Serve hot.

HONEY AND ORANGE ROAST DUCK

The honey and orange used to coat the duck in this recipe gives a dark, lacquered glaze similar to that of Chinese roast duck. As honey can burn at high temperatures, this duck is roasted at a relatively low temperature with regular basting. Serve with a simple green salad and pan-fried potatoes, or a salad of orange segments and watercress.

SERVES: 4

1 (4^1/$_2$ LB) DUCK
1 TBSP SEA SALT
3 TBSP HONEY
^{1}/$_2$ TSP GROUND ALLSPICE
FINELY GRATED ZEST AND 1 TBSP JUICE
   FROM 1 ORANGE

1 Preheat the oven to 325°F.

2 Using a fork, prick the skin of the duck all over and wipe the duck dry with a paper towel. Rub the skin with the salt and then place the duck on a rack in a roasting pan.

3 Roast for 1 hour, then drain off the fat collected in the roasting pan.

4 In a small bowl, whisk together 2 tablespoons of the honey, the allspice, and orange juice and zest and spread it over the duck. Roast for 1 hour, basting every 15 minutes.

5 Spoon the remaining honey over the duck and roast for a final 15 minutes. Remove the duck from the oven and let it rest for 5 to 10 minutes before carving. Serve hot.

LEMON AND HONEY–GLAZED CHICKEN BREASTS

Lemon and honey make a delicious glaze for poultry, and these chicken breasts are equally good hot or cold. Serve cold, sliced, with lemon mayonnaise and a simple bean salad with green olives, wrapped in warmed flatbread. If you are serving it hot, add a couple of slices of fresh or preserved lemon before cooking and then serve with baked rice and green beans, or little roast potatoes and a crisp salad.

SERVES: 4

FINELY GRATED ZEST AND JUICE OF
   2 LEMONS

1 TBSP HONEY

1 CLOVE GARLIC, PEELED AND CRUSHED

1 TBSP CHOPPED FRESH TARRAGON

1 TBSP BROWN SUGAR

4 (5-OZ) SKIN-ON BONELESS CHICKEN
   BREASTS

2/3 CUP DRY SHERRY OR WHITE WINE

SEA SALT AND BLACK PEPPER, FRESHLY
   GROUND

1 TBSP OLIVE OIL

TO SERVE:

1 TBSP CHOPPED FRESH TARRAGON

1 Preheat the oven to 375°F.

2 In a small bowl, whisk together the juice and zest of the lemons with the honey and garlic, then stir in the brown sugar and 1 tablespoon tarragon.

3 Put the chicken breasts in a large bowl. Pour the lemon mixture over the chicken and make sure they are well coated. Cover and chill for 30 minutes.

4 Add the sherry to the chicken and season with salt and pepper; turn the chicken in the marinade.

5 Brush a roasting pan with the oil and place the chicken, skin side up, in the pan. Pour over any remaining marinade, then roast for 35 to 40 minutes, until cooked through. Scatter the rest of the tarragon over the chicken and serve.

MOROCCAN HONEY-CHICKEN TAGINE WITH PRUNES

Moroccan cuisine uses honey extensively in both sweet and savory dishes, and honey and prunes are a classic combination. Tagines (covered clay cooking pots) are used all over North Africa to cook meat and vegetarian dishes, but a traditional casserole dish works just as well. This tagine can be made ahead of time and reheated over low heat for 20 to 30 minutes.

SERVES: 4

1/2 CUP FLAT-LEAF PARSLEY, CHOPPED

2 CINNAMON STICKS, HALVED

1 MEDIUM ONION, CHOPPED

1 PRESERVED LEMON, CHOPPED

2 TBSP HONEY

1/3 CUP OLIVE OIL

SEA SALT AND BLACK PEPPER, FRESHLY GROUND

4 (10-OZ) SKIN-ON CHICKEN PIECES

2 CUPS PITTED PRUNES

1 CUP CHICKEN STOCK

1 1/2 CUPS COUSCOUS

TO SERVE:

2 TBSP PINE NUTS, TOASTED

1 Place half of the parsley, the cinnamon sticks, onion, preserved lemon, honey, and half of the oil in a large bowl and season with salt and pepper. Mix together. Add the chicken and turn to coat well.

2 Cover and chill for 1 hour to allow the meat to marinate.

3 Heat 2 tbsp of the remaining oil in a nonstick skillet over medium heat. Remove the chicken from the marinade, brushing off any onion and lemon; reserve the marinade.

4 Brown the chicken in the pan, 2 pieces at a time, turning to brown all sides. Remove to paper towels to drain; cover and keep warm.

5 Discard the cinnamon sticks and add the remaining marinade to the pan and turn the heat to low. Cook gently for 5 to 8 minutes, until the onion is soft and translucent.

6 Place the chicken and the onion mixture in a large casserole dish or tagine, add the prunes and the stock, and season with salt and pepper. Cover and simmer gently over medium heat for 1 hour, until the chicken is tender.

7 Meanwhile, place the couscous in a large heatproof bowl with the remaining oil and a pinch of salt. Add 1 cup boiling water, cover the bowl with a clean towel, and let soak for 8 to 10 minutes, then fluff up the grains with a fork.

8 Serve the chicken with the couscous, and scatter the remaining parsley and pine nuts over.

Pan-Fried Scallops with Honey Lime Marinade

In this delicious shellfish dish inspired by the flavors of Thai cooking, honey is used to replace the traditional palm sugar. Choose sea scallops, which are firm and white and have a sweet aroma.

SERVES: 4

3 TBSP HONEY, SUCH AS CHESTNUT
FINELY GRATED ZEST AND JUICE OF 2 LIMES
3 TSP PEANUT OIL
1 RED CHILE, SEEDED AND FINELY CHOPPED
12 SEA SCALLOPS, CORALS REMOVED,
    SLICED IN HALF HORIZONTALLY IF LARGE
9 OZ RICE NOODLES
1 TBSP LIGHT SOY SAUCE

TO SERVE:
2 TBSP FRESH CILANTRO LEAVES

1 In a small bowl, whisk the honey with the lime juice and zest and 1 teaspoon of the oil. Add the chile and stir together.

2 Place the scallops in a shallow dish. Pour the marinade over. Cover and chill for 15 minutes.

3 Bring a large pot of water to a boil and add the rice noodles; remove from the heat and let the noodles soak for 5 minutes, then drain and keep warm.

4 Heat the remaining oil in a nonstick skillet over medium heat. When hot, add the scallops, reserving the marinade. Sear for 1 minute, or until the honey begins to caramelize, then flip the scallops over and cook for another 1 to 2 minutes.

5 Pour in the marinade and allow the juice to bubble and reduce down. Add the soy sauce and stir.

6 Divide the noodles among 4 warm serving bowls and top with the scallops and sauce. Scatter the cilantro over, then serve.

HONEY BALSAMIC SALMON FILLETS

Using honey with oily fish such as salmon is not only delicious, it is also a great nutritional combination. The omega fish oils combine with the vitamins and minerals in honey to make a heart-healthy and tasty supper dish.

SERVES: 4

4 (7-OZ) SKIN-ON SALMON FILLETS
3 TBSP HONEY
2 TBSP AGED BALSAMIC VINEGAR
1 TSP SUPERFINE SUGAR
1 TBSP PEANUT OIL

TO SERVE:
4 SCALLIONS, SLICED

1 Wipe the salmon dry with a paper towel and place in a shallow dish, skin side down.

2 Whisk the honey with the vinegar and superfine sugar and spoon over the fish.

3 Heat the oil in a large nonstick skillet over high heat. When the oil is hot, add the fish, skin side down, and cook for 4 to 5 minutes, until the skin is crisp and the salmon is nearly cooked through.

4 Preheat the broiler to high. Spoon any remaining honey mixture over the salmon and place the pan under the grill for 2 minutes, or until the fish is cooked through.

5 Serve the salmon with the scallions sprinkled over.

TERIYAKI HONEY SEAFOOD

Teriyaki is a traditional Japanese sauce used as a marinade and dip for a variety of fish and meat dishes. Made with Japanese rice wine—sake—teriyaki has a pungent sweet/salty flavor and honey adds a rich consistency to the sauce. Teriyaki can be used as a sauce for barbecued meat, fish, or vegetables and is delicious as a dip for tempura vegetables or shrimp.

SERVES: 4

3 TBSP SOY SAUCE

5 TBSP MIRIN

3 TBSP HONEY

2 TBSP SAKE

1 TBSP THAI FISH SAUCE

1-IN PIECE GINGER, PEELED AND GRATED

1 CLOVE GARLIC, PEELED AND CRUSHED

7 OZ FRESH TUNA, CUBED

16 RAW KING PRAWNS, PEELED

9 OZ MONKFISH TAIL OR SALMON, CUBED

TO SERVE:

2 TBSP FRESH CILANTRO, ROUGHLY CHOPPED

2 LIMES, HALVED

1 Whisk together the soy sauce, mirin, honey, sake, and fish sauce and stir in the ginger and garlic.

2 Place the seafood in a large bowl and pour the marinade over. Cover and chill for 30 minutes.

3 Soak 8 wooden skewers in cold water for 30 minutes. Preheat the broiler or prepare a grill.

4 Lift the seafood from the marinade and thread onto the drained skewers, alternating the tuna and monkfish with the shrimp. Strain the marinade into a small saucepan and bring to a simmer over low heat.

5 Broil the seafood for 3 to 4 minutes, then turn the skewers over and cook 2 to 3 minutes longer.

6 Serve each person 2 skewers, with the hot teriyaki sauce poured over. Sprinkle cilantro over the seafood and serve with the limes.

RACK OF LAMB WITH THYME AND HONEY GLAZE

Tender young lamb coated in dark, rich honey works well with thyme, but rosemary would also be suitable. Make sure the racks of lamb are well trimmed or "Frenched," so there is not too much fat on the bones, and remember to rest the meat before carving.

SERVES: 4

1 TBSP OLIVE OIL

3 RED ONIONS, PEELED AND THINLY SLICED

SEA SALT AND BLACK PEPPER,
    FRESHLY GROUND

2 (1 TO 1¼-LB) RACKS OF LAMB

1 TBSP FRESH THYME LEAVES
    (OR ROSEMARY)

4 TBSP DARK HONEY

¼ CUP MARSALA OR VEGETABLE STOCK,
    PLUS MORE AS NEEDED

1 Preheat the oven to 400°F.

2 Heat the oil in a small nonstick skillet over medium heat until it is sizzling. Add the onions and cook gently for 10 minutes, or until they are softened but not brown. Season with salt and pepper.

3 Spoon the onions into a roasting pan and place the lamb on top. Score the fatty side of the lamb with a sharp knife.

4 Place the thyme and 3 tablespoons of the honey in a small saucepan and heat gently for 1 to 2 minutes, then brush the mixture over the lamb. Sprinkle with 1 tsp salt.

5 Roast for 20 minutes, then add the ¼ cup Marsala and roast for 10 to 15 minutes longer.

6 Transfer the lamb to a board and cover with foil. Rest for 5 minutes.

7 Pour the warm pan juices and onions from the roasting pan into a small saucepan. Stir together and heat over low heat, adding a little more Marsala or stock and the remaining honey to create a sauce.

8 Carve the lamb into chops, spoon the sauce over, and serve.

CHINESE SESAME AND HONEY BEEF STIR-FRY

Honey is an essential ingredient in Chinese cooking and is used extensively in sweet and sour recipes. Marinating the steak tenderizes it, and the honey adds a subtle flavor to the aromatic mix. This dish can also be eaten cold—leave out the rice and serve with cold soba noodles (available from supermarkets or Asian food stores) dressed in a little soy sauce and sesame oil for a summer lunch.

SERVES: 4

1¼ LB BEEF STEAK, CUT INTO STRIPS

2 TBSP DARK SOY SAUCE

3 TBSP HONEY

2 TBSP PEANUT OIL

9 OZ BROCCOLINI FLORETS

1⅓ CUPS LONG-GRAIN RICE

1 YELLOW BELL PEPPER, SEEDED AND CUT INTO THIN STRIPS

1 RED BELL PEPPER, SEEDED AND CUT INTO THIN STRIPS

3 SCALLIONS, SLICED

2 TBSP SESAME SEEDS, TOASTED

1 TBSP TOASTED SESAME OIL

1 Place the beef in a shallow bowl. Whisk the soy sauce with the honey and 1 tablespoon of the peanut oil and pour the mixture over the beef. Stir, cover, and chill.

2 Cook the broccoli in plenty of boiling water for 4 minutes, or until tender, then drain well. Set aside.

3 Bring a large pot of water to a boil and cook the rice for 10 to 12 minutes, until tender; drain and keep warm.

4 Heat a wok or heavy skillet over high heat until hot and add the remaining peanut oil. Add the bell peppers and scallions and stir-fry for 2 to 3 minutes.

5 Lift the beef from the marinade and add to the wok; stir-fry for 3 minutes. Reserve the marinade.

6 Add the sesame seeds and broccoli and stir-fry for 1 to 2 minutes longer.

7 Divide the beef and vegetables among 4 warmed serving bowls and drizzle with the sesame oil.

8 Transfer the reserved marinade to the wok and simmer for 2 minutes, then pour over the finished stir-fry. Serve with the rice.

HONEY AND GARLIC ROAST LAMB

A dark honey is most appropriate with this slow-cooked lamb; it gives the potatoes a wonderful golden color and makes the whole dish rich, sticky, and delicious. In Greece, from where this dish takes its inspiration, lamb is often cooked with honey, garlic, and mountain herbs. To cook a butterflied leg of lamb (cut and spread open) on the grill or barbecue, crush the garlic with the honey and olive oil and spread the mixture onto the skin; cook until slightly charred on the outside and with a hint of pink inside.

SERVES: 4 TO 6

3^1/$_2$-LB PART-BONED LEG OF LAMB

3 CLOVES GARLIC, PEELED AND CRUSHED,
 PLUS 5 WHOLE CLOVES GARLIC, WITH
 SKIN ON

4 TSP DARK HONEY

3 TBSP OLIVE OIL

SEA SALT AND BLACK PEPPER, FRESHLY
 GROUND

2^1/$_4$ LBS POTATOES, PEELED AND THICKLY
 SLICED

2 SPRIGS FRESH OREGANO

^{3}/$_4$ CUP HOT VEGETABLE STOCK

1 Preheat the oven to 350°F.

2 Score the fat on the lamb with a sharp knife in a diamond pattern.

3 Mix the 3 crushed garlic cloves with the honey, 1 tablespoon of the oil, and ½ teaspoon salt, and use this mixture to coat the lamb. Season with pepper.

4 Toss the potatoes with the remaining oil, the oregano and salt and pepper. Spread the potatoes in a large roasting pan and add the 5 whole garlic cloves.

5 Place the lamb on top of the potatoes and roast for 1 hour.

6 Add the hot stock to the roasting pan and roast for 35 to 40 minutes longer.

7 Remove the lamb and turn off the oven, but return the roasting pan with the potatoes to the oven to keep warm. Cover the lamb with foil and rest for 15 minutes.

8 Slice the lamb and serve with the potatoes and garlic.

PHYLLO, GOAT CHEESE, AND HONEY PARCELS

In Italy cheeses such as pecorino are served simply drizzled with a delicate light honey, such as rosemary or lavender. These crisp pastry parcels filled with melting goat cheese are equally good—try a single-flower honey such as acacia or clover. Ideal as a starter or for picnics, they can also be made with a spoon of basil pesto or olive tapenade instead of the onion marmalade.

SERVES: 4

VEGETARIAN

1 TSP PEANUT OIL

8 SHEETS FROZEN PHYLLO DOUGH, THAWED

6 TBSP UNSALTED BUTTER, MELTED

3 TBSP FLOWER HONEY

4 TBSP ONION MARMALADE (OR 1 TBSP BASIL PESTO OR OLIVE TAPENADE)

4 SLICES GOAT CHEESE LOG, ABOUT 2 OZ EACH

1 TBSP FRESH MARJORAM LEAVES

5 OZ ARUGULA

1 TSP EXTRA-VIRGIN OLIVE OIL

1 Preheat the oven to 375°F and brush a baking sheet with the peanut oil.

2 Brush 4 of the pastry sheets with melted butter and top each one with another sheet of pastry.

3 Stir 1 tablespoon of the honey and the onion marmalade in a small bowl. Spoon 1 tablespoon of this mixture into the center of each pastry sheet.

4 Top the honey-marmalade mixture with a slice of goat cheese and scatter the marjoram over.

5 Gather up the corners of the pastry to form a parcel and squeeze the pastry edges together to seal. Repeat this until you have 4 pastry parcels.

6 Place the parcels on the prepared baking sheet and brush the parcels with the remaining butter. Bake in the center of the oven for 10 to 15 minutes, until golden.

7 Place each parcel on a serving plate and drizzle with the remaining honey. Toss the arugula with the olive oil and divide among the plates.

Honey Roast Squash with Wild Rice Stuffing

This rich vegetarian dish makes a substantial lunch or supper served with a green salad. The squash can also be served without the stuffing: Roast with the honey and olive oil and a few sprigs of thyme or rosemary. For a less elaborate stuffing you can sauté mixed, chopped mushrooms in butter, season, and spoon these into the cavity, top with a grating of Parmesan cheese and bake for 10 minutes, or until golden. A little extra honey could also be spooned over the baked squash.

SERVES: 4

VEGETARIAN

2 (1¹/₂-LB) BUTTERNUT SQUASH,
    CUT IN HALF LENGTHWISE

5 TBSP OLIVE OIL

3 TBSP HONEY

¹/₂ CUP WILD RICE

1 RED ONION, FINELY CHOPPED

1 RIB CELERY, FINELY CHOPPED

1 RED BELL PEPPER, SEEDED AND FINELY
    CHOPPED

2 CLOVES GARLIC, PEELED AND CRUSHED

1 RED CHILE, SEEDED AND FINELY CHOPPED

¹/₃ CUP DRIED CRANBERRIES

²/₃ CUP MASCARPONE

1 TBSP CHOPPED FRESH PARSLEY

1 TSP FRESH THYME LEAVES

4 TBSP FRESH BROWN BREADCRUMBS

2 TBSP UNSALTED BUTTER

SEA SALT AND BLACK PEPPER,
    FRESHLY GROUND

1 Preheat the oven to 400°F.

2 Using a spoon, scoop out and discard the seeds from the squash. Place the four halves cut side up on a heavy baking sheet.

3 Whisk together 3 tablespoons of the oil with the honey and brush this mixture over the cut sides of the squash. Spoon any remaining mixture into the hollow left by the seeds.

4 Roast in the center of the oven for 30 to 35 minutes, until tender.

5 Meanwhile, bring a large saucepan of water to a boil and add the rice. Cook for 35 to 40 minutes, until tender, then drain well.

6 Heat the remaining oil in a large nonstick skillet over medium heat until sizzling, then add the onion, celery, bell pepper, garlic, and chile, and fry over medium heat, stirring occasionally, for 15 to 20 minutes, until the vegetables are tender.

7 Remove from the heat and add the cranberries, rice, mascarpone, and herbs. Season well and stir together.

8 Spoon the filling into the cavities and top with the breadcrumbs. Dot with the butter and return to the oven for 15 minutes, or until golden brown.

Caramelized Shallot and Honey Tart

This savory tart combines the rich caramel flavor of honey, the sweetness of slow-roasted shallots, and the creamy and nutty taste of Gruyère cheese. It can also be made with sliced white or red onions or whole banana shallots, and can be served warm or cold.

MAKES: 1 (8-INCH) TART
VEGETARIAN

FOR THE PASTRY DOUGH:
1¹/4 CUPS ALL-PURPOSE FLOUR,
    PLUS EXTRA FOR DUSTING
PINCH SALT
6 TBSP CHILLED UNSALTED BUTTER, CUBED
1 MEDIUM EGG, BEATEN

FOR THE FILLING:
6 TBSP UNSALTED BUTTER,
    PLUS EXTRA FOR GREASING
2 TBSP HONEY
12 SHALLOTS, PEELED
3 LARGE EGGS, BEATEN
²/3 CUP CRÈME FRAÎCHE
SEA SALT AND BLACK PEPPER,
    FRESHLY GROUND
2 FRESH SAGE LEAVES, SHREDDED
¹/3 CUP GRATED GRUYÈRE CHEESE

1 Preheat the oven to 375°F.

2 To make the pastry dough, sift the flour into a large bowl and add the salt, then rub in the butter until the mixture resembles breadcrumbs (or process together in a food processor).

3 Add the beaten egg and bring the ingredients together into a dough. If it is too crumbly add a little ice water until the mixture comes together. Knead briefly, then wrap in plastic wrap and chill for 30 minutes.

4 To make the filling, melt the butter with the honey in an ovenproof skillet over low heat and add the shallots. Make sure the shallots are well coated in the butter and honey, then transfer the dish to the oven and roast for 30 minutes, or until they are soft and beginning to caramelize.

5 Remove from the oven and allow the shallots to cool slightly. Lower the oven temperature to 350°F.

6 Roll the pastry dough thinly on a floured board and use the extra butter to grease an 8-inch tart pan with a removable bottom. Line the bottom with the dough and prick the bottom a few times with a fork, then let it chill for 20 minutes.

7 Cover the pastry with parchment paper, fill with dried beans, and bake in the oven for 15 minutes, before carefully removing the beans and paper. Return the pastry shell to the oven for 10 minutes longer, or until it is crisp.

8 Increase the oven temperature to 375°F.

9 Arrange the shallots in the bottom of the pastry shell and make sure they are distributed evenly.

10 Whisk together the eggs and crème fraîche and season well with salt and pepper. Add the sage and mix again.

11 Pour the mixture over the shallots, then sprinkle with the cheese.

12 Return the tart to the oven and bake for 25 minutes, or until set. Serve warm or at room temperature.

Desserts, Cookies, and Sweets
Honey and Cinnamon Baked Apples with Honey Yogurt

Baked apples can be very straightforward, but the addition of honey and brandy makes these rather special. The golden raisins can be soaked in liquid for up to 24 hours before stuffing the apples. Bee pollen has a remarkable flavor and is available from health-food suppliers. A little scattered over the honeyed yogurt makes this a honey lovers' treat.

SERVES: 4

1/2 CUP GOLDEN RAISINS

1 TBSP BRANDY OR CALVADOS (OPTIONAL)

1/2 TSP GROUND CINNAMON

FINELY GRATED JUICE AND ZEST OF
    1 ORANGE

2 TBSP SLICED ALMONDS

6 TBSP HONEY, SUCH AS HEATHER,
    PLUS EXTRA FOR DRIZZLING

4 COOKING APPLES, SUCH AS ROME,
    2 LBS TOTAL

2 TBSP UNSALTED BUTTER, MELTED

1/2 CUP PLAIN GREEK YOGURT

2 TBSP BEE POLLEN (OPTIONAL)

1 Preheat the oven to 375°F.

2 Place the raisins in a bowl with the brandy, if using, and the cinnamon, orange zest, and almonds and stir in 1 tablespoon of the honey.

3 Core the apples and place them in a shallow ovenproof baking dish. Stuff the cavities of the apples with the raisin mixture and pour the orange juice into the dish around the apples.

4 Pour the melted butter and 3 tablespoons of the honey over the apples. Bake for 30 to 35 minutes, until the apples are tender.

5 While the apples are cooking, whisk the yogurt with 2 tablespoons honey and chill.

6 Serve the apples hot with a spoonful of the yogurt and drizzled with more honey. Scatter the pollen, if using, over the top.

CRANACHAN

A traditional Scottish dessert usually served on Burns' Night, cranachan or "crowdie cream" uses oatmeal and Scottish heather honey, rich amber in color and with a caramel flavor. Raspberries or loganberries are the traditional fruits, but any soft fruit can be used. Likewise, Scottish pinhead oats are best here, but the similar steel-cut oats will work too. Serve with a dram of whisky and a piece of shortbread for extra authenticity.

SERVES: 4

¹/₃ CUP PINHEAD OR STEEL-CUT OATS
2 CUPS HEAVY CREAM
3 TBSP MALT WHISKY OR DRAMBUIE
4 TBSP SCOTTISH HEATHER HONEY
4 OZ FRESH RASPBERRIES

1 Heat a nonstick skillet over medium heat and add the oats. Toast for 3 to 4 minutes, stirring all the time, until the oats are beginning to turn golden. Transfer to a plate and cool.

2 Whisk the cream with the whisky and 2 tablespoons of the honey until soft peaks form.

3 Spoon the mixture into 4 dessert glasses. Cover and chill for 3 hours.

4 When ready to serve, sprinkle with the toasted oats and drizzle with the remaining honey. Top with raspberries and serve.

HONEY, LEMON, AND MASCARPONE TART

This is a rich and sophisticated tart with the classic creamy filling of lemon and honey. Drizzling it with lavender honey when the tart is cold adds a delicious flower aroma. You can make your own lavender-infused honey by gently heating honey with a few heads of fresh lavender flowers (pesticide-free); leave to infuse for 6 hours before removing and discarding the lavender. Use the honey to drizzle over cakes and cookies or to pour over vanilla panna cotta.

MAKES 1 (8-INCH) TART

FOR THE PASTRY DOUGH:

1$^1/_4$ CUPS ALL-PURPOSE FLOUR,
    PLUS EXTRA FOR DUSTING
PINCH OF SALT
6 TBSP CHILLED UNSALTED BUTTER, CUBED,
    PLUS EXTRA FOR GREASING
1 MEDIUM EGG, BEATEN
$^1/_4$ TSP LEMON OIL (OPTIONAL)

FOR THE FILLING:
5 MEDIUM EGGS
1 CUP SUPERFINE SUGAR
$^1/_4$ CUP HONEY, SUCH AS LEMON THYME
FINELY GRATED ZEST AND JUICE OF
    4 LEMONS
1 CUP MASCARPONE
2 TBSP LAVENDER HONEY

1 To make the pastry dough, sift the flour into a large bowl and add the salt. Rub in the butter until the mixture resembles breadcrumbs (or process together in a food processor).

2 Add the beaten egg and lemon oil, if using, to bring the pastry together; if it is too crumbly add a little iced water. Knead the dough briefly and form it into a ball, then wrap it in plastic wrap and chill for 30 minutes.

3 Preheat the oven to 350°F. Grease an 8-inch loose-bottomed tart pan with a removable bottom with 1 teaspoon butter. Roll out the dough on a floured surface and line the pan; trim the edges, prick the base with a fork, and chill for 20 minutes.

4 Line the pastry shell with baking parchment and fill it with baking beans. Bake in the center of the oven for 15 minutes, or until dry, then carefully remove the parchment and beans. Return the pastry shell to the oven for 10 minutes longer, or until the shell is crisp. Remove from the oven and allow to cool.

5 Reduce the oven temperature to 325°F.

6 To make the filling, in a large bowl, whisk the eggs with the superfine sugar, honey, and lemon juice until smooth. Add the lemon zest and mascarpone and beat again until smooth.

7 Pour the filling into the pastry shell and bake for 30 to 40 minutes, until the filling is set. Cool completely in the pan.

8 Remove the tart to a serving plate, drizzle with the lavender honey, and serve.

BAKLAVA

A Middle Eastern "sweetmeat" or confectionary from the time of the Ottoman Sultans, baklava is popular all over the eastern Mediterranean, in particular in Greece and Turkey. It is delicious made with Greek flower honey, but any floral, fragrant honey would be suitable. The baklava will keep well in an airtight container at room temperature.

MAKES: 24 PIECES

FOR THE SYRUP:
1¹/₂ CUPS SUPERFINE SUGAR,
    PREFERABLY UNREFINED
FINELY GRATED ZEST AND JUICE OF 1 LEMON
1 CINNAMON STICK
²/₃ CUP GREEK FLOWER HONEY
1 TSP ROSEWATER (OPTIONAL)

FOR THE BAKLAVA:
²/₃ CUP UNSALTED BUTTER, MELTED
18 SHEETS FROZEN PHYLLO DOUGH, THAWED
2 CUPS SALTED SHELLED PISTACHIOS,
    CHOPPED
2 CUPS WALNUTS, CHOPPED
2 TBSP SUPERFINE SUGAR,
    PREFERABLY UNREFINED
1 TBSP GREEK FLOWER HONEY
¹/₂ TSP GROUND CARDAMOM
¹/₂ TSP GROUND CLOVES

TO SERVE:
¹/₃ CUP SALTED PISTACHIOS,
    SHELLED AND CHOPPED

1 To make the syrup, place 1½ cups water, the superfine sugar, and lemon zest and juice in a saucepan with the cinnamon stick. Stir together and heat over medium heat until the sugar has dissolved.

2 Turn up the heat and boil gently for about 15 minutes, until thick and syrupy.

3 Add the honey and cook over low heat for 2 minutes longer.

4 Cool, remove the cinnamon stick, stir in the rosewater (if using), cover, and chill.

5 To make the baklava, preheat the oven to 350°F.

6 Brush a 7-by-11-inch baking pan with a little of the melted butter. Line the pan with a sheet of phyllo dough and brush it with melted butter. Top with another sheet of dough and butter, and repeat, alternating dough and butter until you have used 10 sheets of dough.

7 In a small bowl, mix the pistachios and walnuts with the superfine sugar, honey, cardamom, and cloves and stir well. Spoon the nuts over the dough and spread out evenly. Top the nuts with another piece of dough, brush with butter, and continue until all the dough is used up.

8 Brush the remaining butter over the top and then, using a small, sharp knife, cut halfway through the layers in a diamond pattern.

9 Bake in the center of the oven for 20 minutes. Decrease the oven temperature to 300°F and bake for 35 to 40 minutes longer, until golden on top.

10 Remove from the oven and, using a sharp knife, cut right through to the bottom of the pan. While the baklava is still hot, pour the chilled syrup over and then let cool completely. Sprinkle a few chopped pistachios on top before serving.

WALNUT AND HONEY COOKIES

Honey is hygroscopic—it absorbs moisture—so it is good for keeping cakes and cookies moist for longer. These cookies could also be made with hazelnuts, macadamia nuts, or pecans, or you could replace the nuts with chocolate chips. For a special treat, sandwich two cookies together with ginger preserves or vanilla ice cream.

MAKES: 24 COOKIES

1 TSP PEANUT OIL
1/2 CUP PLUS 1 TBSP UNSALTED BUTTER
1/2 CUP DEMERARA SUGAR
4 TBSP HONEY
1 CUP SELF-RISING FLOUR
1 TSP MIXED GROUND SWEET SPICES
1/4 TSP BAKING SODA
PINCH OF SALT
1/2 CUP ROLLED OATS
1 3/4 CUPS WALNUTS, CHOPPED

1 Preheat the oven to 325°F and lightly oil 2 large baking sheets with the oil.

2 Place the butter, demerara sugar, and 2 tablespoons of the honey in a small saucepan and melt together over low heat. Stir until well combined.

3 Sift the flour, spices, and baking soda into a large bowl and stir in the salt, oats, and walnuts.

4 Pour the butter and honey mixture into the flour mixture and mix well with a wooden spoon to form a dough. Take a teaspoon of dough and roll it into a ball about the size of a walnut, then flatten it slightly to a 2½-inch disk. Repeat until you have 12 and place them all, well spaced, on a baking sheet.

5 Bake in the center of the oven for 15 minutes.

6 Meanwhile, shape 12 more cookies with the remaining dough.

7 Remove the first batch from the oven and bake the second batch for 15 minutes.

8 While the cookies are still warm, brush them with the remaining honey and then cool them on the baking sheet before transferring to a wire rack.

Melomakarona and Honey Syrup

Traditional Greek Christmas cookies drenched in honey, melomakarona are also known as phoenikia and are often taken as gifts to celebrations. Once dipped in the honey syrup, the cookies can also be rolled in chopped nuts. Use Greek honey for an authentic flavor.

MAKES: 38 COOKIES

FOR THE COOKIES:
1 CUP PLUS 2 TBSP SUNFLOWER OIL
1¹/₄ CUPS SUPERFINE SUGAR,
    PREFERABLY UNREFINED
¹/₄ CUP FRESH ORANGE JUICE
JUICE OF ¹/₂ LEMON
1 MEDIUM EGG YOLK
1 TBSP SWEET SHERRY OR OUZO
3 CUPS PLUS 2 TBSP SELF-RISING FLOUR
¹/₂ TSP GROUND CINNAMON

FOR THE SYRUP:
2¹/₄ CUPS SUPERFINE SUGAR,
    PREFERABLY UNREFINED
¹/₂ CUP BOILING WATER
1 CINNAMON STICK
JUICE OF ¹/₂ LEMON
¹/₂ CUP GREEK HONEY

1 To make the cookies, preheat the oven to 350°F and line two large baking sheets with parchment paper.

2 In a large bowl, beat together the oil, superfine sugar, and orange and lemon juices with an electric mixer for 10 minutes. Add the egg yolk and sherry, and beat for 3 minutes longer.

3 Sift in the flour and cinnamon and use a wooden spoon to mix it in to form a light dough.

4 Using a teaspoon, take walnut-sized pieces of the dough and shape them into ovals. Place on the prepared baking sheets, well spaced, and use a fork to press a pattern on the top of each cookie. Repeat until you have used all the dough.

5 Bake the first batch in the center of the oven for 25 minutes or until risen and golden. Cool on the baking sheet while you bake the second batch, also for 25 minutes.

6 To make the syrup, place all the ingredients in a heavy-bottomed saucepan and stir over medium heat for 4 to 5 minutes, until the sugar has dissolved.

7 Bring the mixture to a boil and then simmer for 3 to 4 minutes, until the syrup thickens. Remove the cinnamon stick and discard.

8 Keeping the syrup hot, use a slotted spoon to lower the cooled cookies, 3 or 4 at a time, into the syrup for 10 to 15 seconds. Turn each once carefully, then remove to cool on a baking sheet lined with parchment paper.

HONEY AND GINGER RHUBARB FOOL

Rhubarb fool is a classic English summer dessert, and the addition of ginger and honey makes it a more elegant dish. You might also replace the ginger with a little orange flower water and use orange blossom honey; this also makes a good filling for a sponge cake. Or you can add a little ginger wine when you are whisking the cream, and the fool can be used to make a trifle.

SERVES: 4

1 LB RHUBARB, TRIMMED AND CHOPPED

1 TSP WATER

2 OZ STEM GINGER OR GINGER PRESERVED
    IN SYRUP, CHOPPED

$1/4$ CUP LIGHT MUSCOVADO SUGAR

3 TBSP HONEY, SUCH AS ACACIA

$1^1/4$ CUPS HEAVY CREAM

1 TBSP STEM GINGER SYRUP

TO SERVE:

2 TBSP HONEY, SUCH AS ACACIA

4 THIN GINGER COOKIES

1 Place the rhubarb, water, ginger pieces, muscovado sugar, and 3 tablespoons honey in a saucepan and simmer over low heat for 15 minutes, stirring once, until the rhubarb is soft. Allow the mixture to cool.

2 Whisk the cream with the stem ginger syrup, then fold in the rhubarb mixture. Spoon the mixture into glasses and chill for 2 hours.

3 Just before serving, drizzle with a little more honey. Serve with ginger cookies.

CAKES AND BREADS
HONEY BANANA CAKE

Using honey in cakes to replace some of the refined sugar is a good way of lowering the amount of sugar needed, as honey tastes sweeter than refined sugar. Try using a fragrant flower honey, such as lavender, for the topping of this impressive cake, with some soft dried banana slices or banana chips to decorate. Mascarpone can be used in place of the cream cheese topping: Simply beat with a little superfine sugar and vanilla bean seeds and spread on the cake.

MAKES: 1 (9-BY-5-IN) CAKE

FOR THE CAKE:

¹/₃ CUP SUNFLOWER OIL, PLUS EXTRA
    FOR GREASING

¹/₂ CUP PLUS 1 TBSP UNSALTED BUTTER,
    SOFTENED

¹/₂ CUP LIGHT BROWN SUGAR

¹/₄ CUP DARK HONEY, SUCH AS EUCALYPTUS

2 RIPE BANANAS, 14 OZ IN TOTAL,
    PEELED AND MASHED

2 CUPS PLUS 1 TBSP ALL-PURPOSE FLOUR

2 TSP BAKING POWDER

¹/₂ TSP BAKING SODA

¹/₂ TSP GROUND CINNAMON

2 MEDIUM EGGS, BEATEN

FOR THE TOPPING:

4 TBSP UNSALTED BUTTER, SOFTENED

¹/₂ CUP CREAM CHEESE

1 CUP CONFECTIONERS' SUGAR,
    PREFERABLY UNREFINED

1 VANILLA BEAN, SEEDS ONLY

3 TBSP FLOWER HONEY, SUCH AS LAVENDER

1 To make the cake, preheat the oven to 350°F. Line a nonstick 9-by-5-inch loaf pan with parchment paper and brush the paper with 1 teaspoon oil.

2 In a large mixing bowl, cream the butter and brown sugar together until pale and fluffy, then beat in the dark honey. Add the bananas and stir well.

3 Sift the flour with the baking powder, baking soda, and cinnamon. Add alternate spoonfuls of the flour mixture and beaten egg to the banana mixture beating well to combine before adding the next, until it is all added. Add the oil and mix thoroughly.

4 Spoon the batter into the prepared pan and bake in the center of the oven for 30 minutes, or until golden and risen, and then cover the top with parchment paper and bake for 30 to 35 minutes longer, until a metal skewer inserted into the center comes out clean. Cool in the pan, then turn out onto a wire rack to cool.

5 To make the topping, use a wooden spoon to beat together the butter, cream cheese, and confectioners' sugar in a large mixing bowl until smooth.

6 Add the vanilla and 1 tablespoon of the flower honey, and mix again until combined.

7 Use a palette knife to spread the topping over the cold cake. Drizzle with the remaining 2 tablespoons honey and serve.

HONEY AND APRICOT SPICE CAKE

It's always best to use a medium blended (amber rather than dark-colored) honey for baking, unless you are making gingerbread or chocolate-based recipes. This is a lovely dark, rich teatime cake topped with icing and extra chopped apricots.

MAKES: 1 (8-IN) ROUND CAKE

FOR THE CAKE:

1/2 CUP UNSALTED BUTTER, SOFTENED,
    PLUS EXTRA FOR GREASING

1/2 CUP SUPERFINE SUGAR,
    PREFERABLY UNREFINED

2 TBSP HONEY, SUCH AS ORANGE BLOSSOM

1 1/2 CUPS PLUS 1 TBSP ALL-PURPOSE FLOUR

1 TSP GROUND CINNAMON

1 TSP GROUND MIXED SWEET SPICES

1/2 TSP GROUND CLOVES

1 TSP BAKING SODA

1 TSP BAKING POWDER

1/2 TSP ORANGE BLOSSOM WATER

2 MEDIUM EGGS, BEATEN

1 CUP SOFT DRIED APRICOTS, CHOPPED

FOR THE ICING:

1 3/4 CUPS CONFECTIONERS' SUGAR,
    PREFERABLY UNREFINED

1/3 CUP SOFT DRIED APRICOTS, CHOPPED

1 To make the cake, preheat the oven to 325°F. Grease an 8-inch round spring-form cake pan.

2 In a large bowl, cream the butter with the superfine sugar until pale and light. Add the honey and stir well.

3 Sift the flour, cinnamon, spices, baking soda, and baking powder, then add 1 tablespoon of the flour mixture to the butter mixture.

4 In a small bowl, stir the orange blossom water into the beaten egg and add half to the butter mixture. Beat well, then add the remaining flour mixture and the rest of the egg and beat well.

5 Stir in the apricots and beat until smooth. Spoon into the prepared pan. Level the top and bake in the center of the oven for 35 to 40 minutes, until a metal skewer inserted in the center comes out clean.

6 Cool in the pan for 15 minutes, then turn out onto a wire rack to cool completely.

7 To make the icing, sift the confectioners' sugar into a bowl and stir in 2 to 3 tablespoons water until you have a smooth icing thick enough to coat the back of a spoon. Spread this over the cooled cake and let the icing run down the sides. Top with the chopped apricots and let set before serving.

HONEY POLENTA CAKE

Essentially a crumbly Italian cake, this makes a fabulous dessert when served with crème fraîche or mascarpone. Accompany it with a glass of Italian almond liqueur. The polenta adds an interesting texture to the cake and it's delicious served with sliced strawberries—or try black cherries soaked in kirsch or poached plums in the autumn. This cake keeps well; wrap it in foil and store in an airtight container for up to five days.

MAKES: 1 (8-INCH) ROUND CAKE

1 CUP UNSALTED BUTTER, SOFTENED,
    PLUS EXTRA FOR GREASING
3/4 CUP SUPERFINE SUGAR,
    PREFERABLY UNREFINED
3 TBSP HERB HONEY, SUCH AS THYME
1 1/4 CUPS FINE POLENTA
2 CUPS GROUND ALMONDS
1 TSP BAKING POWDER
3 LARGE EGGS, BEATEN
3 TBSP FLOWER HONEY
1 SPRIG FRESH ROSEMARY

TO SERVE:
FRESH ROSEMARY SPRIGS
CRÈME FRAÎCHE

1 Preheat the oven to 350°F.

2 Use 1 teaspoon butter to grease an 8-inch round spring-form cake pan, and line the bottom with parchment paper.

3 In a large bowl, beat the butter with the superfine sugar until light and fluffy. Add the herb honey and beat again until pale.

4 Mix the polenta with the ground almonds and baking powder, then add a large tablespoon of this mixture to the butter and sugar mixture and beat well.

5 Add one third of the beaten eggs and beat in. Repeat, adding polenta mixture and eggs alternately until you have a thick batter. Spoon this into the prepared pan and smooth the top.

6 Bake for 30 minutes, then reduce the oven temperature to 325°F and bake for 20 minutes longer, or until the cake is golden and risen and a metal skewer inserted in the center comes out clean. Cover the cake with parchment paper toward the end of baking if it is browning too much. Remove from the oven.

7 Heat the flower honey with the rosemary sprig in a small saucepan over low heat until the honey begins to thin.

8 Place the warm cake still in its pan on a wire rack and, using a thin metal skewer, make holes in the top. Spoon the warm honey over the cake so that it soaks into the cake, and then let the cake and honey cool to room temperature. Garnish with rosemary sprigs and serve with crème fraîche.

SPICED CINNAMON AND HONEY-SWEETENED DATE MUFFINS

These muffins are perfect for breakfast or as a mid-morning snack and are best eaten on the day they are made. Chopped dried apricots or soft dried banana can be used in place of the dates. Use a fragrant wildflower honey, such as borage, mixed with chopped nuts to top the warm muffins or, for a nut-free version, use a mixture of toasted seeds, such as sunflower and pumpkin, mixed with the honey.

MAKES: 12 MUFFINS

1^1/$_2$ CUPS WHOLE-WHEAT FLOUR

1^1/$_2$ TBSP BAKING POWDER

^{1}/$_2$ TSP BAKING SODA

^{1}/$_4$ CUP BROWN SUGAR

1 TSP GROUND CINNAMON

1 TSP GROUND MIXED SWEET SPICES

1^1/$_2$ CUPS MEDJOOL DATES,
     PITTED AND CHOPPED

^{1}/$_3$ CUP RAISINS

^{2}/$_3$ CUP WHOLE MILK

^{1}/$_4$ CUP HONEY, SUCH AS CHESTNUT

4 TBSP UNSALTED BUTTER,
     MELTED AND COOLED

1 LARGE EGG

^{1}/$_3$ CUP WALNUTS OR PECANS, CHOPPED

6 TBSP LIGHT WILDFLOWER HONEY,
     SUCH AS BORAGE

1 Preheat the oven to 400°F and place 12 paper liners in a 12-cup muffin pan.

2 In a large mixing bowl, stir together the flour, baking powder, baking soda, brown sugar, cinnamon, and spices.

3 Add the dates and raisins and stir.

4 In a separate bowl, whisk together the milk, chestnut honey, butter, and egg until well combined.

5 Make a well in the center of the dry ingredients and pour in the milk mixture, stirring until the ingredients are just mixed—the mixture should be lumpy rather than smooth.

6 Spoon it into the muffin cups and bake in the center of the oven for 20 minutes, until well risen.

7 Meanwhile, stir the nuts into the light honey and then, using a teaspoon, divide the honey nuts among the warm muffins. Transfer the muffins to a wire rack to cool.

HONEY SCONES

These scones are very easy to make, and are a good recipe to do with children. Replacing some of the sugar with honey keeps the scones light and moist. You can also try brushing the scones with a little milk and scattering over sesame seeds mixed with a little bee pollen (available at health-food stores) before baking. Serve the scones warm with butter and extra honey.

MAKES: 12 SCONES

1³/₄ CUPS SELF-RISING FLOUR,
    PLUS EXTRA FOR DUSTING
PINCH OF SALT
4 TBSP UNSALTED BUTTER, SOFTENED,
    PLUS EXTRA FOR GREASING
1 MEDIUM EGG, BEATEN
2 TBSP SUPERFINE SUGAR
2 TBSP HONEY, SUCH AS CHESTNUT
¹/₄ CUP WHOLE MILK

TO SERVE:
1 TBSP BUTTER
HONEY

1 Preheat the oven to 400°F.

2 In a large mixing bowl, sift the flour with the salt, then add the butter and lightly rub it in until the mixture resembles breadcrumbs.

3 Using a fork, mix in the egg, superfine sugar, and honey, then gradually add the milk until you have a stiff dough.

4 Using 1 teaspoon each of butter and flour, grease and then flour a heavy baking sheet.

5 Roll the dough out on a floured surface to 1 inch thick and then, using a 2½-inch cutter dipped in flour, stamp out the scones; you may have to re-roll the dough for the last couple.

6 Place on the prepared baking sheet and bake in the center of the oven for 12 to 15 minutes, until risen and golden. Remove the scones immediately to a wire rack and cover with a clean towel.

7 To serve, split each scone in half and spread with butter and your favorite honey.

TRADITIONAL HONEY TEA BREAD

Eucalyptus honey has a slightly caramel flavor, which works well in this recipe. To serve, slice the tea bread and spread with butter and extra honey. The bread will keep wrapped in an airtight container for five days and is good for picnics and lunchboxes.

MAKES: 1 (9-BY-5-IN) LOAF

1 CUP RAISINS
$^1/_3$ CUP CANDIED MIXED PEEL, CHOPPED
3 ENGLISH BREAKFAST TEA BAGS
1$^1/_4$ CUPS BOILING WATER
$^1/_2$ CUP HONEY, SUCH AS EUCALYPTUS
1 TBSP UNSALTED BUTTER, FOR GREASING
2 MEDIUM EGGS, BEATEN
2 CUPS ALL-PURPOSE FLOUR
1$^1/_2$ TSP BAKING POWDER
$^1/_2$ TSP GRATED NUTMEG

1 Grease a 9-by-5-inch loaf pan. Place the raisins and candied peel in a large bowl.

2 Make the tea in a heatproof pot using the tea bags and the boiling water; steep for 3 minutes. Discard the tea bags and stir the honey into the tea, making sure it has dissolved before pouring the sweetened tea over the raisins and peel.

3 Stir together, cover the bowl, and let the fruit steep for 3 to 4 hours.

4 Preheat the oven to 350°F.

5 Stir the egg into the fruit mixture, then add the flour, baking powder, and nutmeg and mix well.

6 Spoon the batter into the prepared pan and bake in the center of the oven for 1 hour, 15 minutes, or until a metal skewer inserted in the center comes out clean.

7 Cool the tea bread in the pan for 5 minutes, then turn it out onto a wire rack to cool completely.

DRESSINGS AND CHUTNEYS

WARM HONEY AND WHOLE-GRAIN MUSTARD DRESSING

Honey and whole-grain mustard are a popular combination, and this warm dressing is good on potato salad, spooned over salad greens, or as a sauce for roast cod or haddock. Use it to dress robust winter salads or coleslaw, and as a lighter alternative to creamy mayonnaise or crème fraîche dressings.

SERVES: 4

1 TBSP HONEY, SUCH AS LIME BLOSSOM
1 TSP WHOLE-GRAIN MUSTARD
$^1/_4$ CUP PEANUT OIL
1 TBSP WHITE WINE VINEGAR
SEA SALT AND BLACK PEPPER,
   FRESHLY GROUND

1 Spoon the honey and mustard into a small saucepan and stir together. Warm over low heat for 2 minutes.

2 Gradually add the oil, then remove from the heat and whisk in the vinegar.

3 Season to taste and use as desired. The dressing will keep for three days in a screw-top jar in the refrigerator.

HONEY VINAIGRETTE

Make a batch of this useful dressing and keep it in a screw-top jar in the refrigerator. The ingredients may separate, so take it from the fridge before use and allow it to come up to room temperature, then shake the jar vigorously. Try it on a crisp salad of mixed leaves, sliced avocado, and chopped walnuts or use it to dress peppery arugula and scatter thin shavings of Parmesan cheese over the salad.

SERVES: 4

1 TBSP RED WINE VINEGAR
$^1/_4$ CUP EXTRA-VIRGIN OLIVE OIL
2 TBSP HONEY, SUCH AS WILD THYME
$^1/_2$ TSP DIJON MUSTARD
SEA SALT AND BLACK PEPPER,
   FRESHLY GROUND
1 SPRIG FRESH ROSEMARY

1 Place all the ingredients except the rosemary in a clean screw-top jar and shake vigorously to combine.

2 Add the rosemary to the jar, re-seal, and let sit overnight in the refrigerator. Shake again to emulsify before using. The vinaigrette will keep for 1 week in an airtight jar in the refrigerator.

APPLE, HONEY, AND CHILE CHUTNEY

Chutneys were originally developed by the British during the colonial era, when they adapted Indian chatni—thick pastes made with fresh fruit or vegetables, vinegar, and herbs and spices— to preserve fruits and vegetables and add interest to plain English foods. Apple chutney is a classic recipe, and this tasty sweet-and-hot one will keep for up to three months if stored in a cool, dark cupboard. It is delicious with cold meats (such as honey roast ham) or cheese.

MAKES: 2 PINT JARS

2$\frac{1}{2}$ LBS COOKING APPLES, PEELED, CORED,
    AND DICED
3$\frac{1}{4}$ CUPS LIGHT MUSCOVADO SUGAR
$\frac{1}{4}$ CUP BLENDED HONEY
3 CUPS GOLDEN RAISINS
3 ONIONS, PEELED AND CHOPPED
2 TSP MUSTARD SEEDS
2 TSP GROUND GINGER
1 TSP SEA SALT
2 RED CHILES, SEEDED AND FINELY CHOPPED
3 CUPS CIDER VINEGAR

1 Put all the ingredients in a large stainless-steel saucepan or jam pan and stir well with a wooden spoon.

2 Place the pan over medium heat and gradually bring to a boil. Simmer over medium heat for 1½ to 2 hours, stirring occasionally, until the chutney is thick and all the liquid has been absorbed and evaporated. Keep a close eye on the chutney for the last 30 minutes of cooking, stirring frequently to keep it from sticking. It is done when a spoon drawn across the bottom of the pan leaves a line.

3 Sterilize 2 pint jars by putting them through a dishwasher cycle (or washing thoroughly) and dry them in an oven at 250°F.

4 Remove the saucepan from the heat and spoon the chutney into the warm, sterilized jars. Seal and label.

5 Set the chutney aside for at least 2 weeks before opening.

DRINKS

TRADITIONAL MEAD

Believed to be the oldest alcoholic drink known to mankind, mead is made by the simple fermentation of honey with spices and other flavorings. There are a bewildering number of variations and recipes for mead—including spiced mead (metheglin), fruit mead (melomel), and mead with mulberries (morat) or hops (sack)—but this is a basic home recipe. Once you have mastered the technique, you can begin to experiment with your own flavors. The mead will reflect the flavor of the honey you use, so bear this in mind.

Note: *Only use wine (or champagne) yeast—available from health-food suppliers—not brewer's or baking yeast.*

MAKES: 5½ PINTS

1 GALLON BOTTLED (OR FILTERED) WATER

5 CUPS MILD HONEY, SUCH AS
    ORANGE BLOSSOM

JUICE AND PEEL OF 1 LEMON

2 WHOLE CLOVES

1 CINNAMON STICK

1 PACKET ALL-PURPOSE WINE OR
    CHAMPAGNE YEAST

1 In a large stainless-steel pan, bring the water to a boil, then allow it to simmer. Add the honey, stir, and simmer, skimming off the scum when it appears, until no more scum forms (this can take up to 45 minutes). Turn down the heat and add the lemon juice and peel to the mixture, along with the cloves and cinnamon stick. Simmer for 5 minutes, then cover and set aside to cool to room temperature.

2 When cool, remove the cloves, lemon peel, and cinnamon stick and skim off any more scum completely.

3 Add the yeast and stir, then transfer into a sterilized 1-gallon fermentation jug with a fermentation lock.

4 Allow to ferment for 48 to 60 hours, until bubbling ceases, and then siphon into sterilized bottles, carefully avoiding the layer of yeast at the bottom of the jug.

5 Seal tightly and store upright in the fridge for 5 to 7 days before opening. Drink within 2 weeks.

HONEY EGGNOG

A traditional Christmas drink, this eggnog is not cooked, so you can use your best single-flower honey. If you wish to make it nonalcoholic, just leave out the rum and add a little extra honey or half a teaspoon of vanilla extract. For an alternative version, place a few pitted black cherries in the bottom of each glass before adding the eggnog, and top with cocoa powder rather than nutmeg.

Note: *Recipes using raw egg should not be given to infants, the elderly, pregnant women, and other vulnerable groups.*

SERVES: 4

2 LARGE EGGS, SEPARATED

2 TBSP SUPERFINE SUGAR,
    PREFERABLY UNREFINED

3 TBSP HONEY, SUCH AS ORANGE BLOSSOM

1/3 CUP WHOLE MILK

1/3 CUP DARK RUM OR MADEIRA

3/4 CUP HEAVY CREAM

1/2 TSP FRESHLY GRATED NUTMEG, TO DUST

TO SERVE:
8 TO 12 LADYFINGER COOKIES

1 In a large bowl, whisk the egg yolks with the superfine sugar until pale and light. Add the honey and whisk again until fluffy.

2 Stir the milk in a pitcher with the rum then gradually whisk the liquid into the egg yolk mixture.

3 In another large bowl, whisk the egg whites until stiff peaks form, then fold them into the egg yolk mixture.

4 Whisk the cream in a large bowl until soft peaks form, then fold it into the egg mixture.

5 Spoon the eggnog into 4 glasses, top with nutmeg, and serve immediately with cookies for dipping.

HONEY MINT JULEP

This classic Southern cocktail is traditionally served over crushed ice and made with sugar syrup; honey syrup is a lovely alternative. This will make more syrup than you need for two drinks but is very useful for other cocktails too. Keep in the refrigerator in a screw-top jar for up to a week.

SERVES: 2

1/4 CUP HONEY, SUCH AS SUNFLOWER
    OR CLOVER
2 TBSP BOILING WATER
6 SPRIGS FRESH MINT
CRUSHED ICE
1/4 CUP BOURBON
CLUB SODA, TO TASTE

1 In a heatproof bowl, stir the honey into the boiled water until the honey has dissolved and you have a syrup.

2 Strip the mint leaves from 4 of the sprigs. Crush them slightly into the syrup. Leave until completely cooled.

3 Strain into a large pitcher.

4 For each julep, fill a glass with crushed ice and pour 2 tablespoons bourbon and 2 tablespoons mint-honey syrup over.

5 Top up with soda and garnish each with a mint sprig.

HONEY AND RED BERRY SMOOTHIE

Smoothies are great for children, as you can pack them with all the things they think they don't like, and this one, with nutrient-rich berries and honey, makes a delicious drink as a good summer breakfast. Alternatively, make an almond and banana smoothie by using almond milk, 1 banana, honey, a drop of almond extract, and a few chopped raw almonds to top the smoothie.

Note: *Honey should not be given to children under 18 months old.*

SERVES: 2

¹/₄ CUP HONEY, SUCH AS ACACIA

1¹/₄ CUPS REDUCED-FAT MILK, OR MORE
   AS NEEDED

³/₄ CUP PLAIN YOGURT

1 CUP FRESH RASPBERRIES

³/₄ CUP BLUEBERRIES

1¹/₂ CUPS STRAWBERRIES, HULLED
   AND ROUGHLY CHOPPED

¹/₂ BANANA, SLICED

6 ICE CUBES

1 Place all the ingredients except the ice in a food processor or blender and process for 20 to 30 seconds, until smooth.

2 Add 2 ice cubes and blend again, adding more milk if you like.

3 Place 2 ice cubes in each glass and pour the smoothie over. Serve immediately.

HOT HONEY CIDER

Perfect on Halloween, this warming drink can also be made with apple juice (instead of cider) for children, or made more lethal with the addition of a shot of Calvados for adults. To serve cold, leave out the apple rings; when the cider mix is cold, strain over ice and add a few slices of fresh orange.

MAKES: 4¼ CUPS

4¼ CUPS APPLE CIDER
1 ORANGE
10 WHOLE CLOVES
2 TBSP HONEY, SUCH AS CHESTNUT
1 STAR ANISE
2 CINNAMON STICKS

TO SERVE:
DRIED APPLE RINGS

1 Pour the cider into a medium saucepan, stud the orange with the cloves, and add it to the pan, along with the honey, star anise, and cinnamon sticks. Stir. Warm gently over low heat for 5 minutes, stirring once or twice.

2 Place 2 apple rings in the bottom of each heatproof glass or mug.

3 Remove the orange from the saucepan and cut it in half; squeeze the juice from one half into the cider.

4 Strain the cider over the apple slices and serve.

Iced Honey Earl Grey Tea

A nonalcoholic alternative to Pimm's, this iced tea makes an elegant summer party drink. Use a pale fragrant honey, such as lavender, to complement the bergamot in the tea. Assam or fruit teas also work well for this recipe: Vary the fruits to suit the type of tea—sliced stone fruits, melon, and dark berries are all fine additions. Keep covered in the refrigerator for up to 24 hours.

SERVES: 4

6 EARL GREY TEA BAGS

6 CUPS BOILING WATER

3 TBSP HONEY, SUCH AS LAVENDER

6 SPRIGS FRESH MINT

2 PEACHES, PITTED AND SLICED

1 CUP PEACH OR GRAPEFRUIT JUICE

TO SERVE:
ICE CUBES

1 Place the tea bags in a large heatproof pitcher and pour the boiling water over. Stir in the honey until dissolved and add half of the mint. Let steep for 10 minutes, then strain, cool, and chill.

2 When you are ready to serve, place the ice in a large pitcher and add the peach. Pour the chilled tea and the fruit juice over. Stir well and add the remaining mint.

SPICED HONEY COFFEE

This coffee is especially good when made with hazelnut or mocha coffee. The honey melts into the hot coffee, and a shot of Kahlúa coffee liqueur makes it a real heart starter. For extra luxury, stir the coffee with a solid chocolate stirrer.

SERVES: 2

2 TBSP HONEY, SUCH AS CLOVER
1 TSP GROUND CINNAMON
1/2 TSP GROUND CARDAMOM
1/2 TSP VANILLA EXTRACT
11/4 CUPS STEAMED WHOLE MILK
2 SHOTS STRONG BREWED ESPRESSO
2 CINNAMON STICKS

1 Place 1 tablespoon of the honey in each of the 2 coffee cups and swirl to cover the bottom.

2 Whisk the ground cinnamon, cardamom, and vanilla into the hot milk.

3 Pour the espresso into the cups (over the honey) and top up with the hot, frothy milk. Use a cinnamon stick to stir.

Beauty Recipes

PURE HONEY is renowned for its skin-soothing and moisturizing properties, and is one of the oldest natural cosmetics known to man. Honey is a humectant (it attracts moisture from the atmosphere), meaning that, if you use honey-based beauty products on your skin or hair, the thin honey coating helps to attract and retain moisture. It is suitable for sensitive skin, and its antibacterial properties (see page 379–88) make it an ideal skin-cleansing agent.

Beeswax has many applications in natural beauty cosmetics and is an effective barrier cream for protecting skin in all weathers. If you don't have bees for producing your own beeswax you can buy it in granule or pellet form from health-food shops, or in blocks of wax from beekeepers and craft shops. Granules dissolve faster than a large block of wax, but you can grate a block using a food grater or cut it into slivers with a sharp knife. The best way to melt beeswax is over low heat using a double boiler or water bath by placing the container of wax—for example a small steel saucepan—inside a larger pan of water.

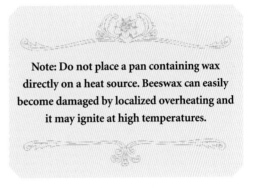

Note: Do not place a pan containing wax directly on a heat source. Beeswax can easily become damaged by localized overheating and it may ignite at high temperatures.

MILK AND HONEY BATH

This luxurious soak is an adaptation of Cleopatra's milk bath. The addition of milk to a bath helps to revitalize the skin by dissolving the proteins that hold together dead skin cells, and the honey provides a natural moisturizer. The mixture may be stored in the refrigerator for up to five days.

MAKES: ENOUGH FOR 2 OR 3 BATHS

2 CUPS WHOLE MILK
3 TBSP HONEY

1 Whisk together the milk and honey in a mixing bowl until the honey has dissolved, then pour the mixture into a pint jar with a secure lid.

2 Shake the jar before use. Pour half or one third of the mixture into the filled bath and swirl it around.

3 Soak in the bath for up to 10 minutes. The bath will need a thorough clean afterward to remove the sticky residue.

CREAMY HAIR CONDITIONER

Honey can smooth and moisturize damaged hair, making it an excellent conditioner. This recipe is based on a concoction favored by British Queen Anne, who wore her hair long in the Romantic style of the early 1700s. Use the mixture as soon as it is made.

Note: *Thick hair responds well to the deep conditioning properties of olive oil, but for finer hair types use lighter oils such as sesame or almond oil.*

MAKES: 1 APPLICATION

2 TBSP CREAMED HONEY (LIGHT COLORED
    AND THICK)
1 TBSP OLIVE, SESAME, OR ALMOND OIL

1 Prepare the conditioner just before you are ready to use it to keep the honey and oil from separating. Wash hair as normal, and towel dry.

2 Mix the honey and oil together in a bowl using a spoon, then apply liberally to damp hair, gently massaging through to the ends.

3 Pile the hair on top and wrap your head in a warm towel or place a shower cap over the hair. Wait for 20 minutes to optimize the conditioning effect of the honey, then wash off using a small amount of mild shampoo to remove excess oils, and rinse thoroughly.

SIMPLE HONEY WATER CLEANSER

Honey water has been used as a face freshener and tonic for centuries. The antiseptic properties of honey make it perfect for skin health; poured over the face and neck, this cleanser leaves the skin clear and revitalized. Use the mixture immediately once made.

MAKES: 1 APPLICATION

2 TBSP HONEY
2 PINTS RAINWATER OR DISTILLED WATER, WARMED

1 In a mixing bowl, dissolve the honey in the warm water by stirring gently.

2 Splash the honey water over the face and neck for 5 to 10 minutes, then rinse with water and pat dry.

COLD CREAM

The inventor of cold cream was a physician of second-century Greece named Galen, and in France the cream is still known as Galen's wax (cérat de Galien). This recipe uses borax, a natural ingredient that acts as a simple preservative, and, stored in the refrigerator, the cream should keep for at least a month. Cold cream can be used to remove makeup and dirt, to soften skin and relieve dryness as a night treatment, or as a cooling and soothing treatment after sunburn.

MAKES: 1 TO 2 (4-OZ) JARS OR POTS

2 OZ BEESWAX
1 CUP SWEET ALMOND OIL
$1/2$ TSP BORAX POWDER (SODIUM BORATE)
4 TBSP RAINWATER OR DISTILLED WATER
2 OR 3 DROPS ESSENTIAL OIL OF CHOICE (OPTIONAL)

1 Prepare clean glass jars or ceramic pots with lids.

2 In a double boiler, combine the beeswax and sweet almond oil over low heat, stirring until well blended.

3 In a separate pan, dissolve the borax in the distilled water over low heat (do not boil).

4 Slowly pour the borax solution into the wax and oil mixture, stirring constantly. When combined, whisk briskly until the mixture has cooled.

5 Add the essential oil, if using, and continue to whisk until the cream is light and airy.

6 Spoon into jars or pots, cover, and store in the refrigerator.

TO USE FOR CLEANSING: Massage a small amount into your skin and either wipe off with a tissue or rinse off with warm water.

LEMON AND HONEY FACE PACK
(FOR OILY SKIN)

The ancient Egyptians used lemon in beauty treatments, and today we know that the fruit acids it contains help to revitalize the skin naturally. This recipe combines egg white—an effective cleanser and pore-reducer—with lemon and honey, which tone and soften the skin. Apply the face pack while the mixture is fresh.

MAKES: 1 FACE PACK

1 TBSP LEMON JUICE
1 TSP HONEY
2 EGG WHITES
1 TSP RAINWATER OR DISTILLED WATER,
    WARMED
1 TSP CHICKPEA FLOUR

1 Whisk together the lemon juice, honey, egg whites, and water thoroughly, then stir in the chickpea flour.

2 Spread the mixture evenly over the face. Leave it on for 15 to 20 minutes, allowing the ingredients to cleanse and soothe the skin.

3 Rinse with warm water. Pat dry with a clean towel, then moisturize.

CHAMOMILE AND HONEY FACE PACK
(FOR DRY SKIN)

This recipe combines the soothing effects of chamomile, the hydrating and skin-softening properties of honey, and the exfoliating action of bran and oatmeal to help rejuvenate a tired complexion. Use the mixture while it is still warm and fragrant.

MAKES: 1 FACE PACK

1 TBSP DRIED CHAMOMILE FLOWERS
2/3 CUP BOILING WATER
1 TSP HONEY
1 TBSP OATMEAL, FINELY GROUND
1 TBSP WHEAT BRAN

1 Place the chamomile flowers in a heatproof bowl and pour the boiling water over. Allow to cool until just warm, then strain.

2 Combine the honey, oatmeal, and bran, then add the chamomile infusion in small amounts, blending together using a spoon until it forms a sticky paste.

3 Apply evenly to the face, leave on for 30 minutes, then rinse with lukewarm water. Pat dry with a warm towel, and apply moisturizer or cold cream.

HONEY AND CUCUMBER EYE TREATMENT

The fresh pulp and juice of the cucumber have soothing properties, which cool and refresh the skin around the eyes, reducing puffiness and dark circles. Store the cream in a cool, dry place, preferably in the refrigerator, and use within 1 week.

Note: *Always be careful applying new creams to the face, particularly to the sensitive areas around the eyes; test a small amount and check the reaction before applying more.*

MAKES: 1 (1/2-OZ) POT

1/2 TSP DRIED CHAMOMILE FLOWERS
2/3 CUP BOILING WATER
1 TSP ALOE VERA GEL
1 TSP HONEY
1 TSP PEELED, SEEDED, AND MINCED
    CUCUMBER

1 Prepare a clean glass or ceramic covered pot.

2 Place the chamomile flowers in a heatproof bowl and pour the boiling water over. Allow to cool for 30 minutes, then strain.

3 Meanwhile, blend the aloe vera, honey, and cucumber until smooth.

4 Add 1/2 to 1 teaspoon of the cooled chamomile infusion and stir until fully combined. The consistency should be firm but easily spreadable. Scoop into the pot and chill. Apply gently to the area around the eyes.

LAVENDER AND BEESWAX HAND CREAM

This soothing beeswax moisturizer is great for hardworking gardeners' hands. Lavender is an effective healer and antiseptic for chapped and damaged skin, and adds a pleasing fragrance. Stored in a cool, dry place, this cream should keep for several months.

MAKES: 1 TO 2 (4-OZ) POTS

4 TBSP SWEET ALMOND OIL
4 TBSP COCONUT OIL
3 TBSP WHITE (BLEACHED) BEESWAX
    GRANULES
6 TBSP GLYCERINE
5 OR 6 DROPS LAVENDER ESSENTIAL OIL

1 Prepare clean glass or ceramic pots with lids.

2 Combine the sweet almond oil, coconut oil, and beeswax in a double boiler over low heat, stirring continuously to blend the ingredients. Add the glycerine very gradually, while stirring.

3 When the ingredients have fully combined, remove the pan from the heat and continue stirring until it is a creamy consistency.

4 Add the lavender oil, mix well, then immediately spoon the mixture into the pots.

PEPPERMINT AND BEESWAX FOOT TREATMENT

This thick moisturizer is ideal for soothing tired feet. Peppermint oil helps restore cracked skin, and its primary constituent, menthol, creates a pleasant cooling sensation on the skin. Stored in a cool, dry place, this cream should keep for several months.

MAKES: 1 TO 2 (4-OZ) POTS

4 TBSP SWEET ALMOND OIL

4 TBSP COCONUT OIL

3 TBSP WHITE (BLEACHED) BEESWAX GRANULES

6 TBSP GLYCERINE

5 OR 6 DROPS PEPPERMINT ESSENTIAL OIL

This recipe is similar to the recipe for Lavender Beeswax Hand Cream (page 374). Combine the oils, beeswax, and glycerine in the same way and, when they have blended together and the mixture has become a creamy consistency, add the peppermint oil. Mix well, then immediately spoon the mixture into the pots.

SOLID PERFUME

This is an inexpensive and simple way to make perfume—simply add your fragrance oil of choice. Stored in a cool, dry place, the solid perfume should keep for several months.

MAKES: 1 TO 2 (2-OZ) POTS

1 TBSP GRATED BEESWAX

1 TBSP SHEA BUTTER

2 TBSP SWEET ALMOND OIL

8 TO 10 DROPS FRAGRANCE OIL
  OF CHOICE

1 Prepare clean containers, preferably glass or ceramic with lids (an ice-cube tray can also be used).

2 Place the beeswax, shea butter, and sweet almond oil in a double boiler over low heat, and stir until fully combined.

3 Remove from the heat and add the fragrance oil, stirring as the mixture cools and thickens. Pour immediately into the containers.

4 Let cool for 30 minutes, and the perfume will be solid and ready to use.

HONEY SOAP

The Arabs began making cold-press soaps in the seventh century, and the same basic process is used today. Honey in soap helps to keep skin from drying out and provides a wonderful sweet aroma. This mild recipe is suitable for sensitive skin. The recipe may take an hour or so, but the full process of converting the fats into soap (known as "saponification") can take up to six weeks to complete. **Read the safety instructions opposite before you start.**

MAKES: 5 OR 6 BARS OF SOAP

5 OZ RAINWATER OR DISTILLED WATER

2¹/₄ OZ LYE (SODIUM HYDROXIDE) PELLETS, FLAKES, OR GRANULES

5 OZ COCONUT OIL

6 OZ OLIVE OIL

5 OZ HEMPSEED OIL

1 TBSP HONEY

1 Prepare suitable soap molds, such as tough (heat-proof) plastic or metal molds from cake or jelly making.

2 Working in a well-ventilated place and wearing safety goggles, gloves, and long sleeves, place the (tepid) water in a large, heatproof bowl (not aluminium) and add the lye (always add lye to water and not the other way around), taking care not to breathe in the fumes.

3 Use a steel mixing spoon and stir well. Lye produces heat when it dissolves in water, so set the mixture aside and allow it to cool to approximately 100 to 122°F.

4 Heat all the oils in the top of a double boiler over low heat, stirring until the oils melt and combine. Allow the mixture to cool to the same temperature as the lye mixture (100 to 122°F).

5 Carefully add the lye solution to the melted oils, stirring vigorously (taking care not to splash) until "trace" occurs, which is when the mixture produces a slight thickening and color change and is often likened to the moment when custard thickens. The mixture will leave a trace when run off the back of a spoon. This can take anywhere from a few minutes to over an hour.

6 Stir in the honey and immediately pour the mixture into the prepared molds. Let sit uncovered at room temperature for 3 to 5 days for the chemical reaction to complete.

7 Once hardened, remove the soap from the molds and cut and shape. Leave the bars to cure on a breathable surface, such as a clean towel or muslin cloth, for 4 to 6 weeks to allow the soap to harden up.

SOAP-MAKING SAFETY AND EQUIPMENT

❀ A good set of scales is essential, as all the ingredients must be weighed accurately (including the water as it gives a more accurate measurement). A variance of as little as ¼ ounce could result in a caustic soap that is dangerous to use.

❀ You will need a stainless-steel thermometer for checking the temperature of the fats. Once you have used any piece of equipment for soap-making, do not use it for preparing foods.

❀ Always add solid lye to the liquid (water), not the other way around, which could cause a violent reaction.

❀ The amount of alkali used in soap making, known as lye (caustic soda or sodium hydroxide), is critical, and there are serious safety issues associated with handling it. Follow the product instructions, wear well-fitting protective neoprene gloves, long sleeves, and safety goggles, and work in a well-ventilated place.

❀ Lye comes in pellet, flake, or granule form and must be stored in an airtight container.

❀ Do not substitute one oil for another, as each oil has a different saponification value (the point at which the oil mixes with the alkali to form soap).

LIP BALM

Beeswax provides an effective barrier to harsh conditions, and the olive oil and cocoa butter in this recipe help to soften and smooth dry lips. Almond oil adds a sweet, nutty flavor. Stored in a cool, dry place, this lip balm should keep for several months.

MAKES: 5 TO 6 (½-OZ) POTS

³/₄ OZ BEESWAX

¹/₂ OZ LANOLIN

¹/₂ OZ COCOA BUTTER

¹/₂ OZ OLIVE OIL

1¹/₂ OZ SWEET ALMOND OIL

2 TSP HONEY (ANY VARIETY CHOSEN FOR TASTE, BUT TRY BORAGE OR ORANGE BLOSSOM)

1 Melt the beeswax in a double boiler over low heat.

2 In a separate pan, gently heat the lanolin and cocoa butter, stirring until the mixture is fully liquid.

3 Add the oils, stirring over low heat until fully combined, then stir in the melted beeswax.

4 Remove from the heat and let the mixture cool until just beginning to set, then add the honey, stirring thoroughly until fully incorporated.

5 Scoop into pots and let set at room temperature.

LIPSTICK

Carnauba wax, which comes from the leaves of Brazilian wax palm trees and is known as the "queen of waxes," combines well with beeswax to give lipstick its base and ease of application. You can buy it from specialist suppliers online or from craft shops. Lip-safe mica pigments, also available from craft suppliers, can be ordered in every conceivable shade. Castor oil acts as an emollient, and the grapefruit seed extract is a natural preservative. When stored in a cool, dry place, this lipstick should keep for several months.

MAKES: 6 TO 8 (1-OZ) POTS OR TUBE MOLDS

FOR THE BASE MIXTURE:
1/2 OZ BEESWAX
1 OZ CARNAUBA WAX
4 OZ CASTOR OIL
4 OZ JOJOBA OIL
1/2 OZ WHEAT GERM OIL

FOR THE COLORANT MIXTURE:
6 TSP LIP-SAFE MICA (COLOR OF CHOICE)
1 OZ CASTOR OIL (FOR SHINY FINISH) OR
   WHEAT GERM OIL (FOR MATTE FINISH)

TO FINISH:
1 OR 2 DROPS GRAPEFRUIT SEED EXTRACT

1 Prepare clean glass pots with lids, or lipstick tube molds.

2 To make the base mixture, melt the beeswax and carnauba wax in a double boiler over low heat, stirring until combined.

3 Add the oils, and continue stirring over low heat until all the oils have combined.

4 To make the colorant mixture, in a separate bowl combine the mica and the castor oil and mix well.

5 Add the colorant mixture to the base mixture in the double boiler and stir over low heat (do not overheat).

6 Once combined, remove from the heat, stir in the grapefruit seed extract, and allow to cool and thicken. Pour into pots or tubes and leave for several hours in the refrigerator for the lipstick to become solid.

Health Recipes

HONEY HAS BEEN recognized for its healing properties for thousands of years—both the ancient Greeks and the Mayans used it. In recent years, its medicinal value has been confirmed by science as research has demonstrated that the enzymes in honey do indeed have antiseptic and antibacterial properties. It inhibits the growth of bacteria and contains anti-oxidants that may help in fighting disease. Raw honey—untreated and uncooked—contains a higher number of beneficial enzymes than processed honey that has been treated at high temperatures. Darker honeys, such as Manuka, are particularly potent, as they contain a higher mineral count and more iron than lighter honeys.

Honey has a wide range of applications as a natural remedy for treating ailments from sore throats to skin problems: it can aid digestion, increase energy levels, and promote all-around good health. The unique properties of honey, which support the body's own healing process, have led some holistic practitioners to believe that honey and bee pollen may also help protect against hay fever and breathing difficulties associated with allergies.

Note: The following remedies should not be treated as a substitute for the medical advice of your doctor or nutritionist. Always seek qualified professional medical advice before trying these remedies.

GENERAL HEALTH

Honey is used to treat many ailments but is also particularly effective simply in supporting the body's natural defenses. Its combination of natural vitamins and minerals and disease-fighting antioxidants makes it a valuable addition to any healthy diet.

HONEY AND APPLE CIDER VINEGAR TEA

MAKES: 1 MUG OF TEA

1 APPLE, PEELED, HALVED,
   AND CORED
1 TBSP APPLE CIDER VINEGAR
2 TSP HONEY, SUCH AS EUCALYPTUS
1 HERBAL TEA BAG, SUCH AS ECHINACEA
BOILING WATER

1 Slice the apple thinly into a mug. Add the vinegar, honey, and tea bag. Top up with boiling water.

2 Let steep for 1 minute, then remove the tea bag and serve.

BLADDER INFECTIONS

Honey's antibacterial properties are thought to be especially effective in treating mild bladder infections. Although all types of honey contain hydrogen peroxide (which is antibacterial), Manuka honey is thought to be the most effective honey for this.

WARM HONEY AND CINNAMON DRINK

MAKES: 1 DOSE

1 TBSP MANUKA HONEY
¾ CUP RECENTLY BOILED WATER
1 TSP GROUND CINNAMON

Add the honey to the water and stir until dissolved. Add the cinnamon and stir again.

DOSE: *Repeat every 6 hours.*

CHOLESTEROL

Happily, both honey and cinnamon are traditionally used to lower cholesterol, and this simple topping for toast is a delicious way to help do this. The antioxidants in darker honeys, such as chestnut or eucalyptus, in particular, may help to protect against the effects of low-density lipoprotein (LDL) cholesterol—sometimes known as "bad" cholesterol because of where it is deposited in the arteries and its serious long-term effects.

HONEY AND CINNAMON TOAST

1 For each piece of toast, make a paste of 1 tablespoon honey and 1 teaspoon ground cinnamon.

2 Spread on whole-grain toast for a nutritious and healthy breakfast or snack.

COLDS

Taking some raw honey daily will support your body's immune system and boost general well-being. But if you do succumb to a cold, honey is a fine natural remedy. A soothing, warm drink of honey and lemon will relieve symptoms and aid sleep. Using eucalyptus honey may also help to ease congestion.

HONEY AND LEMON HERB TEA

MAKES: 1 MUG OF TEA

2 FRESH SAGE LEAVES
SMALL SPRIG FRESH ROSEMARY
2 SLICES LEMON, PLUS 1 FOR SERVING
1¼ CUPS BOILING WATER
1 TO 2 TEASPOONS ROBUST HONEY,
   SUCH AS EUCALYPTUS

1 In a mug, lightly crush the sage, rosemary, and 2 lemon slices, then cover with the boiling water. Stir in the honey and steep for 2 minutes.

2 Strain into a clean mug, add a fresh slice of lemon, and serve.

CONSTIPATION

A mild laxative, honey has been used since Roman times to help with digestive disorders. The addition of dates and linseed oil makes this a natural and effective treatment for constipation.

HONEY, DATE, AND LINSEED SMOOTHIE

MAKES: 1 SMOOTHIE

2 CUPS REDUCED-FAT MILK
2 DATES, PITTED AND ROUGHLY CHOPPED
½ TSP LINSEED OIL
1–2 TABLESPOONS HONEY, SUCH AS APPLE
   BLOSSOM
LINSEEDS OR PUMPKIN SEEDS, TOASTED AND
   CHOPPED

1 In a blender or food processor, blend the milk with the dates, linseed oil, and honey.

2 Pour into a glass and top with the linseeds or pumpkin seeds.

DOSE: *Take once a day.*

HANGOVER CURE

As honey is high in carbohydrates and natural glucose, it is a fast way to deliver energy. Easily absorbed and kind to the internal organs, the fructose sugars in honey also help the liver to break down alcohol. The orange juice in this remedy helps with rehydration, but drink plenty of water, too. Try a spoonful of honey dissolved in water throughout the day or take a spoonful of honey before you start drinking as a preventive measure.

HONEY, ORANGE, AND EGG HANGOVER CURE

MAKES: 1 DRINK

2 EGGS, RAW
1¼ CUPS ORANGE JUICE
1 TBSP HONEY, SUCH AS ORANGE BLOSSOM

Place all the ingredients in a food processor or blender and blend to combine. Pour into a glass and drink immediately.

COUGHS

Manuka honey from New Zealand is particularly effective in easing coughs and soothing sore throats, but any dark honey such as chestnut or eucalyptus could be used. A teaspoon of honey on its own will help with a sore throat, but this syrup is particularly good for a tickly cough. Licorice root is available at health-food stores.

HONEY, THYME, AND LICORICE COUGH SYRUP

MAKES: 1¼ CUPS

2 TBSP LICORICE ROOT,
    PEELED AND SHREDDED
2 TBSP DRIED THYME
1 CUP MANUKA HONEY

1 Put the liquorice root in a medium saucepan, add 2 cups water, and stir over medium heat. Bring slowly to a boil. Cover and simmer gently for 20 minutes.

2 Remove the pan from the heat, add the thyme, and stir. Replace the lid and let infuse for 20 minutes.

3 Strain the mixture into a clean pan and return to the heat. Once warm, add the honey. Stir with a wooden spoon to dissolve the honey, then simmer for a few minutes, stirring all the time, until the mixture is a syrupy consistency. Let the mixture bubble for a moment, but not overheat, then turn off the heat and let cool.

4 Pour the syrup into a sterilized bottle (put the bottle through a dishwasher cycle to sterilize). Label with the name, date, and dose. Keep in a cool, dark place and use within 6 months.

DOSE: *Adult: 2 to 3 teaspoons 3 to 6 times daily.*
Child over 18 months: 1 teaspoon 3 to 4 times daily.

HAY FEVER

It has been suggested that honey and honey products such as pollen and honeycomb may have anti-allergy properties, which are said to be especially effective if they are locally produced. The theory is based on the idea that if honey made from pollen from local plants is ingested, the body will become tolerant to those pollens (see page 300). Try any of the methods below.

LOCAL HONEY AND POLLEN

❀ Take 1 tablespoon local honey (ideally produced within about 20 miles of your home) after each meal and well before the pollen count begins to rise. Continue to do this until the pollen season subsides.

❀ Bee pollen (available locally in health-food stores) should be eaten raw, so scatter it over cooked dishes or stir it into muesli or cereals. Include pollen in smoothies and shakes (yogurt is particularly good at extracting the nutrients from pollen) or scatter over crunchy salads.

❀ If you find local comb honey, cut it up and mix it into Greek or plain yogurt or melt it onto toast. Chew a small piece of comb (discarding the resulting wax) to relieve the symptoms of hay fever.

MORNING TONIC

Green tea is renowned for its antioxidants, and the addition of honey and cinnamon makes it hard to beat as a morning kick-start. A healthy alternative to black tea, green tea has a subtle and delicate aroma, which means you can choose your favorite honey and really appreciate its flavor. The natural glucose in honey is quickly absorbed and provides an instant energy boost.

HONEY AND CINNAMON GREEN TEA

MAKES: 1 MUG OF TEA

1 GREEN TEA BAG
BOILING WATER
1 TO 2 TSP HONEY
1 CINNAMON STICK

1 Brew a cup of green tea and, while still hot, stir in the honey.

2 Let steep for 2 minutes, then add a cinnamon stick and stir gently to make sure the honey is dissolved. Leave the cinnamon stick in the tea to impart its flavor.

RHEUMATISM AND BACK PAIN

Honey and wax have long been used in massage treatments to relieve the symptoms of arthritis and other aches and pains, and are used in many cultures for treating aching joints—by ingesting the honey or including the wax as an ingredient in salves and poultices.

GREEN MOUNTAIN SALVE

MAKES: 1 (8-OZ) POT

½ CUP BALSAM FIR ESSENTIAL OIL

2 TBSP ROSEMARY ESSENTIAL OIL

2 TBSP BAY LEAF ESSENTIAL OIL

4½ OZ BEESWAX

1 In a medium saucepan over low heat, stir the essential oils together and add the beeswax. Heat gently, stirring, for about 5 minutes until the wax has melted.

2 Pour the liquid salve carefully into a clean jar or tin and let cool.

Note: *Take care with essential oils on sensitive skin. Do not use on sore, broken, or inflamed skin, do not use on babies, and keep away from the eyes. Seek advice before using if pregnant.*

SLEEP ENTICER

Hot toddies are a traditional sleep enticer; the combination of warm, sweet honey and a little alcohol is believed to encourage relaxation and promote deep sleep. Scottish heather honey is traditionally used to complement the whisky. To make a child's toddy, simply replace the whisky with 1 teaspoon lemon juice.

HONEY AND WHISKY TODDY

MAKES: 1 TODDY

2 TSP HONEY
²/₃ CUP BOILING WATER
2 TBSP WHISKY
1 CINNAMON STICK
1 THICK SLICE LEMON
FRESHLY GRATED NUTMEG

1 Stir the honey into the boiling water until it dissolves.

2 Add the whisky and cinnamon stick.

3 Leave for 2 to 3 minutes, then remove the cinnamon stick and the lemon slice. Sprinkle with a little freshly grated nutmeg and serve.

SORE THROAT

The combination of honey and lemon is perhaps the best-known treatment for sore throats, as the antibacterial properties in honey help to fight the infection and the bacteria that cause the problems. If you are unable to make the toddy, for instant relief take 1 tablespoon Manuka honey and let it melt on the back of your tongue and trickle down the throat.

HONEY AND LEMON TODDY

MAKES: 1 TODDY

1 TBSP CIDER VINEGAR
1 TSP LEMON JUICE
1 TO 2 TBSP HONEY, SUCH AS LEMON THYME
²/₃ CUP HOT WATER

1 Stir the vinegar, lemon juice, and honey into the hot water.

2 Stir until the honey is melted.

3 Either sip or use as a gargle.

ULCERS

Manuka honey has been used by the Maori in New Zealand as a natural remedy for generations. Its strong antibacterial properties make it particularly effective in destroying certain bacteria in the gut and in treating stomach disorders, including ulcers.

MANUKA HONEY

The honey will be more effective on an empty stomach, so take 1 tablespoon raw Manuka honey three times a day one hour before eating. Continue until symptoms abate.

WOUNDS

As a mild antiseptic, honey has long been used to treat wounds and minor burns, keeping them clean and infection-free. Honey absorbs moisture and discourages the growth of bacteria and other harmful micro-organisms around a skin lesion. It is also used to keep dressings from drying out on a wound, allowing the skin to heal faster and with less scarring.

FOR CUTS AND SCRAPES: Thoroughly clean the area and make sure it is dry. Cover with raw (unprocessed) honey and repeat, cleaning and covering with honey twice a day until completely healed. Alternatively, for minor cuts, clean the cut thoroughly and spread a little honey onto a waterproof dressing. Apply to the cut and change the dressing two or three times a day as necessary.

FOR MINOR BURNS OR SCALDS: If applied directly to a minor burn or scald, honey should prevent the formation of blisters and aid natural healing.

*Note: These remedies are for minor cuts and scalds. **Always seek appropriate medical attention.***

Domestic Crafts

P URE BEESWAX has many applications around the home, from easing a stiff-opening drawer (by rubbing a block of wax along the runners) to treating a saw to prevent it from rusting and sticking in new, green wood (the wax lubricant helps the blade to run smoothly inside the moist, fresh wood). Some of these uses have been known since ancient times, such as the waxing of thread to make it stiffer and smoother; in quilting and heavy sewing, seamstresses will often draw a thread over a small block of beeswax to help the thread pass through tough materials.

ONE OF THE MOST COMMON DOMESTIC USES FOR WAX IS FOR MAKING CANDLES.

Note: When melting beeswax use a double boiler or water bath. Do not place a pan of wax directly on a heat source. Beeswax can easily become damaged by localized overheating and may ignite.

ROLLED BEESWAX CANDLES

Candles made of rolled beeswax foundation are the easiest and safest candles to make, as the wax doesn't need to be melted.

MAKES: 1 CANDLE

SHEET OF BEESWAX FOUNDATION—USUALLY
   8 BY 16 IN
CANDLEWICK—THE LENGTH OF THE SHORT
   SIDE OF THE BEESWAX SHEET, PLUS 1½ IN

1 Place the beeswax sheet on a clean, flat surface. Lay the candlewick just inside the short edge of the foundation so the wick extends beyond the sheet at both ends.

2 Bend the edge of the wax sheet over the wick (about ¼ inch of wax) and press down firmly. Start rolling the candle slowly, pressing gently to avoid compressing the honeycomb pattern on the wax. Make sure the ends are even, and continue rolling until you reach the edge of the sheet.

3 Gently press down the final edge with your thumb or thumbnail. Choose the best end for the top and trim the wick to about ½ inch. Cut the wick off at the bottom.

HAND-DIPPED BEESWAX CANDLES

Beeswax candles are the most natural candles. Scented with the smell of honey, they burn longer and more cleanly than commercial candles made from paraffin wax.

MAKES: 2 (3½-IN) CANDLES

18-OZ BLOCK OF BEESWAX, NATURAL
   YELLOW OR BLEACHED WHITE
CAN OR HEAT-RESISTANT CONTAINER
   TALLER THAN THE FINAL CANDLE LENGTH
   (AN EMPTY STANDARD-SIZED FOOD CAN
   CAN BE USED FOR THIS SIZE CANDLE)
STAINLESS-STEEL THERMOMETER
CANDLEWICK—THE LENGTH OF 2 CANDLES
   PLUS EXTRA
PIECE OF WOOD OR RULER—ABOUT 1½ IN
   WIDE BY 1 FT LONG

1 Melt the beeswax in a double boiler over low heat, then carefully pour the hot wax into the can away from the heat source.

2 Place the can in a heated water bath to keep the wax at a temperature 160 to 170°F. Do not overheat.

3 Cut the wick to about twice the height of the can plus 2 inches and place it around the piece of wood, so the wick hangs down on both sides.

4 To dip two candles at once, lower both ends of the wick into the heated wax for about 3 seconds, so the wick is coated along its full length. Allow the wax to cool on the wick for 1 minute before dipping again.

5 After the first dip, when the wax has cooled a little but is still soft, gently pull the wick to straighten it. After the second or third dip, if the candle is not straight, roll it carefully on a smooth, clean surface to straighten it before dipping again. Repeat until each candle reaches the desired diameter.

6 Hang the candles over a rail and let cool, then trim off the candle bottoms to make them flat and cut the top wick to about ½ inch.

BEESWAX FURNITURE POLISH

One of the oldest wood polishes known, beeswax is still recommended for hardwood furniture care by many furniture conservators and museums. The polish can be applied to most hardwood furniture that does not already have an oil finish and on unsealed wood floors. To apply the polish, put a little onto a clean rag and rub in a circular motion before finishing with strokes along the grain of the wood. Allow to dry for 30 minutes to let the turpentine evaporate, then wipe in the direction of the wood grain with another clean, soft cloth to bring out the shine.

MAKES: 2 (14-OZ) CANS

9 OZ BEESWAX
LARGE GLASS JAR WITH LID
  (ABOUT 1½ PINTS)
1 PINT PURE TURPENTINE

1 Shave the beeswax and melt it in a double boiler over low heat.

2 Place the melted wax in the jar and add the turpentine. Stir well with a wooden spoon.

3 Pour into cans, put on the lids, and set aside for 1 to 2 hours to cool before using.

WAX FURNITURE FILLER

Wax filler is used to repair cracks, splits, and small to medium-sized holes within moldings, furniture joints, or carvings. No sanding is required.

MAKES: ENOUGH TO FILL A CRACK
ABOUT 8 IN LONG, 1/16 IN WIDE, AND
1/2 IN DEEP

3/4-OZ NATURAL BEESWAX STICK

1 Select the appropriate color of beeswax to match your furniture—colors can be mixed for a better match. Take the approximate amount required to fill the gap you wish to repair and melt it in a double boiler over low heat.

2 Allow the wax to cool to a malleable consistency, then fill in the crevices.

WAX VARNISH

Used on furniture or wooden fittings, as with other forms of varnish, as it dries this forms a film that is tough and resilient, keeping out moisture and retaining a shine.

MAKES: ABOUT 1 PINT

1 OZ GRATED BEESWAX
1 PINT LINSEED OIL

In a double boiler, heat the beeswax and the linseed oil over low heat until the mixture forms "strings." The clear portion can then be poured off and painted on as a varnish.

SEALING WAX

The use of seals to secure documents is a practice as old as writing itself. Today they are used to personalize letters and invitations. To melt sealing wax and make a seal, hold the stick of wax at an angle over a flame and drop a small amount onto the flap of the envelope. A stamp to identify the sender can be applied to the wax before it cools and hardens.

MAKES: ABOUT 6 (6-BY-$\frac{1}{2}$-IN) MOLDS

ALUMINUM FOIL
$\frac{1}{2}$ TSP COOKING OIL, TO GREASE MOLDS
$\frac{1}{2}$ OZ GRATED BEESWAX
$\frac{1}{2}$ OZ LARD
$7\frac{1}{2}$ OZ RESIN, SUCH AS PINE RESIN (AVAILABLE FROM HARDWARE AND CRAFT SUPPLIERS)
$\frac{1}{2}$ OZ RED, BLACK OR GREEN ARTIST'S PIGMENT

1 Shape the foil into several molds about $\frac{1}{2}$ inch wide, $\frac{1}{2}$ inch deep, and 6 inches long, and oil the inside with cooking oil.

2 In a double boiler, melt the beeswax and lard over low heat, stirring to combine.

3 Add the resin and pigment and mix well.

4 Remove from the heat, cool slightly, and pour into the molds. Let cool and harden, then remove from the molds.

GRAFTING WAX

This useful wax is essential for horticultural grafting—the method used for propagating plants where the tissues of one plant are fused with those of another. Grafting wax seals the cut surfaces, keeping out water and preventing drying. It can be applied at once or kept for months, even years.

MAKES: ABOUT 13 OZ

9 OZ RESIN, SUCH AS PINE RESIN (AVAILABLE FROM HARDWARE AND CRAFT SUPPLIERS)
$2\frac{1}{2}$ OZ TALLOW (RENDERED BEEF OR MUTTON FAT, AVAILABLE FROM FOOD SUPPLY COMPANIES)
$2\frac{1}{2}$ OZ BEESWAX

1 Melt the ingredients in a double boiler over low heat.

2 Mix well and pour into a heatproof container large enough to hold 14 oz.

3 After cooling, and before use, knead until pliable.

BEESWAX CRAYONS

These pastel crayons are perfect for young children, as the pigment is non-toxic food-coloring paste, available from cake-decorating suppliers.

MAKES: 1–2 SMALL CRAYONS

ALUMINUM FOIL
$^1/_2$ TSP COOKING OIL, TO GREASE MOLDS
2 TBSP GRATED BEESWAX
2 TBSP GRATED SOAP
$^1/_2$–1 TSP FOOD-COLORING PASTE,
   COLOR OF CHOICE

1 Prepare 1 to 2 molds by shaping aluminum foil into small, rectangular block molds and lubricating them with oil.

2 Melt the beeswax in a double boiler over low heat. Add the soap and melt, keeping the heat low and stirring until the mixture is smooth.

3 Add the food-coloring paste and stir thoroughly to combine.

4 Pour the mixture into the molds and let cool. The crayon(s) may be melted again after testing if more coloring is required.

BUMBLEBEE COSTUME

Dress your child up as a bumblebee in this made-to-measure bumblebee costume.

YELLOW AND BLACK FLEECE (AMOUNT
   ACCORDING TO SIZE OF CHILD/ADULT)
ELASTIC
2 BLACK KNEE-HIGH NYLONS
1 PAIR BLACK NYLONS
BLACK PIPE CLEANERS

1 Measure the shoulder width of the child (or adult). Cut the fleece (in both colors) into strips 5 inches long by the width you have just measured, plus an extra 12 inches.

2 To make the front of the costume, sew the strips of black and yellow together, alternately, along the long edges, until the total length matches the desired length of the costume measured from the shoulders to below the waist or knees.

3 Repeat the process for the back of the costume.

4 Lay the front and back of the costume on top of one another (seam side out) and sew the left and right sides, leaving room for the arms.

5 Next, sew the top, leaving room for the head.

6 Sew a piece of elastic inside a hem along the bottom for a better fit around the hips.

7 Use the knee-highs for the arms and the full-length tights for the legs.

8 With the black pipe cleaners, create two antennae to fit on the head.

FRESHLY CUT WAXED FLOWERS

This recipe produces long-lasting preserved flowers that retain the appearance and fragrance of freshly cut flowers.

MAKES: 4 TO 8 WAXED FLOWERS,
DEPENDING ON SIZE AND VARIETY

ABOUT **18** OZ BEESWAX
FRESHLY CUT FLOWERS (TRY ROSES, DAISIES,
   TULIPS, AND HYACINTHS)
ALUMINUM FOIL

1 Line a baking sheet with foil. Melt the beeswax in a double boiler over low heat, and keep the wax hot throughout the following process.

2 Cut a flower stem to 2 inches.

3 Dip the flower head into the melted wax for a very short time (about 1 second) and lift it out, allowing the excess wax to drip back into the pot. For a flower with many petals, spoon wax into the flower center after dipping to ensure good coverage.

4 Let the wax on the flower cool for 30 seconds, then place the flower on the prepared baking sheet for about 5 minutes.

5 When the wax is hard, gently hold the flower head and dip the stem in the wax. Let cool and harden.

6 Repeat the process once more, allowing the wax to cool and harden between each step. Repeat with the remaining flowers.

Bibliography and Further Reading

Aston, D., and Bucknall, S., *Plants and Honey Bees: Their Relationships* (Northern Bee Books, 2004)

Bailey, L., and Ball, B.V., *Honey Bee Pathology* (Academic Press, 2nd ed., 1991)

Beeswax Crafts (Search Press, 1996)

Benjamin, A., and McCallum, B., *A World Without Bees* (Guardian Newspapers Ltd, 2008)

Benton, T., *Bumblebees* (Collins, New Naturalist Series, 2006)

Berthold, R., Jr., *Beeswax Crafting* (Wicwas Press, 1993)

Briggs, M., *Honey and Its Many Health Benefits* (Abbeydale Press, 2007)

Brown, R., *Beeswax* (Bee Books New and Old, 1981)

Butler, C.G., *The World of the Honey Bee* (Collins, 1976)

Caron, D., *Africanized Honeybees in the Americas* (A.I. Root Co., 2001)

Charlton, J., and Newdick, J., *A Taste of Honey* (Reader's Digest, 1995)

Coggeshall, W.L., and Morse, R.A., *Beeswax: Production, Harvesting, Processing and Products* (Wicwas Press, 1984)

Collision, C.H., *What Do You Know?* (A.I. Root Co., 2003)

Cooper, B.A., *The Honeybees of the British Isles* (British Isles Bee Breeders' Association, 1986)

Cramp, D., *A Practical Manual of Beekeeping* (How to Books, 2008)

Crane, E., *Bees and Beekeeping: Science, Practice and World Resources* (Heinemann, 1990)

Crane, E., *A Book of Honey* (IBRA, 1980)

Crane, E., *Honey: A Comprehensive Survey* (Heinemann, 1975)

Crane, E., *The World History of Beekeeping and Honey Hunting* (Duckworth, 1999)

Dade, H.A., *Anatomy and Dissection of the Honeybee* (IBRA, 2009)

Davies, A., *Beekeeping* (Collins & Brown, 2007)

Davis, C., *The Honey Bee Around and About* (Bee Craft Ltd, 2007)

Davis, C., *The Honey Bee Inside Out* (Bee Craft Ltd, 2004)

de Bruyn, C., *Practical Beekeeping* (Crowood Press, 1997)

Delaplane, K., *First Lessons in Beekeeping* (Dadant and Sons, 2007)

Diemer, I., *Bees and Beekeeping* (Merehurst Press, 1988)

Digges, J.G., *The Practical Bee Guide* (Talbot Press)

Edwards, M., *Field Guide to the Bumblebees of Great Britain and Ireland* (Ocelli Ltd, 2005)

Edwards, R., *Social Wasps: Their Biology and Control* (Rentokil, 1980)

Fearnley, J., *Bee Propolis: Natural Healing from the Hive* (Souvenir Press, 2001, reprinted 2005)

Fernandez. N. and Coineau, Y., *Varroa: The Serial Bee Killer Mite* (Atlantica, 2007)

Flottum, K., *The Backyard Beekeeper* (Quarry, 2010)

Flottum, K., *Honey Handbook* (Quarry, 2009)

Free, J.B., *Pheromones of Social Bees* (Chapman and Hall, 1987)

Free, J.B., *The Social Organization of Honeybees* (Edward Arnold, Studies in Biology No. 81, 1977)

Furness, C., *How To Make Beeswax Candles* (British Bee Journal, n.d.)

Goodman, L., *Form and Function in the Honey Bee* (IBRA, 2003)

Gould, J.L, and Gould, C.G., *The Honey Bee* (Scientific American, 1988)

Graham, J.M. ed., *The Hive and the Honey Bee* (Dadant and Sons, 2005)

Hands, P., *Beekeeping for Beginners* (The Kitchen Garden, 2000)

Hanssen, M., *The Healing Power of Pollen and other Products from the Beehive* (Thorsons, 1979)

Herrod-Hempsall, W., *Beekeeping New and Old Described with Pen and Camera* (BBJ, 1930 [Vol. 1], 1937 [Vol. 2])

Hodges, D., *The Pollen Loads of the Honeybee* (IBRA, 1974)

Hooper, T., *Guide to Bees and Honey* (Northern Bee Books, 4th ed., 2008)

Hooper, T., and Taylor, M., *The Beekeeper's Garden* (A. & C. Black, 1988)

Howes, F.N., *Plants and Beekeeping* (Faber and Faber, rev. ed., 1979)

IBRA, *Garden Plants Valuable to Bees* (IBRA, 1981)

Johansson, T.S.K., and Johansson, M.P., *Some Important Operations in Bee Management* (IBRA, 1974)

Kesseler, R., and Harley, M., *Pollen: The Hidden Sexuality of Flowers* (Papadakis Publishers, 2006)

Kevan, P.G., *The Asiatic Hive Bee* (Enviroquest Ltd, 1995)

Kirk, W., *A Colour Guide to Pollen Loads of the Honey Bee* (IBRA, 2nd rev. ed., 2006)

Langstroth, L.L., *Langstroth's Hive and the Honeybee: The Classic Beekeeper's Manual* (Dover Publications, 2004)

Macfie, D.T., *Practical Bee Keeping and Honey Production* (Collingridge, 1936)

Matheson, A., *Practical Beekeeping in New Zealand* (GP Publishing, 1993)

Meyer, O., *Basic Beekeeping: Everything a Beginner Should Know to Ensure Honey from Home* (Thorsons, 1978)

Meyer, O., *The Beekeeper's Handbook: A Practical Manual of Bee Management* (Thorsons, 1981)

Morse, R.A., and Flottum, K., *Honey Bee Pests, Predators and Diseases* (A.I. Root Co., 3rd ed., 1997)

Morse, R.A., and Hooper, T., *The Illustrated Encyclopaedia of Beekeeping* (Blandford Press, 1985)

Oldroyd, B.P., and Wongsiri, S., *Asian Honey Bees: Biology, Conservation and Human Interactions* (Harvard University Press, 2006)

O'Toole, C., and Raw, A., *Bees of the World* (Blandford Press, 1991)

Procter, M., Yeo, P., and Lack, A., *The Natural History of Pollination* (Collins, New Naturalist Series, 1996)

Ribbands, R., *The Behaviour and Social Life of Honey Bees* (Bee Research Association, 1953)

Riches, H., *Insect Bites and Stings: A Guide to Prevention and Treatment* (IBRA, 2003)

Riches, H., *Mead—Making, Exhibiting and Judging* (Riches, H.R.C., 1997)

Ruttner, F., *Biogeography and Taxonomy of Honeybees* (Springer-Verlag, 1988)

Schramm, K., *The Compleat Meadmaker* (Brewers, 2003)

Seeley, T.D., *The Wisdom of the Hive* (Harvard University Press, 1995)

Shimanuki, H., Flottum, K., and Harman, A., *The ABC & XYZ of Bee Culture* (A.I. Root Co., 2007)

Snodgrass, R.E., *Anatomy of the Honey Bee* (Comstock Publishing Associates, 1985)

Stephens-Potter, L.A., *The Beekeeper's Manual* (David & Charles, 1984)

Tautz, J., *The Buzz About Bees: Biology of the Superorganism* (Springer-Verlag, 2008)

Tonsley, C.C., *Honey for Health* (Tandem, 1969)

Traynor, J., *Honey, the Gourmet Medicine* (Kovak Books, 2002)

Verma, L., *Beekeeping in Integrated Mountain Development* (Aspect Publications Ltd, 1991)

von Frisch, K., *The Dance Language and Orientation of Bees* (Oxford University Press, 1967)

von Frisch, K., *The Dancing Bees* (Methuen, 1966)

Vosnjak, M., *The Miracle of Propolis* (Thorsons, 1980)

Waring, A.C., *Better Beginnings for Beekeepers* (BIBBA, 2nd ed., 2004)

Waring, A.C., and Waring, I.C., *Teach Yourself Beekeeping* (Hodder Education, 2006)

Wedmore, E.B., *A Manual of Beekeeping* (Bee Books New and Old, 1979)

White, E.C., *Super Formulas: Arts and Crafts* (Valley Hills Press, 1993)

Winston, M., *The Biology of the Honey Bee* (Harvard University Press, 1987)

Wood, Rev. J.G., *Bees: Their Habits, Management and Treatment* (Routledge, 1853)

Zahradnik, J., *Bees, Wasps and Ants* (Hamlyn, 1991)

Index

abdomen, honeybee:
  drone 97
  effect of low temperatures upon 117
  egg laying, role in 84
  evolution of 84
  flexibility of 84
  hairs on 81, 96
  identifying honeybee and 81, 97
  pollen collection, role in 96
  queen 97, 98
  wax secretion and 57, 97, 126
  worker 94, 97, 126
acacia (wattle, mimosa) 134
Achilles 25
Aculeata 80
Adam, Brother 92
Africa:
  beekeeping in 37
  hive beetle in 144–5
  honey in ritual 23
  killer bees 37, 91
  mythology, bees in 13
African honeybee (*Apis mellifera scutellata*) 91
agricultural economy, importance of honeybee to 7, 122
Ah Mucan Cab 13
Alexander the Great 32
alfalfa 120, 135
almond (*Prunus dulcis*) 132
Amalthea 20
Ambrose, Saint 14, 16, 17
ambrosia 20, 25
America, beekeeping in:
  advent of commercial beekeeping in 49
  bees that saved America 66
  Colony Collapse Disorder in 149–50
  disease in *see* disease
  early beekeeping in 26
  European honeybee in 64–6
  migratory beekeeping in 72, 77
American Beekeeping Federation 263
American foulbrood (AFB):
  attacks old larvae and younger pupae 149, 237
  causes of 149, 237
  controlling 149
  history of 74
  identifying 237, 238
  law and 236
  prevention of 221

spread of 74
  treatment of 237
American Honey Producers Association 263
Amoeba 238, 239
Anaitis 13
Ancient beekeeping 12, 28–34
antennae, honeybee:
  description of 83
  drone 98
  early work on 58
  flagellum and 95
  function of 95–6
  hearing and 95, 113
  Johnston's organ and 113
  sensillae and 95
  smell, prime organ of 95–6
  touch, organ of 95
ants 80, 126, 144
Apiaceae, carrot family 142
Apiaries Act, 1906 (New Zealand) 74
Apiary Inspectors of America 246
apiculture *see* beekeeping
Apiguard 240, 241
Apilife Var 239, 241
Apis cerana (Eastern/Oriental honeybee) 32, 89, 92, 93, 100, 146, 238, 240, 245
Apis dorsata (Giant honeybee) 88, 92, 93, 125, 245
Apis florae and *Apis andreniformis* (Dwarf honeybee) 87, 88, 92, 93, 126, 245
Apis laboriosa (Himalayan honeybee) 26, 88, 245
Apis mellifera (European/Western honeybee) 32, 90, 92, 93, 100, 240, 245
Apis mellifera capensis (Cape honeybee) 91
Apis mellifera carnica (Carniolan honeybee) 91
Apis mellifera lamarcki (Egyptian honeybee) 91
Apis mellifera ligustica (Italian honeybee) 91
Apis mellifera mellifera (Dark European or German black honeybee) 91
Apis mellifera scutellata (African honeybee) 91
Apis nigrocinta (close relative of the Oriental honeybee) 89
Apistan 240, 241
Apitherapy:
  history of 20, 43, 44

honey, medicinal properties of 20, 43, 44, 283–4
pollen, medicinal properties of 300, 301
propolis, medicinal properties of 297–8
recipes 379–88
see also health and healing
Apollo 13, 14
appearance of honeybee 80–3
apples 132
Apples with Honey Yogurt, Honey and Cinnamon, Baked 338, 339
Aristaeus 14
Aristotle 14, 33, 45, 50, 62
"Arrival of the Bee Box, The" (Plath) 21
Artemis 13
Asteraceae, daisy family 142
Atergatis 13
attacks on bees from animals and other insects 144–6 *see also* predators
Australia 68–9
Autumn tasks, beekeeping 234
avocado (*Persea americana*) 132, 274
Ayurvedic medicine 293

Babilos 13
back, beekeeper's 223–4
Baklava 342–3
Balché 13
Bath, Milk and Honey 370
batik, beeswax in 294–5
Baugi 26
Bedouin, Syrian Desert 26
"Bee Meeting, The" (Plath) 21
Bee Research Association 30
bee-eaters 145
beekeeping:
  ancient 12, 28–34
  commercial 250–63
  future of 264–5
  hives *see* hives
  honey and other bee products 267–304 *see also* honey
  hygiene and safety 221–5
  law and 40–2, 167, 190, 252, 257–8
  Middle Ages and beyond 34–40
  migratory 70–2, 76, 77, 122
  practical 153–247
  recipes and home crafts 305–88
  records, keeping 187–90
  scientific advances in 50–63
  year 226–35
"Bees and the Beekeeper, The" (Aesop) 16

beeswax 7, 42, 44, 72, 161
in batik 294–5
candles 46, 48–9, 281, 285, 289–92, 390–1
coating for cheese 296
composition of 285
in cosmetics and pharmaceuticals 280, 285, 292
crayons 394
to decorate eggs 296
dubbin 286
in encaustic painting 295
as a filler 48, 392
as food 48
foundation 286–8
grafting wax 286, 393
harvesting 259–61
lost-wax casting 46, 294
as a lubricant 48, 283
in modeling and casting 294
in musical instruments 48, 296
in polishes 285, 292–3, 391
processing 261, 285
production 124–6
recipes 261
reclaiming from comb 285
as a sealant 46–8, 285–8, 293, 393
secretion of 57, 97, 126
sheets, production of 63
shoe shine 296
uses of 46–9
varnish 392
in watermarks, detailed 296
wax for hair 296
wax tablets 46
waxed flowers 395
BeeVital Ltd 298
"beo-ceorl" 41, 45
Beowulf 45
Best, Henry 50, 52
Bible 34, 268
birds 145–6, 242
birth from carcasses of animals, (ox-born) bee 14, 53, 82
black bees 69
black queen cell virus 149
bladder infections, remedies for 380
blueberry (Vaccinium) 134
boles, bee 39–40, 41
Bonaparte, Napoleon 15, 18
Book of Mormon, The 18
Book of Nature, The (Swammerdam) 55
borage (Borago officinalis) 134
Boraginaceae, borage family 142
boxes, bee 40, 176
bracket/bragget/bragot (beer mixed with honey or mead) 46
brain, bee 55
bramble (Rubus fruticosus) 134, 139
Brassicaceae, cabbage family 142
Bread, Traditional Honey Tea 356
breeding bees 170

Britain:
bee boles 39–40
beekeeping in 49, 92, 165, 166, 258
beekeeping law 41–2, 190, 236, 241, 250, 257–8
beeswax, uses of 47–8
disease see disease
early beekeeping 36, 40
migratory beekeeping 71
British Beekeepers' Association (BBKA) 49, 165, 258
British black bee 92
brood/brood cells/brood box:
Cape honeybee 91
checking the brood pattern 197
disease in 74, 149, 181, 221, 236–8
Dwarf honeybee 87
egg to adult development 99–104
eggs/larvae, evolution of 84, 86, 122
giant honeybee 88
hive layout, position in 155, 157, 161, 163, 165
hygiene and safety of 221–2
nursing/feeding of 87, 94, 99–105, 106, 107, 108, 109, 111, 217–20
records 188–9
replacement of 178, 182, 184, 221–2, 224
swarm prevention/control and 198, 204
transferring new frames into hive 178, 182, 184, 221–2, 224
Buckfast bee 92
buckwheat (Fagopyrum esculentum) 34, 274
bugonia 53
Bulgaria 23
bumblebee:
costume 394–5
identifying 80, 81, 83, 85, 86
pollination 98, 120
Bumby, Mary 69
Burnens, François 58, 59
busy as a bee 12
Butler, Reverend Charles 53–4

cake:
Honey and Apricot Spice 349
Honey Banana 348
Honey Polenta 350, 351
Campanulaceae, bellflower family 142
candles, beeswax 281, 285
church 46, 48–9
dipping 291, 390–1
molded 291–2
pouring 289
rolled 292, 390
Cannabaceae, hemp family 142
canola 135, 272, 277
Cape honeybee (Apis mellifera capensis) 91
caps, wax cell 100, 102, 189, 237, 244, 247, 252–3, 259

capsicumel 46
Carniolan honeybee (Apis mellifera carnica) 91
carpenter bee (Xylocopa) 81, 82
Carr, William Broughton 158
Cartland, Dame Barbara 280
Casas, Bartolomé de las 31, 32
Celsus 298
Central America 66
Chalkbrood 149, 181, 237, 238
characteristics, colony see temperament of bees
Charles II, King 57
Charter of the Forest 42
CheckMite+ 241, 245
cheese, beeswax as a coating for 296
cherry (Prunus) 139
Chicken Breasts, Lemon and Honey–Glazed 320, 321
Childeric I 18
China:
beekeeping in ancient 32, 90
honey used in healing within 44
mead in 45
medicinal use of bee produce 303
royal jelly, largest consumer of 303
chouchenn 46
Christian symbol, bee as 16, 17, 21
Chronic Bee Paralysis Virus (CBPV) 77
Chutney, Apple, Honey, and Chile 358, 359
Cider, Hot Honey 366
Cleanser, Simple Honey Water 371
Cleopatra 280
clothing, protective 171, 172–3, 209, 221
clovers (Trifolium species) 132
coconut palm (Cocos nucifera) 132, 273
Coffee, Spiced Honey 368
Cold Cream 371
colds, remedies for 381
Collin, Abbé 63
colonies, honeybee see hives and nests
Colony Collapse Disorder (CCD) 7, 77, 149–50, 246, 264
columella 15, 26
comb:
building/formation of 46, 97, 106, 109, 115–16, 125–6, 154, 159, 163
early use of 43, 44
extraction of beeswax from 259–61
extraction of honey from 34, 52, 63, 231, 247, 250–3, 254, 262
frames, foundation and 159, 163 see also frames and foundation
fumigating 221
honey direct from 275
organization of within hive 88, 89
removal/replacement of 33, 92, 178, 182, 184, 221–2, 224
temperature control 100, 109, 123, 125

commercial beekeeping 255
  beeswax, extracting and harvesting
    259–61
  beeswax products 263
  brand name, choosing 263
  bulk contract with wholesaler 263
  cleaning up 257
  equipment, basic 250–2
  help with 263
  honey extraction 250–4
  honey filtering or straining 254–6
  honey harvest 250–8
  labeling 263
  law and 252, 257–8
  number of hives 177
  pollination fees 177, 262–3
  presenting honey 258
  regulations 257–8
  selling products 177, 257–8, 262–3
  setting up a commercial apiary
    262–3
  small-scale 33–4
  storage and bottling, bulk 256–7
  supers, reusing empty 254
commonwealth, bee 14–15
communication, bee 28, 54, 87, 92, 95,
  110, 112–13
*Complete Guide for the Management of
  Bees Throughout the Year, A*
  (Wildman) 60
constipation, remedies for 382
Cookies, Walnut and Honey 344
Cooperative Movement, Lancashire 18
cosmetics and pharmaceuticals, beeswax
  in 280, 285, 292
Cotton, Reverend William Charles 69
cotton (*Gossypium hirsutum*) 136
Cough Syrup, Honey, Thyme, and
  Licorice 383
coughs, remedies for 383
Coumaphos 241
coursing 26
Cowan, T.W. 63
Cranachan 340
Cranbrook, Lord 136
Crane, Dr. Eva 30, 32
crayons, beeswax in 394
Cybele 13
cyser 46, 279
czwornniak 46

Dadant hive:
  frames and foundation 160, 162,
    164
  invention of 61
  single-walled hive 158
dances, honeybee:
  differ between type and species of
    honeybee 54
  dwarf honeybee 87, 92
  flagellum and 95
  Johnston's organ and 95, 113

new forager recruit learns signals of
  110
observation of 112, 113
optic flow, measurement of distance
  and 112
round dance 112
significance of 112–13
use of sound vibration in 95, 113
waggle 28, 112–13
dandelion (*Taraxacum officinale*) 132,
  273
dark bees 50, 52, 91
dark European or German black honey
  bee (*Apis mellifera mellifera*) 91
Darwin, Charles 104
De Clementia (Seneca) 15
De Hruschka, Francesca 63
De re rustica (Columella) 15, 26
de Réaumur, René-Antoine Ferchault
  58, 63
death's head hawkmoth (*Acherontia
  atropos*) 144, 145
defecation, bee 109, 117, 168, 227, 230
deformed wing virus 149
Delphi 13
Delphic Bee 13
Demeter 12
democracy, bee colony as model for 16
demonstrations, bee 60
Department of Agriculture, US 246, 248
Description of the bar-and-frame hive, A
  (Munn) 61
Deseret Times 18
destroying bees 33, 59, 201
Dipsacaceae, teasel family 142
disease and pests:
  American Foulbrood (AFB) 74,
    149, 221, 236, 237, 238
  Amoeba 238, 239
  black queen cell virus 149
  Chalkbrood 149, 181, 237, 238
  Chronic Bee Paralysis Virus (CBPV)
    77
  Colony Collapse Disorder (CCD) 7,
    77, 149–50, 246, 264
  death's head hawkmoth (*Acherontia
    atropos*) 144, 145
  deformed wing virus 149
  dysentery 74, 229–30, 238
  European Foulbrood (EFB) 74, 149,
    221, 236, 237
  Isle of Wight 74, 77, 91, 92, 149
  Israeli Acute Paralysis Virus 246
  large hive beetle (*Hyplostoma
    fuligineus*) 145
  law and 42
  louse, bee (*Braula coeca*) 144
  Nosema 74, 77, 149, 229–30, 238–9,
    246
  Sacbrood 149, 236, 238
  small hive beetle (*Aethina tumida*)
    144–5, 236, 245, 264

Stonebrood 149
tracheal mite (*Acarapis woodi*) 74,
  77, 146, 239
Tropilaelaps mites 236, 245, 265
varroa 68, 74, 146, 149, 150, 151,
  155, 156, 166, 180, 181–2, 222,
  240–1, 246, 264
wax moths (*Achroia grisella* and
  Galleria mellonella) 144, 145, 244,
  246 see also predators
distance, honeybee judgment of 112
Domesday Book 41
Dressing, Warm Honey and
  Whole-Grain Mustard 359
Drink, Warm Honey and Cinnamon
  380
drone honeybee:
  abdomen 97, 98
  antennae 98
  cells 103, 110
  culling of 15
  emergence from cell 100
  evolution of 85, 105
  eyes 97, 98
  function/description of 50, 53–4, 94,
    97
  jaws 98
  larvae 103
  life-span 103
  mating 28, 116
  retention of within colony 232, 234
  sting 97, 98
  tongue 97, 98
dubbin, beeswax in 296
Duck, Honey and Orange Roast 318, 319
dwarf honeybee (*Apis florae* and *Apis
  andreniformis*) 87, 88, 92, 93, 126, 245
dwojniak 46
Dyce, Elton J. 277, 279
dysentery, honeybee 74, 229–30, 238
Dzierzon, Jan 36, 63, 154–5

Eastern honeybee (*Apis cerana*) 32, 146,
  238, 245
edible honey:
  early spring/long flowering 132–5
  late flowering 136
Edward VI, King 15
Eggnog, Honey 361
eggs, honeybee:
  evolution of 84, 86, 122
  laying and hatching of 99–105 see
    also larvae
  nursing/feeding of 87, 94, 99–105,
    106, 107, 108, 109, 111, 217–20
    see also brood/brood box/
    brood cells
Egypt, ancient 37, 53, 90, 91
  bee hieroglyphs 30, 31, 34
  bee as sacred emblem of royalty 18, 30
  beeswax, uses in 47–8
  Ebers papyrus 43, 44

first beekeepers 12, 30–1
honey in healing 44
honey in ritual 21, 23, 25
opening of the mouth ceremony 21, 25
sacred significance and mythical power of bees 12, 13, 18, 30, 31, 34
Egyptian honeybee (*Apis mellifera lamarcki*) 91
Elaeagnaceae, Elaeagnus family 142
El-Beithar, Ibn 300
Elyot, Thomas 15
Elysium Britannicum (Evelyn) 62
encaustic painting, beeswax in 295
environment, importance of honeybee to 119–22
Environmental Protection Agency (EPA) 241
equipment, buying beekeeping 178–9
Eros 25
eucalyptus 274, 306
Eucharist 21
Euphora sepulchralis (scarab beetle) 145
European Foulbrood (EFB) 74, 149, 221, 236, 237
European social wasps (*Vespula*) 86, 144
European/Western Honeybee (*Apis mellifera*) 32, 100
    description of 90
    global spread of 64–77
    nests 92, 93
    threats to *see* disease, pests, and predators
Eurydice 14
"Eve of St. Agnes, The" (Keats) 25
evolution of bees:
    colony 104–5
    flowers and 120, 122, 123
    group selection 104–5
    kin selection 105
    social bees 84–6
excluders, queen 63, 195, 196, 198, 212, 222, 228
eye, honeybee compound:
    discovery of 55, 56
    drone 98
    identifying honeybee and 83
    light polarization and 95
    ocelli and 95, 98
    ommatidia and 94
    optic flow and 112
    optic nerve and 55
    "pol" 95
    queen 98
    Swammerdam's drawing of 56
    worker 94–5
Eye Treatment, Honey and Cucumber 374

Fabaceae, pea family 142
Fable of the Bees: or, Private Vices, Publick Benefits, The (Mandeville) 16

face pack:
    Chamomile and Honey 372
    Lemon and Honey 372
false acacia (*Robinia pseudoacacia*) 134
Fearnley, James 298
feeding bees:
    amount of feed 217
    ancient instruction on 33
    candy 220
    feeders 218–20
    queen 106
    robbers and 217
    spring/summer 232
    syrup 216, 217–18
    winter and 228
    worker 106, 107, 108
Feminine Monarchie or a Treatise Concerning Bees, and the Due Ordering of Them, The (Butler) 52, 53–4
feral bees 65–6
fireweed (rosebay, willowherbs) 136
flight, honeybee:
    birth to early 108–10
    flight hole 30
    maiden 109
    nuptial 99, 116
    reconnaissance 109
    see also forage *and* pollination
flowers:
    pollination of 119–22, 132–43
    see also plants and flowers
    waxed 395
fly-catchers 145
folklore, beekeeping 43
Food and Drug Administration (FDA) 258
Food Standards Agency (FSA) 250
food:
    bee processing of 80, 84, 85, 86, 96, 98, 100, 102–3, 106, 108–9, 110–11, 112–13, 119–22, 123–7, 154, 189
    beeswax as a 48
    brood, feeding of 87, 94, 99–105, 106, 107, 108, 109, 111, 217–20
    feeding bees 33, 106, 107, 108, 215–20, 228, 229, 232
    pollen as a food supplement 298
    recipes 305–68
    see also honey, pollination *and under individual recipe*
Fool, Honey and Ginger Rhubarb 346, 347
Foot Treatment, Peppermint and Beeswax 375
forage 111
    availability of 54, 178
    communication during 112–13
    dances and 112–13 *see also* dances, honeybee
    effects of water upon 113
    location of 112–13

location of hive during 179
nectar 111, 123 *see also* nectar
plant evolution and 122
pollen 111 *see also* pollen
propolis 111 *see also* propolis
reception of incoming forager bees 110, 123
scouting for 87, 106, 110, 115, 154
forests, bee 34, 36
foundation, wax 214
    commercially produced 286–8
    first manufacture of 63
    frames and 161–3, 165, 182
    making 161–3, 261, 286–8
    matrix, using a 287–8
    ordering 178–9
    press, using a 286–7
    wiring 288
frame spacers 157, 161, 165, 166
frames, hive 159–60
    bottom bars 159, 161
    British Standard (BS) 160–1
    cracking apart 182
    foundation and 161–3, 165, 182
    in hive layout 156, 157, 158, 159–61
    Hoffman 161, 251
    inspecting a 193, 194, 195
    making up 162–3, 165, 166
    movable-frame hives 36, 61, 63, 74, 114, 154–5, 156
    nucleus box 181, 182, 183, 184, 193, 200–4, 215
    in a radial extractor 252
    side bars 159, 160–1
    top bars 159, 160
    transferring bees and 182, 183, 184
    turning the 193
France, migratory beekeeping in 71, 76
Freemasons 18, 19
Frisch, Karl von 112, 113
fuchsia (*Fuchsia*) 136
Fumagil-B 238
fumagillin 238
future of beekeeping 264–5
Galen 302
gall wasps 80
Gallaland 23
Gard Star 245
garden:
    bee-friendly 139–43
    keeping bees in 167, 168, 169
Georgics (Virgil) 15, 33–4, 45, 50, 51, 53
Germany, beekeeping in 34, 36, 37
giant honeybee (*Apis dorsata*) 88, 92, 93, 125, 245
globe thistle (*Echinops*) 139
goldenrod (*Solidago*) 136, 139
gooseberry (*Ribes uva-crispa*) 132
grafting wax, beeswax in 286, 393
Greece, Ancient:
    Attic honey 43
    beekeeping in 32, 33, 90

beeswax uses in 46
honey in ritual 23, 24, 279
mythology 13, 14
Grossulariaceae, gooseberry family 143
guards, hive 106, 109, 110, 144
gum tree (*Eucalyptus* species) 133
Gurung people, Nepal 26
gut parasite (*Nosema apis*):
  causes of 74, 149, 238
  CCD and 246
  checking for 229–30
  defecation in hive and 74
  diagnosing 239
  spread of 238
  symptoms of 149
  treatment of 149, 238

Hackenberg, Dave 246
hair:
  honeybee 80, 81, 84, 96, 98
  wax for 296
Hair Conditioner, Creamy 370
Ham, Honey Roast 314, 315
Hamilton, William D. 105
Hand Cream, Lavender and Beeswax 374
Hangover Cure, Honey, Orange, and Egg 382
hangovers, remedies for 283–4, 382
Harbison, John 66
Hartlib, Samuel 52–3
hawthorn (*Crataegus monogyna*) 134, 274
hay fever, remedies for 384
health and healing, bee produce and:
  history of 20, 43, 44
  honey, medicinal properties of 43, 44, 283–4
  pollen, medicinal properties of 300, 301
  propolis, medicinal properties of 297–8
  recipes 379–88
hearing, honeybee 95
heather (bell) 136, 274, 306
heather (ling) 136, 274
Henry III, King 42
Henry V (Shakespeare) 15
heraldry, bees in 18
Himalayan honeybee (*Apis laboriosa*) 26, 88, 245
Hinduism 13, 21, 23, 32
Hippocras 279
Hippocrates 44, 298, 302
Historia Animalium (Aristotle) 14, 33, 57
history of bees and beekeeping 11–77
  bees in myth and symbol 12–25
  European honeybee, global spread of the 64–77
  honey hunting and early beekeeping 26–49

scientific advances in beekeeping 50–63
History of Beekeeping in Britain (Fraser) 39
Hittite Empire 40
Hive and the Honey-Bee, The (Langstroth) 61
hive beetle:
  large (*Hyplostoma fuligineus*) 145
  small (*Aethina tumida*) 144–5, 236, 245, 264
hives:
  ancient 29, 30–1, 32, 33, 34
  bar 59, 61
  basket-weave "skeps" *see* skeps
  bee space 61, 156, 158, 178
  bee-tight 154, 165, 243
  bottoms 156, 166
  brood box/cell 94, 155, 157, 161, 163, 165, 178, 182, 184, 188–9, 197, 198, 204, 221–2, 224
  caps, wax cell 100, 102, 189, 237, 244, 247, 252–3, 259
  choosing suitable 154–64
  clean, keeping 33, 226
  coating "cloomed" 36
  Commercial 158, 160, 164, 177, 254
  creating and keeping your 165–6
  Dadant 61, 158, 164
  debris, clearing out 106, 108, 110, 240
  designs, innovations in 59–63
  dividing 30–1
  division of labor within 106, 107, 108
  early English 36
  earthenware 33
  entrance 34, 54, 106, 155, 156, 178, 179, 228, 232, 234
  ferula 33
  flight hole 30
  foundation 63, 161–3, 165, 178, 213
  frames 159–61, 162–3, 165, 166, 181, 182, 184, 193, 194, 195, 251, 252
  frame spacers 157, 161, 165, 166
  garden 167, 168, 169
  guards 106, 109, 110, 144
  how many? 177–8
  industry, as symbol of 12
  inspections, colony 155, 158–9, 181–2, 190, 191, 192–7, 227
  Langstroth 61–2, 63, 160, 161, 164, 165
  layout 155–8
  leaf hive (Huber) 58–9, 62
  location of 167–71
  log 36
  Middle Ages 36
  Modified National 164
  movable-frame 36, 61, 63, 74, 114, 154–5, 156
  neighbors and 42, 141, 167–8, 198

nucleus box 181, 182, 183, 184, 193, 200–4, 215
number of bees in 94
observation/glass 58, 60, 62–3
occupants of 94–8
Omlet's Beehaus 264, 265
out-apiaries 171
paint 165
placing on site 179
plastic 265
Porter bee escape 158, 247
queen excluder 156–7, 181, 196, 198–9, 212, 222, 228
registering 167, 177
roof 156, 158
runners 155–6, 157, 165
screen, bee-proof 154, 155, 156, 158, 166
shading entrance of in snowy conditions 228
shed, keeping in 168
single- or double-walled 158–9
skeps *see* skeps
Smith 164, 165
social life in 57, 106–11
solar 264, 265
stand 156, 179
suitable bees for, finding 179
supers 156, 157–8, 165, 181, 195, 196, 199, 203, 231, 247, 250, 252, 254, 259
swarms, hiving 39, 50, 54, 159, 198, 208–14
temperament of bees within 54, 170, 171, 180, 182, 189, 190, 222
temperature within 100, 109, 123, 125
tool 170, 176, 196, 201, 221
transferring bees to 182–4
types of 29, 30–1, 32, 33, 34, 36, 59–63, 74, 114, 154–64, 221, 234, 250–1, 264, 265
urban 170
ventilation 86, 168, 228, 229, 230
water evaporation 86, 168
water source and 170–1
waterproofing 36, 226
WBC 158, 159, 160, 164
wood, thickness of 165–6
wooden 33, 36, 59
holly (*Ilex aquifolium*) 133
honey 266–84
  acacia 271, 308
  alfalfa 271
  antimicrobial 24
  apple 271
  beauty benefits of 280
  birth and 21
  blended 278, 306
  blossom 277
  blueberry 271
  borage 134, 142, 271, 277

bulk storage and bottling 256–7, 262
canola (*Brassica napus* subsp.
   oleifera) 135, 272, 277
chestnut 274, 306
cleaning up 257
clover 271
cold-pressed 278
color, aroma, and flavor 269–74
composition and quality 268
consumption and production 124
cooking with 306
cotton 271
coughs, colds and 283
crystallization 275, 276, 277, 279,
   281
cut comb and sections 275
dark 269, 274, 281
death and 24–5
deposited in comb storage cells 125
different types of 33
drinks 279–80 *see also* mead
embalming and 24–5, 43
exhibiting products 281
extraction 34, 52, 63, 231, 247,
   250–3, 254, 262
false acacia 271
fireweed 271
as food 124
frosting 258, 276
fuchsia 271
by geographic origin 278
goldenrod 271
granulated, naturally 276
great eloquence and (honey-
   tongued) 12, 14, 16, 20, 25
hangovers and 283–4, 382
harvest 54, 161, 250–1
heat and 269
holly 272
honeydew and 268, 269
hygiene and preparations 252
immortality and 25
ivy 272
jarrah 283
knapweed 272
labeling 258, 270, 278, 281
leatherwood 272
light-colored 271–2
ling heather 136, 274, 277
liquid 275–6, 281
local 384
longevity of 25, 44
love and 25
manuka 274, 283, 388
maple 272
marriage and 23–4
mead *see* mead
measuring and storing 306
medicinal properties of 43, 44,
   283–4
medium-colored 273
melilot 272

mesquite 272
monofloral and multifloral 277, 306
mythology of 12
nectar to 123–4
nectar, ambrosia and 20, 25
organic and fair-trade 278
presenting 258
preservative 24, 43
regulations for selling 257–8
remedies 20, 43, 44
rewarewa 274
in ritual 21–5
selling 177, 257–8, 263
soft set 275, 276–7
storage 86, 124, 125, 279
straining 254, 255, 256
symbolism of 20–5
taxes and tributes, use as 42, 43
toxic/poison 45, 137, 138, 278–9
tulip tree 274
types of 33, 270–9, 306
ulcers and 284
uncapping 252–3
uses of 43–6
warming 256
wax production and 126
who shouldn't eat? 270
honey badger or ratel (*Mellivora
   capensis*) 144, 147
honeybee, understanding the 78–151
   communication 112–13
   evolution 84–6
   from egg to adult 99–105
   hive *see* hive
   honey *see* honey
   identifying 80–3
   man's best friend 119–27
   nests 84–5, 90–2, 93
   occupants of the hive 94–8
   pollination 130–42
   social life in the hive 106–11
   species 87–91
   sting *see* sting
   swarming *see* swarming
   threats to 144–51
   winter survival *see* winter, honeybee
      survival during
honeydew 268, 269
honeymoon, etymology of 25
Hooke, Robert 54–5
Hopkins, Isaac 74
hornets (*Vespa* species) 81, 82, 83, 144,
   146
houses, bee 40
hover flies 80, 82, 83
How to Acquire Wealth (Fan-Li) 32
Huber, François 57, 58–9, 62
Huber, Maria 58
Hughes, Ted 21
Hummus, Spiced Honey 313
Hungary 24
Hunter, John 57

hunting, honey 26–8
hydromel 46
hygiene and safety 221–5, 250
Hymenoptera 80, 105

ichneumons 80
identifying the honeybee 80–3
India 13, 14, 20
Indicator indicator (Greater
   Honeyguide) 26, 28, 144
industrious nature, bee 17–18
inspections, colony:
   before buying 181–2
   brood pattern, checking 197
   checking for colonies retaining their
      drones 232, 234
   checking hives are stable and
      waterproof 227
   checking the colony isn't starving
      229
   checklist 190
   closing the hive 195–6
   cover cloths 196
   early spring 229–30
   effect of different types of hive on
      155, 158–9
   finding the queen 194
   inspecting a frame 193
   late spring/summer 231, 232, 234
   moving to the next frame 194
   opening the colony 192
   to prevent swarming 231
   propolis and 193
   queen excluder and 196
   records of 190–7
   turning the frame 193
   using the smoker during 192
inspector, bee 181, 237
Integrated Pest Management (IPM)
   program 166, 240, 241, 264
International Bee Research
   Association (IBRA) 30, 40, 283
Isaiah 21
Islamic tradition 20
Isle of Wight disease 74, 77, 91, 92, 149
Israeli Acute Paralysis Virus 246
Istar 13
Italian honeybee (*Apis mellifera
   ligustica*) 91
ivy (*Hedera helix*) 136

Janscha, Anton 59
Johnston's organ 95, 113
Jupiter (Zeus) 20

Kai Si (Pu Me) 32
Kama 13, 25
Kashmir 32
Kattunaikkar tribe, India 26
Kecskemét 24
keps, bee 40
killer bees 37, 91

Kingdon-Ward, Frank 137
knapweed (hardheads, blackhead) 134
Kung Bushmen of Kalahari 13

La Cueva de la Araña (Cave of the
 Spider), Spain 13, 26
Lamb, Honey and Roast Garlic Roast
 332
Lamb with Thyme and Honey Glaze,
 Rack of 328, 329
Lamiaceae, mint family 143
Langstroth, Reverend Lorenzo L. 61–2,
 63, 74, 155, 158, 162, 164, 303
Langstroth hive:
  dimensions and features 164
  frames 160, 161
  invention of 61, 62, 63
  layout of 61
  self-made 165
larvae:
  evolution of 84, 86. 122
  from egg to adult 99–104
  nursing/feeding of 87, 94, 99–105,
   106, 107, 108, 109, 111, 217–20
  see also brood/brood box/
   brood cell
lavender (*Lavendula*) 135, 139, 273
law, beekeeping and the:
  ancient 40–1, 42
  English 41–2
  Hittite Empire 40
  honey 252, 257–8
  hygiene 252
  Irish 42
  legal records 190
  Middle Ages 41–2
  Roman 40–1
  selling bee products and 247–8
  where to keep your bees and 167
leaf-cutter bees (*Megachile*) 80, 81
leaf hive (Huber) 58–9, 62
leatherwood (*Eucryphia lucida*) 133
Leeuwenhoek, Anthony van 54
life cycle, honeybee 57, 58
life spans, honeybee 103
Liliaceae, lily family 143
lime tree (linden, basswood) 135, 273
Limnanthaceae, meadowfoam family
 143
lip balm 377
lipstick 378
literature, bees in 21, 25
location of bees/hive:
  bee shed 168
  bees, flight path and 168
  damp locations, avoid 168
  fields 167
  gardens 167, 168, 169
  law and 167
  neighbors and 167–8, 170
  out-apiaries 171
  planning 167

registering 167
swarming and 167, 168
temperament of bees and 170, 171
urban 167, 170
vandalism and 171
water sources 170–1
"log hive" 36
Lost Gardens of Heligan, Cornwall,
 England 40
louse, bee (*Braula coeca*) 144
Lythraceae, loosestrife family 143

Ma 13
Madame Tussaud's 48, 294
Madhava 13
Madheri 13
Magna Carta 41–2
maiden flights, honeybee 109
Maimonides 300
Malvaceae, mallow family 143
Manuka (*Leptospermum scoparium*) 133,
 284
maple tree (*Acer* species) 133
Maraldi, Giacomo 62–3
Marsden, Reverend Samuel 68
Marx, Karl 16
material benefits of honeybee 119–22
mating, honeybee 58, 59, 85, 116
"Maud" (Tennyson) 25
Maya, Ancient 13, 31, 44, 46
McCain, J. N. 66
mead 7, 360
  exhibiting 281
  history of 20, 23, 32, 44–5
  making of 52, 360
  in Rig-Veda 20, 32, 45
  traditional 360
  types of 46
  Vikings and 23, 45
meadow flower 273
medieval apiary 22
Mehring, Jan 63
melilots (sweet clover) 135
Melipona beecheii 13, 66
Melittosphex burmensis 84
mellissae 13
Melomakarona and Honey Syrup 345
melomel 46, 279, 281
*Mémoires pour servir à l'historie des
 insectes* (de Réaumur) 63
Merovingians 18
Mesolithic era 26
mesquite (*Prosopis glandulosa*) 133
Meteorologica (Aristotle) 45
metheglin 46, 279, 281
mice 242–3
Micrographia (Hooke) 55
microscope, effect of upon knowledge
 of bees 54–7
Middle Ages, beekeeping in 34–40
migratory beekeeping 70–2, 77, 122
Miller, Nephi Ephraim 72

mining bee (*Andrena*) 81
Mint Julep, Honey 362, 363
MiteAway II 241
mites and parasites:
  tracheal mite (*Acarapis woodi*) 74,
   77, 146, 239
  Tropilaelaps 236, 245, 265
  Nosema apis 74, 77, 149, 221,
   229–30, 238–9, 246
  varroa 68, 74, 146, 149, 150, 151,
   155, 156, 166, 180, 181–2, 222,
   240–1, 246, 264
  *see also under individual mite and
   parasite name*
model of human society, bee colonies
 as a 14–15
modeling and casting, beeswax in 294
Modified National hive:
  dimensions and features 164
Mormonism 18
Morocco 23
Moroccan Honey-Chicken Tagine with
 Prunes 322
mountain laurel (lambkill) 138
mouthparts, honeybee 55, 83, 96, 98,
 108
movable-frame hives:
  Dadant hive *see* Dadant hive
  design 36, 154–8
  hive layout 154–8
  invention of 36, 61, 63, 74
  Langstroth hive *see* Langstroth hive
   61
  swarming and 114
Muffins, Spiced Cinnamon and Honey-
 Sweetened Date 352, 353
mulsum 46
musical instruments, beeswax in 48, 296
My Bee Book (Cotton) 69
mythology, bee:
  African 13, 23
  bee as bridge between world and
   underworld 12
  Christian 16, 17, 21
  commonwealth, bee 14–15
  Egyptian 12, 13
  as emblem of thrift and industry
   17–18, 19
  as embodiment of soul 12
  Greek 12, 13, 14
  heraldry 18
  India (Hinduism) 13, 20, 23
  Mayan, Ancient 13
  as model of human society 14–16
  Northern Europe 13
  of the origin of bees 14, 53, 82
  symbolism 14–19
  tombs, appearance of bees on 12

Nailsworth, Gloucestershire 40
Nanosvelta 13
Nasonov pheromone/glands 115, 210

National Bee Unit (NBU) 245
National Honey Board 263
National Honey Show 49, 281
National Show of Bees and Honey,
　Crystal Palace, 1934 281
Naturalis Historia (Pliny the Elder) 45,
　50, 62, 302
nectar 85, 268
　　in classical mythology 20, 25
　　foraging for 111
　　to honey, process of 123–4
　　plant production of 120
　　see also pollen *and* pollination
needle bush (silky oak) 133
neighbors, effect of beekeeping upon
　42, 141, 167–8, 198
neonicotinoids 246
Nero, Emperor 15, 280
nests 92–3, 154
　　Cape honeybee 91
　　comb organization 88, 89
　　dark cavity 89, 90
　　dwarf honeybee 87, 93
　　evolution of 84–6
　　giant honeybee 88, 93
　　intruders 92
　　Oriental honeybee 89, 93
　　propolis in 92
　　social nesting 105
　　wasps 84–5
　　Western honeybee 90, 93
New Zealand 69, 74, 86
niches, bee 40
Nosema apis:
　　causes of 74, 149, 238
　　CCD and 246
　　checking colonies for 229–30
　　defecation in hive and 74
　　diagnosing 239
　　effects of 149, 238
　　spread of 74, 238
　　symptoms of 149
　　treatment of 149, 221, 238–9
　　in UK 238
　　viruses associated with 77
Nouvelles observations sur les abeilles
　(Huber) 58

ocelli, honeybee 95, 98
Odin 27, 45
odors, bees, dislike of certain 54
OIE (World Organisation for Animal
　Health) 236
Okrap 23
Omlet's Beehive 264, 265
omphacomel 46
Onagraceae, willowherb family 143
optic flow, honeybee 112
optic nerve, honeybee 55
orange blossom 273
oranges 133
Oriental honeybee (*Apis carana*) 88, 92,

93, 100, 240
origin of bees 14, 53, 82, 84–6
Orpheus 14
out-apiaries 171
ovaries, honeybee 55, 100
Owl and the Pussycat, The 270
oxalic acid 240, 241
ox-born bee 14, 53, 82
oxytetracycline 237

Pantopolion (Georg Pictorius of
　Villingen) 52
parasites *see* mites and parasites
paschal candle 34, 49
Pepys, Samuel 55
perennial colonies 85
perfume, solid 375
pesticides 141, 246
pests:
　　ants 80, 126, 144
　　bee louse 144
　　death's head hawkmoth 144, 145
　　hive beetles 144–5, 236, 245, 264
　　honey badger 144, 147
　　honeyguide birds 26, 28, 144
　　mice 242–3
　　rats 243
　　wasps *see* wasps
　　wax moths 144, 145, 244, 246
　　see also disease *and* mites and
　　　parasites
pheromone secretion:
　　alarm 96, 189
　　queen 55, 98, 114, 115, 116
　　sex 55, 116
　　swarming 210
Phyllo, Goat Cheese, and Honey Parcels
　333
Physiologus 50
Plant-Hunter's Paradise (Kingdon-
　Ward) 136
plants and flowers 120, 122, 123, 132–43
　　acacia (wattle, mimosa) 134
　　alfalfa 122, 135
　　almond (*Prunus dulcis*) 132
　　Apiaceae, carrot family 142
　　apples 132
　　Asteraceae, daisy family 142
　　to attract honeybees 139–41
　　avocado (*Persea americana*) 132, 274
　　blackberry 273
　　blueberry (*Vaccinium*) 134
　　borage (*Borago officinalis*) 134
　　Boraginaceae, borage family 142
　　bramble (*Rubus fruticosus*) 134, 139
　　Brassicaceae, cabbage family 142
　　buckwheat (*Fagopyrum*
　　　esculentum) 134, 274
　　Campanulaceae, bellflower family
　　　142
　　Cannabaceae, hemp family 142
　　canola (*Brassica napus* subsp.

oleifera) 135, 272, 277
cherry (*Prunus*) 139
chestnut 274, 306
clovers (*Trifolium* species) 132
coconut palm (*Cocos nucifera*)
　132, 273
common garden 142–3
cotton (*Gossypium hirsutum*) 136
dandelion (*Taraxacum officinale*)
　132, 273
Dipsacaceae, teasel family 142
edible honey: early spring/long
　flowering 132–5
edible honey: late flowering 136
Elaeagnaceae, Elaeagnus family 142
eucalyptus 274, 306
Fabaceae, pea family 142
false acacia (*Robinia pseudoacacia*)
　134
fireweed (rosebay, willowherbs) 136
fuchsia (*Fuchsia*) 136
globe thistle (*Echinops*) 139
goldenrod (*Solidago*) 136, 139
gooseberry (*Ribes uva-crispa*) 132
Grossulariaceae, gooseberry family
　143
gum tree (*Eucalyptus* species) 133
hawthorn (*Crataegus monogyna*)
　134, 274
heather (bell) 136, 274, 306
heather (ling) 136, 274, 277
holly (*Ilex aquifolium*) 133
ivy (*Hedera helix*) 136
knapweed (hardheads, blackhead)
　134
Lamiaceae, mint family 143
lavender (*Lavendula*) 135, 139, 273
leatherwood (*Eucryphia lucida*) 133
Liliaceae, lily family 143
lime tree (linden, basswood) 135,
　273
Limnanthaceae, meadowfoam
　family 143
Lythraceae, loosestrife family 143
Malvaceae, mallow family 143
manuka (*Leptospermum*
　scoparium) 133, 284
maple tree (*Acer* species) 133
meadow flower 273
melilots (sweet clover) 135
mesquite (*Prosopis glandulosa*) 133
mountain laurel (lambkill) 138
native 140–1
needle bush (silky oak) 133
Onagraceae, willowherb family 143
orange blossom 273
oranges 133
poison honey 137, 138
Polemoniaceae, phlox family 143
Primulaceae, primrose family 143
privet (*Ligustrum vulgare*) 138
ragwort (*Senecio*) 138

rata tree (*Metrosideros*) 135, 272
rhododendron (*Rhododendron ponticum*) 137, 138
Rosaceae, rose family 143
rosemary (*Rosmarinus officinalis*) 133, 273
Rutaceae, citrus family 143
sainfoin (*Onobrychis vicifolia*) 135, 272
spurge (*Euphorbia*) 138
sunflower (*Helianthus anuus*) 136, 272
sweet chestnut (*Castanea sativa*) 135
thistle (*Cirsium arvense*) 135
thyme (*Thymus*) 133, 273
traveler's joy (*Clematis*) 139
tulip tree (yellow poplar) 135
tupelo (sour gum) 133, 273
Valerianaceae, valerian family 143
Plath, Sylvia 21
Plato 14
Pliny the Elder 45, 50, 62, 302
Pork Stir-Fry, Citrus Honey and Balsamic 316
Poland 23, 34, 36
Polemoniaceae, phlox family 143
polishes, beeswax in 285, 292–3, 391
political and economic model, bee colony as a 16
pollen:
    collection 96, 98, 120, 122, 123, 126, 132–43
    corbicula (pollen basket) 96, 98, 126
    as food for honeybee 85
    as a food supplement 298
    harvesting 111
    in hive 102
    local 384
    medicinal benefits of 300, 301
    protein, as source of 102, 111
    storage cells 87
pollination of flowers:
    evolution of 120, 123
    financial implications of 122
    flower fidelity 120
    hairs on honeybee's body and 96, 98
    plants for honeybees: edible honey: early spring/long flowering 132–3
    plants for honeybees: edible honey: late flowering 136
    plants for honeybees: edible honey: late spring/summer flowering 134–5
    poison flowers 137, 138
    worker bee and 96, 98
    see also plants and flowers
poltorak 46
Poppea 280
Porter, E.C. 63
Poussin, Nicolas 20
predators, bee:

bee-eaters 145
birds 145, 242
robber bees 54, 84, 144, 232, 242, 243–4
spiders 146
wasps 54, 144, 146, 243
wax moths 144, 145, 244, 246
woodpeckers 145–6
see also pests
Primulaceae, primrose family 143
privet 138
propolis:
    allergies 298
    collection of 111, 126
    function of 92, 111, 126, 155, 196, 242
    hygiene, safety and 222
    nature's miracle cure 297–8
    removing/cleaning 196
    variation in amount of collection between colonies 193
pyment 46, 279

queen, honeybee 12, 94, 97, 98, 99, 102
    absolute rule of 16
    attendant bees and 28, 184
    buying by post 184
    cells 100, 103, 104, 114, 115, 116, 189, 190, 197, 199–200, 201, 202, 203, 204, 205, 206, 208, 214
    eggs, laying of 94, 97
    evolution of 105
    excluders 63, 195, 196, 198, 212, 222, 228
    finding 194
    killing of other queens 116
    legs 98
    life cycle 192
    life span 85–6, 99, 102
    marking and wing clipping 186, 188
    mating 99, 100, 116
    nuptial flight 99, 116
    ovaries 55, 100
    pheromone secretion, queen substance 55, 96, 98, 114, 115, 116, 189, 210
    piping noise 28
    removal/replacement of 33, 84, 103–4, 189, 200–4, 231, 235
    sperm kept in reserve 99
    stinger 28
    swarming and 114–15, 116, 188, 199–208, 212–15
    thorax 98
    transportation of 63
    virgin 54, 180, 199, 200, 208, 215, 232
quilting bee 18
Quinby, Moses 63

ragwort 138
rata tree (*Metrosideros*) 135, 272
rats 243

recipes:
    Apples with Honey Yogurt, Honey and Cinnamon, Baked 338, 339
    Baklava 342–3
    Bath, Milk and Honey 370
    beauty 369–88
    beeswax 261, 293
    Bread, Traditional Honey Tea 356
    Cake, Honey and Apricot Spice 349
    Cake, Honey Banana 348
    cakes and breads 348–56
    Chicken Breasts, Lemon and Honey–Glazed 320, 321
    Chutney, Apple, Honey, and Chile 358, 359
    Cider, Hot Honey 366
    Cleanser, Simple Honey Water 371
    Coffee, Spiced Honey 368
    Cold Cream 371
    for colds 381
    for constipation 382
    Cookies, Walnut and Honey 344
    Cough Syrup, Honey, Thyme, and Licorice 383
    for coughs 383
    Cranachan 340
    desserts, cookies, and sweets 338–47
    Dressing, Warm Honey and Whole-Grain Mustard 359
    dressings and chutneys 357–9
    Drink, Warm Honey and Cinnamon 380
    drinks 360–8
    Duck, Honey and Orange Roast 318, 319
    Eggnog, Honey 361
    Eye Treatment, Honey and Cucumber 374
    Face Pack, Chamomile and Honey 372
    Face Pack, Lemon and Honey 372
    food and drink 305–68
    Fool, Honey and Ginger Rhubarb 346, 347
    Foot Treatment, Peppermint and Beeswax 375
    Hair Conditioner, Creamy 370
    Ham, Honey Roast 314, 315
    Hand Cream, Lavender and Beeswax 374
    Hangover Cure, Honey, Orange, and Egg 382
    for hangovers 283–4, 382
    for hay fever 384
    health 379–88
    Hummus, Spiced Honey 313
    Lamb, Honey and Roast Garlic Roast 332
    Lamb, Rack of, with Thyme and Honey Glaze 328, 329
    lip balm 377

lipstick 378
main dishes 314–37
Mead, Traditional 360
Mint Julep, Honey 362, 363
Muffins, Spiced Cinnamon and
    Honey-Sweetened Date 352, 353
Phyllo, Goat Cheese, and Honey
    Parcels 333
Polenta Cake, Honey 350, 351
for rheumatism and back pain 385
Ribs, Sticky Honey 317
Salad, Endive, Pancetta, and
    Honey-Broiled Fig 310, 311
Salmon Fillets, Honey Balsamic 324,
    325
Salve, Green Mountain 385
Scallops, Honey-Broiled 312
Scallops, Pan-Fried with Honey
    Lime Marinade 323
Scones, Honey 354, 355
Seafood, Teriyaki Honey 326, 327
sleep enticer 386, 387
Smoothie, Honey, Date, and Linseed
    382
Smoothie, Honey and Red Berry
    364, 365
Soap, Honey 376, 377
for sore throat 387
Soup, Curried Honey Sweet Potato
    308, 309
soups and appetizers 308–13
Squash with Wild Rice Stuffing,
    Honey Roast 334, 335
Stir-Fry, Chinese Sesame and
    Honey Beef 330, 331
Stir-Fry, Citrus Honey and
    Balsamic Pork 316
Syrup, Melomakarona and Honey
    345
Tagine, Morrocan Honey-Chicken
    with Prunes 322
Tart, Caramelized Shallot and
    Honey 336, 337
Tart, Honey, Lemon, and
    Mascarpone 341
Tea, Honey and Apple Cider
    Vinegar 380
Tea, Honey and Cinnamon Green
    384
Tea, Honey and Lemon Herb 381
Tea, Iced Honey Earl Grey 367
Toast, Honey and Cinnamon 381
Toddy, Honey and Lemon 387
Toddy, Honey and Whisky 387
Tonic, Morning 384
Vinaigrette, Honey 357
records, keeping 187–90, 226
Rectitudines Singularum Personarum 41
Redi, Francisco 53
Reformed Common-Wealth of Bees, The
    (Hartlib) 52
Rhea 13

rheumatism and back pain, remedies
    for 385
rhododendron 137, 138
rhodomel 46
Ribs, Sticky Honey 317
Richard III, King 49
Rig-Veda 20, 32, 45
robbers/thieves:
    ants 80, 126, 144
    bee louse 144
    bees 54, 243–4
    brood thieves 84
    death's head hawkmoth 144, 145
    helping honeybees defend hive
        against 232
    hive beetles 144–5, 236, 245, 264
    honey badger 144, 147
    honeyguide birds 26, 28, 144
    wasps 144, 242, 243
    wax moths 144, 145, 244, 246
Rome, Ancient, beekeeping in 32–4, 46,
    47, 90
Rosaceae, rose family 143
rosemary (Rosmarinus officinalis) 133,
    273
round dance 112
royal jelly 21, 102, 114, 303
"Royal Jelly" (Dahl) 21
Royal Society 57
Russia 13, 34
Rutaceae, citrus family 143

Sacbrood virus 149, 236, 238
sainfoin (Onobrychis vicifolia) 135, 272
Salad, Endive, Pancetta, and
    Honey-Broiled Fig 310, 311
sales, honey 177, 257–8, 262–3
Salmon Fillets, Honey Balsamic 324, 325
Salve, Green Mountain 385
San (Bushmen) of Southern Africa 12
sawflies 80
Scallops:
    Honey-Broiled 312
    Pan-Fried with Honey Lime
        Marinade 323
scientific advances in beekeeping 50–63
Scones, Honey 354, 355
scout bees 87, 106, 110, 115, 154
Seafood, Teriyaki Honey 326, 327
sealant, beeswax in 46–8, 285–8, 293,
    393
Secret Life of Bees, The (Kidd) 21
Senchus Mor 42
Seneca 15
sensillae, honeybee 95–6
sex determination, honeybee 99, 105
Shakers 65
Shakespeare, William 15
shells, bee 40
shoe shine, beeswax 296
Siberia 12
sideliners 177

Siva 13
skeps:
    boles and 39–40, 41
    driving 37
    hiving a swarm with 50, 159, 209
    medieval 22, 36, 37–8
    multi-layered 59–60
    origin of 36, 37–8
    sulfuring 37
Sleep Enticer 386, 387
small hive beetle (SHB) (Aethina
    tumida) 144–5, 236, 245, 264
Small Hive Beetle: A Serious New Threat
    to European Apiculture 245
smell, honeybee sense of 94–5
Smith, Adam 16
Smith, John Maynard 105
Smith hive 164
smoking honeybees:
    bees' reaction to 174
    collecting a swarm and 209, 210
    history of 30, 54
    smoker and bellows, development
        of 63, 173–4, 176
    smoker, lighting a 175, 182
    smoker, using a 175, 192, 225
Smoothie:
    Honey, Date, and Linseed 382
    Honey and Red Berry 364, 365
Soap, Honey 376, 377
social life in hive 57, 106–11
society, bees as a model for 14–15
solitary bees 80–1, 82, 96
solitary wasps (Philanthus triangulum)
    82, 146
Solomon, King 34, 268
Solon 43
Song of Men of England, A (Shelley) 16
Sophocles 14
sore throat, remedies for 387
Soup, Curried Honey Sweet Potato 308,
    309
South America 12
    European honeybee in 66
    stingless bee (Melipona beecheii) 13,
        66, 114
    see also under individual nation name
space, bee 61, 89, 155, 158
species of honeybee:
    identifying 81–2
    number of 28, 81
    races across the world 87–91
spelling bee 18
spiders 146
spring tasks:
    early 228–30
    late spring/summer 230–4
Squash with Wild Rice Stuffing, Honey
    Roast 334, 335
St. Mary's Churchyard, Hartpury 40
sting, honeybee:
    action of 128

allergic reaction to 171, 222–3, 302
avoiding 222
death after 28, 302
drone 98
evolution of 84, 128
first descriptions of 55
identifying honeybee and 83
queen 98, 99, 116, 128
removing 222
structure 128, 129
venom 45, 128, 129, 302
stingless bee, South American (*Melipona beecheii*) 13, 66, 114
"Stings" (Plath) 21
Stir-Fry:
Chinese Sesame and Honey Beef 330, 331
Citrus Honey and Balsamic Pork 316
Stone Age 12, 26
Stonebrood 149
Strabo 45
suitable bees, finding 180–4
Sumeria 44
sunflower (*Helianthus anuus*) 136, 272
super-colonies 86
supers:
adding 157–8, 196, 199, 231
building your own hive and 165
buying bees and 181
colony inspections and 195, 196
function 157, 158
honey harvest/extraction and 247, 250, 252, 254, 259
place in hive layout 156
removing 203
reusing empty 254
swarming and 199, 231
Swammerdam, Jan 54, 55, 56, 57, 59
"Swarm, The" (Plath) 21
swarming, honeybee:
artificial 204–5, 241
basis of 114
hiving a/collecting 39, 50, 54, 159, 198, 208–14
honey storage and swarming, evolution of 86
location of hive and 167, 168
neighbors and 198
new nest site and 115–16, 154
prevention or control 114, 188, 198–215, 231, 264
queen and 114–15, 116, 188, 199–208, 212–15
signs of 199
triggers 115
when they are most likely to 54
sweat bees (*Halictus*) 81
Sweden 23
sweet chestnut (*Castanea sativa*) 135
symbolism, bee 14–19

T.W. Woodbury, Exeter 61
tablets, wax 46
tagging bees 26
Tagine, Morrocan Honey-Chicken with Prunes, 322
Tart:
Caramelized Shallot and Honey 336, 337
Honey, Lemon, and Mascarpone 341
Tea:
Honey and Apple Cider Vinegar 380
Honey and Cinnamon Green 384
Honey and Lemon Herb 381
Iced Honey Earl Grey 367
tej 46
"Telling the Bees" [poem] (Whittier) 43
Telling the Bees [painting] (Napier Hemy), 73
telling the bees if beekeeper is dead 43, 73
temperament of bees 54, 170, 171, 180, 182, 189, 190, 222
Tempest, The (Shakespeare) 15
Temple of Artemis, Ephesus 13
Temple of Kalabcha, Egypt 31
Temple of the Sun, Abu Gurab, Egypt 30, 34
tent, bee 60
Terramycin 237
Thetis 25
threats to honeybees:
attacks from animals and other insects 92, 144–6, 242–5
Colony Collapse Disorder (CCD) 7, 77, 246, 264
disease 42, 68, 74, 77, 91, 149–51, 181–2, 234–41
migration 77
mites and parasites *see* mites and parasites
pests and predators *see* pests *and* predators
thrift and industry, bee as emblem of 17–18, 19
thistle (*Cirsium arvense*) 135
thyme (*Thymus*) 133, 273
Toast, Honey and Cinnamon 381
Toddy:
Honey and Lemon 387
Honey and Whisky 387
Tolstoy, Leo 15
tombs, appearance of bees on 12
tongue, honeybee:
drone 97, 98
feeding brood and 108
nectar to honey process, role in 123
pollination, use in 120, 122
queen 97, 98
shortness of 120, 122
worker 96, 108
Tonic, Morning 384

tools, beekeeping 61, 176, 182, 190, 195, 196, 201, 221
tracheal mite (*Acarapis woodi*)
checking for 239
effects upon honeybee of 146, 239
Isle of Wight disease and 77
spread of 74, 77, 239
treatment of 239
tracking bees 26
Traité …sur les abeilles (Féburier) 63
transportation of bees:
commercial 262–3
European honeybee throughout the globe 65–9
migratory beekeeping and 70–2, 76, 77, 122
queen 63
traveler's joy (*Clematis*) 139
Treatise Concerning the Right Use and Ordering of Bees, A (Southerne) 50
Treatise on the Management of Bees, A (Wildman) 59–60
trees, keeping bees in 34, 36
trojniak 46
Tropilaelaps mites 236, 245, 265
tulip tree (yellow poplar) 135
tupelo (sour gum) 133, 273
Tussaud, Marie 48, 294
Tylosin 237
tyrant flycatcher 148, 149

ulcers, treating 388
United Kingdom *see* Britain
United States *see* America
uniting colonies 232, 234

varnish, beeswax 392
Varro 15, 33
varroa destructor mite:
Australia, absence of within 68
causes of 149, 240
CCD and 246, 264
close-up of 150, 151, 240
co-existence with some species of honeybee 146, 240
feeds on pupae 155
monitoring and controlling 156, 166, 180, 181–2
spread of 74, 146, 240, 264
treatments 180, 181–2, 240–1, 264
Veddahs, Sri Lanka 26
Veianius 33
venom, honeybee 45, 128, 129, 302
see also sting
Vikings 23, 27, 45
Vinaigrette, Honey 357
Virgil 14–15, 17, 33–4, 45, 50, 51, 53

waggle dance 28, 112–13
wagtails 145
walls, bee 40
War and Peace (Tolstoy) 15

Washington, George 19
wasps:
    evolution of 84, 85, 120
    honeybee, confusing with 80, 81,
      82, 83
    social 81, 82, 85, 86, 144, 146
    threat to honeybee 54, 144, 146, 243
water:
    evaporation from hive 86, 168
    collection of, honeybee 106, 111,
      170–1
    sources for hive 170–1
    waterproofing hive 36, 226
watermarks, beeswax in 296
wax *see* beeswax
Wax Chandlers, The 49
wax moths (*Achroia grisella* and *Galleria mellonella*) 144, 145, 244, 246
WBC hive:
    dimensions and features 164
    invention of 158
    problems with 158–9
Weed, E.B. 63
Weiss, Frederic 63
Western Honeybee *see* European/
    Western honeybee (*Apis mellifera*)
Westphalia 23
where to keep your bees:
    bee shed 168
    bee flight path and 168
    damp locations, avoid 168
    fields 167
    gardens 167, 168, 169
    law and 167
    neighbors and 167–8, 170
    out-apiaries 171
    planning 167
    registering 167
    swarming and 167, 168
    temperament of bees and 170, 171
    urban 167, 170
    vandalism and 171
    water sources 170–1
white wax of commerce 44
White, Shaker Anna 65
Wildman, Daniel 59–60
Wildman, Thomas 59–60, 61
William I, King 41
wings, honeybee:
    identifying honeybee 83
    worker bee 94
Winnie-the-Pooh 21
"Wintering" (Plath) 21
winter, honeybee survival during 54,
    117–18, 124, 163, 227–8, 232
wisdom, bees as symbol of 18
woodpeckers 145–6, 242
worker bees:
    abdomen 97
    antennae 95–6
    birth to early flight 108–10
    body heat regulates hive

    temperature 109
    career path of 106, 107, 108
    comb/cells, construction of 46, 97,
      106, 109, 114–16, 125–6, 154,
      159, 163
    dances *see* dances, bee
    defecation 209
    defense of hive 189
    emergence from cell 100
    evolution of 105
    eye, compound 94–5
    fat reserves 111
    feeding the brood 99, 106, 107,
      108, 109, 217–20
    first identified 57
    flights 109–10
    foraging 110, 111
    guard duties 110
    head 94
    honey sac 98
    larvae 102, 103
    life stages 108–11
    life-span 103
    mating, incapable of 85
    maturation of honey 110
    mouthparts 96
    pollen collection 96, 97, 98
    removing debris from hive 108, 110,
      232, 240
    replacement of 84
    smell, sense of 95–6
    sting 97, 98, 189
    wax production, role in 98, 124–5
    wings 94–5

Xenophon 45

year, beekeeping 226–35
    autumn tasks 234–5
    beekeepers' meetings, attending
      local 227
    if bright, snowy conditions, shading
      the entrance with a board leaning
      against the hive 229
    checking for colonies retaining
      their drones 232, 234
    checking the colony isn't starving
      229
    checking hives are stable and
      weatherproof 227
    cleaning equipment and checking
      for disease 227
    early spring tasks 229–30
    extracting honey from supers 231
    feeding a colony with sufficient
      stores for the winter 234
    feeding if stores are low 228, 232
    inspections 227–30, 231, 234
    late spring/summer tasks 230–3
    late winter tasks 227–9
    pests and predators, protecting
      against 234

    record cards, maintaining 226
    removing ventilation props and
      closing the feed hole 230
    replacing old frames and combs 230
    requeen colonies if necessary 231
    robbers, helping the bees defend
      the hive against 232
    supers, adding extra as the number
      of bees increases 231
    swarming to increase your stocks,
      using 231
    swarming, inspections to control
      231
    swarming preparations, watching
      for signs of 230
    tidy, keeping hives 226
    uniting colonies 235
    varroa, treating against 232
    ventilation, making sure there is
      sufficient within the hive 228
    visiting the apiary to check the bees
      227–8
    winter, preparing colonies for 232
yellow jackets 81, 85, 144
yew 34

Acknowledgements

An enormous amount of time and effort goes into producing a book of this size.

HarperCollins would like to thank the following contributors for their hard work and dedication:
Richard Jones, Sharon Sweeney-Lynch, Claire Waring, Sally Bucknall, Sarah Banberry, Bertrand Saugier, Nicola Charlton, Pascal Thivillon, Paul Carney, Anna Martin, Kirstie Addis, Geoff Hayes, Richard Phipps, Kate Parker, Lesley Robb, Tim Winter, Catherine Hill, Rachel Jukes, Norman Carreck, Liz Miles, Richard Rosenfeld, and Ben Murphy.

Bertrand Saugier would like to thank the following people for giving their time and authorizing access to their apiaries for taking photographs:
Jean-Paul Suc (beekeeper, Saint-Régis du Coin, France), The Suc family, Maurice Chaudière, Apiary Centre (Saint-Etienne), Les Ruchers de Verseau, Nicolas Guitini, Vincent Clair, Jean-Pierre Bastet, Claire and Adrian Waring, Peter Springall, Nicolas Primat, and Olivier Darné.

Claire Waring would like to thank:
Adrian Waring, NDB, and Ann W. Harman.

Richard Jones would like to thank:
My namesake, Richard Jones, of the International Bee Research Association.

Sally Bucknall would like to thank:
David Aston, John Phipps, and Jeremy Burbidge.

Nicola Chalton and Pascal Thivillon would like to thank:
Pauline Aslin, Peter Springall, and John and Ekaterina of Frog on the Green, Nunhead.

Picture Credits

COMMISSIONED IMAGES

Commissioned nature photography by Bertrand Saugier
p37 [top], p76, p80 [bottom], p81, p83 [columns 1 & 2], p99, pp101–3, pp106–7, p108 [top right & bottom], pp110–11, p117, p119, p121, p125, p126 [bottom], pp130–1, pp139–40, p154, p156 [bottom], p157, p159 (hive design © Maurice Chaudière), p163, pp166–7, p169, pp171–5, p176 [top], pp177–8, pp182–6, p187 [right], p188, pp190–2, p193 [middle & bottom], p195, pp197–206, p208, pp212–5, p218, pp220–1, p222 [top], pp223–5, pp228–33, p235 [left], pp241–4, p247, pp250–7, p259, pp262–3, p264 [left] (hive design © Maurice Chaudière), p265 (hive design © Maurice Chaudière), p268, p285, p288, pp297–300

Commissioned food and cosmetics photography © Tim Winter
pp307–39, p343, p347, pp351–65, p373, p386

Cover illustration and chapter opener illustrations © Richard Phipps
p10, p78, p152, p266, p304

Commissioned bee illustrations by Paul Carney
pp87–92, pp96–97, p100, p108 [top left], p112, p114, p116, p120, p128

STOCK IMAGE PERMISSIONS

© *Alamy*
p12, p45, p71, p118, p127, p226, p261, pp269–70, p275, p279, p303

© *Basement Press*
p155, p160, p161 [left], p162, p164, p260, pp271–4

© *Bridgeman Art Library*
pp8–9, pp42–3, p73, p75

© *Corbis*
p21, pp67–8, p196, pp248–9, p302

© *FLPA*
p133 [Leatherwood & Tupelo], p136 [Ivy], p142 [Broom & Wallflower], pp150–1, p179, p207

© *Getty*
p29

© *Heritage Image Partnership*
p51

©*istockphoto*
p133 [Gum & Manuka], p134 [Hawthorn], p135 [Sainfoin & Thistle], p136 [Goldenrod & Sunflower], p138 [Rhododendron], p142 [Teasel], p143 [Currant, Mint, Grape hyacinth, Bergamot & Valerian], p284, p296, p340, p366, p368

© *Mary Evans Picture Library*
p14

© *NHPA*
p115 [left]

© *NPL*
p93 [top & bottom left], p109, p113, p126 [top]

© *Photolibrary*
p6, p24, p31 [top], p49, p133 [Holly & Needlebush], p134 [False acacia], p135 [Lucerne, Rata tree, Sweet chestnut & Tulip tree], p142 [Angelica & Elaeagnus], p146, p176 [bottom], p258, p277, p282, p291, p294 [top], p395

© *Shutterstock*
p132 [all images], p133 [Maple, Mesquite, Orange, Rosemary & Thyme], p134 [Acacia, Blueberry, Borage, Bramble, Buckwheat & Knapweed], p135 [Lavender, Lime, Melilots, Canola], p136 [Cotton, Fireweed, Fuchsia, Heather bell & Heather ling], p138 [Mountain laurel, Privet, Ragwort & Spurge], p142 [Cosmea, Viper's bugloss, Bellflower & Hop], p143 [Poached egg plant, Loosestrife, Mallow, Evening primrose, Phlox, Primula & Rose]

Other © permissions
p30 [bottom] Egyptian and Papyrus Museum, Berlin; p41 Sally Bucknal; p52 University of Guelph Library; p54 Grace Doherty Library; pp55–6 University of Oklahoma Libraries; p70 Sally Bucknal; p82 [top] Richard Jones; p83 [column 4] Susan Ellis, Bugwood.org; p161 [right] Claire Waring; p170 John Howe / Green Brooklyn; p181 Ettamarie Peterson; p187 [left] & p233 Claire Waring; p264 [right] Omlet UK; p281 [right] National Honey Show